AF540553

ANALYTICAL BIOCHEMISTRY

By

Dr. Arvind N. Shukla

School of Studies of Zoology & Biotechnology

Vikram University

Ujjain

(India)

DISCOVERY PUBLISHING HOUSE PVT. LTD.

NEW DELHI-110 002

Published by:

DISCOVERY PUBLISHING HOUSE PVT. LTD.

4383/4B, Ansari Road, Darya Ganj
New Delhi-110 002 (India)
Phone: +91-11-23279245, 43596064-65
Fax: +91-11-23253475
E-mail: discoverypublishinghouse@gmail.com
sales@discoverypublishinggroup.com
web: www.discoverypublishinggroup.com

First Published: **2009**

Reprinted: **2017**

ISBN: 978-81-8356-485-4

Analytical Biochemistry

Printed at:
Infinity Imaging Systems
Delhi

Preface

Biochemistry today has made spectacular progress in unraveling the mysteries of animate nature. This progress has allowed us to gain deeper insight into the principles of vital activity and has to a very significant extent stimulated the development of applied disciplines, especially medicine. The present title *"Analytical Biochemistry"* is intended for those who wish to understand living organisms, especially man. Biochemistry is essential for this purpose, but it would be almost impossible for a student to survey on his own the massive body of existing knowledge, constantly augmented by a remarkable torrent of brilliant discoveries. The purpose of the book, then is to organize our knowledge into something that can be comprehended in a relatively short time and still convey a reasonable complete picture of the chemical structure and function of man. Readability without sacrifice of coverage has been a prime goal. An important device in gaining that goal is to keep attention constantly focused on function, with repeated use of rationalization to show that the chemical facts are not isolated, but part of a whole.

Biochemistry has two major goals as a fundamental science, it treats the vital functions from the standpoint of physical chemistry, and as an applied discipline, it points out practical applications for the wealth of scientific knowledge it has acquired. The dual purpose has been, as far as possible, take into account in the writing of this book. We have also tried to summarize our pedagogical experience in teaching biochemistry to students specializing in medicine and pharmacy. The material of this book has been organized according to the principle of functionally in order to trace the close relationship between the functions of the living organism and the structures of its constituents molecules as well as the chemical and physico-chemical process in which they are involved.

The aim of this book is to present a core of biochemical knowledge that is desirable for undergraduate and postgraduate students and also those involved in the field of medical, microbiology, biotechnology and pharmaceutical. Every attempt has been made to keep abrest of the advances in the subject and at the same time to include the fundamentals.

To make the work more comprehensive and informative, the author has consulted many authoritative books, research journals, abstracts, monographs etc. He is grateful to all those great scholars whose work are cited or substantially reproduced.

There can be nc claim to originality except in the manner of treatment and much of the information has been obtained from the books and scientific journals available in the different libraries.

The author expresses his thanks to his friends and colleagues whose continue inspirations have initiated him to bring out this book.

The author expresses his gratitude to Mr. Wasan and Staff of M/s Discovery Publishing House Pvt. Ltd., for their whole hearted co-operation in the publication of this book.

In the mean time, the author will remain sincerely responsible for any shortcomings of the book and be grateful to the readers for their suggestions and constructive criticism for the continuous betterment of the book. He takes this opportunity to appeal to the readers to send their suggestions straightway to his publisher.

Author

CONTENTS

CHAPTER 1

Auto-Analysers

The increased demand for analyses has led to the introduction of machines that will perform all or part of an analytical procedure. Many of the manipulations in laboratory methods are common to a variety of different tests (*e.g.* pipetting) and lend themselves readily to mechanization. The introduction of instruments into a laboratory that will perform these tasks will result in a reduction in the analysis time and mean that an increased workload can be met without the need for additional staff. This should reduce the overall cost per test even though the expense associated with the purchase and running of the instruments may be high. In addition to cost considerations, the fact that automated instruments are designed to give consistently good analytical precision is a factor that has greatly influenced their widespread acceptance.

It is recognized, however, that reliability depends upon the quality of their design and manufacture and subsequently their effective and regular servicing in the laboratory situation. An automated system, by definition, should perform a required act at a predetermined point in the process and should have a self-regulating action. This implies that intervention by the analyst is not required during the procedure and that only those systems that incorporate a microprocessor or computer to control and monitor their performance can be designated as automated. Some systems may not comply strictly with this definition but are a valuable means of mechanizing laboratory activities. The first so-called automated analysers were developed to carry out analyses by replacing manual pipetting by the mechanical transfer of fixed volumes of sample and reagents. They were designed to mimic the manipulations of a convential manual procedure and sometimes had the capacity to measure the amount of reaction product, usually spectrophotometrically.

The precision of the pipetting was often poor and the instruments were cumbersome and prone to breakdown. Their modern counterparts offer a much higher degree of precision and reliability and a wide range of designs are available, each offering different levels of mechanization. The simplest are the auto-pipettors/diluters, which can be adjusted to measure and expel selected volumes of solution using a syringe system. They relieve the analyst of the pipetting but require constant attendance. Instruments which are more highly mechanized and will pipette samples and reagents sequentially into racks of tubes play an important role in the mechanized preparation of samples. These were the forerunners of what are now known as sample processors.

An operator still may be required to move the racks of tubes to different positions within the instrument so that each of the steps in a particular procedure may be completed. When a method demands other manipulations which the mechanized analyser cannot perform, *e.g.* centrifugation, then it is necessary for the operator to transfer the reaction tubes to the appropriate laboratory instrument. From this approach, a series of automated analysers known as discrete analysers were developed by the incorporation of a microprocessor to control their operation. The design and capabilities of the different analysers vary but they all require only small volumes of sample and reagents and often 'ready-to-use' reagents are supplied by the manufacturer. Many of the newer instruments are designed to use disposable, dry reagent strips or film devices rather than liquid reagents.

The modern analysers show high levels of sensitivity and specificity owing to the development of reliable analytical methods which suffer only minimal interference effects. Of particular importance in this respect is the elimination of the centrifugation stage, which was previously needed to remove the protein which is often present in biological samples. The 1950s saw the introduction of a completely new approach to automation, in the form of continuous flow analysis. This made a significant contribution to the advance of automated analysis and subsequent development has been in the form of flow injection analysis. The original instruments were single channel and capable of measuring only one constituent in each sample.

Multichannel instruments were then developed which could simultaneously carry out several different measurements on each sample. These were useful in laboratories where many samples required the same range of tests. A microprocessor or computer is now incorporated into most discrete and flow analysers and it is an absolute requirement of multichannel instruments. It allows the operator to set the optimum analysis conditions for each different assay and to give instructions regarding the acceptable limits of the results for samples of known concentrations. With this information the computer is capable of continual self-evaluation of the performance of the analyser and the quality of results being produced.

All the necessary calculations will be done before the concentration of each sample is printed out, although sometimes the printer will be instructed to produce data in the form of graphs or histograms. The results from single-channel instruments are usually presented as a list of sample numbers with their corresponding concentrations. Multichannel analysers require all the data for each sample to be collected together before they are printed out and in their case each sample must carry its own identification, *e.g.* barcode, which the computer can recognize. A natural extension of analytical automation is some means of data processing all the results that are generated. This usually takes the form of a central computer which accepts information from different analysers for presentation in a useful manner. The identification of a sample and the tests performed are typed in using a keyboard and the computer collates all the data on each sample. As well as collecting information, computing and statistically assessing results, an important facility of the computer lies in its ability to store information for future recall via a visual display unit. With the introduction of robotics the scope of laboratory automation has been expanded.

These robotic instruments are fully computerized and programmed to perform either single stages or entire analytical procedures. The more complex systems comprise a series of modules which sequentially perform the various steps in an analytical procedure from sample preparation right through to the production of the final report. Although the introduction of a robotic system into a laboratory may require a considerable period of performance evaluation before it can be routinely used with confidence, once functioning it will continue unattended, thus releasing staff to perform other duties; it may also be used overnight?

Other advantages are in the improved precision of complex, labour-intensive procedures and reduced exposure of laboratory workers to hazardous samples and reagents. There is a wide range of mechanized and automated instruments available from a variety of manufacturers and the choice of which one to purchase should involve several considerations. The nature of the expected workload must be assessed with respect to both the numbers of samples and the variety of tests that will be performed. This will indicate the level of mechanization or automation required.

Neither the existing laboratory equipment nor the effects of automation on the current and future staffing policy should be overlooked. The cost is an important aspect and this includes the price of the instrument, its running costs in terms of reagents and personnel and its servicing requirements. The reliability of the instrument and its life expectancy are also important and these facts, together with information regarding the quality of the results that can be obtained, must be sought before a decision can be made. It is advisable to request an instrument on a trial basis so that its performance can be assessed on site.

DISCRETE ANALYSERS

Instruments in which each test is performed in its own container or slide are known as discrete analysers, in contrast to flow analysers in which the samples follow each other through the same system of tubing. All discrete analysers have a common basic design incorporating a pipetting system, a photometric detector and a microprocessor. A development of the single test instrument is the parallel fast analyser, which analyses several samples simultaneously but for only one constituent. However, the change-over from one analytical procedure to another is quick and simple. The multitest analysers simultaneously measure more than one con-stituent in each sample and are often designed to meet the needs of a particular type of laboratory, *e.g.* clinical chemistry departments.

The instruments offer different test combinations (menu) but these are often determined by the manufacturer, leaving little opportunity for modification within the laboratory. Earlier instruments carried out the same menu of tests on each sample but the introduction of random access analysers made it possible for the operator to select individually a test combination for each sample from the total range available. Other automated systems may be purchased for a specific purpose and are called dedicated instruments, *e.g.* glucose analyser. Others have fairly restricted applications, an example being the reaction rate analysers which are specifically designed for the kinetic measurements of enzyme activity. Some of the more recently developed instruments employ individual pre-prepared disposable test packs or strip devices which contain all the reagents for each particular assay in a dry form.

Centrifugal Analysers

A new concept in automated analysis was developed in the late 1960s, resulting in the introduction of centrifugal fast analysers. The development project was sponsored in the United States of America by the Institute of General Medical Sciences and the Atomic Energy Commission, and the prototype was named the GeMS AEC system. Several instruments are now available and they are all based on the original idea in which batches of samples are treated in parallel and centrifugal force is used to transfer and mix the solutions, the reactions being continuously monitored photometrically. These instruments have considerable flexibility and are capable of being used for a variety of different types of analysis simply by changing the reagents in the dispensers and altering the spin programme.

In addition, they use only small volumes of reagents, not only reducing the running costs but also permitting the analysis of very small volumes of sample.

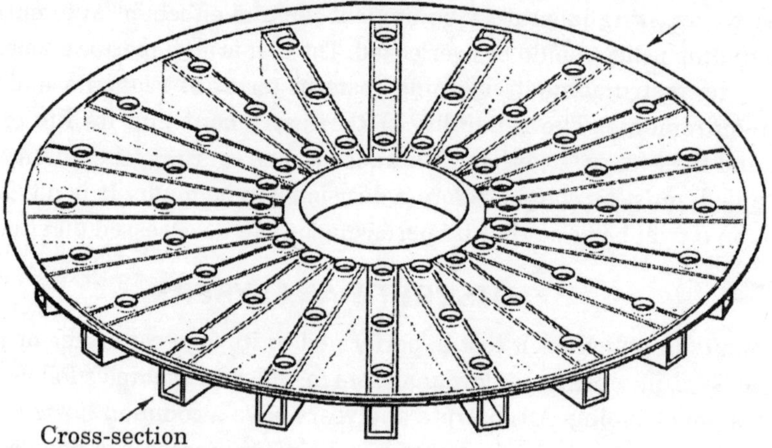

Fig. 1.1. Diagrammatic representation of a typical transfer disc for a centrifugal analyser.

The analysis takes place in a circular disc called the transfer disc or rotor made of an inert material such as Teflon or acrylic whose surface properties allow minimal adhesion of liquid. It contains radially arranged sets of cavities, each set composed of two or more interconnected cavities, one for the sample and the others for the reagents.

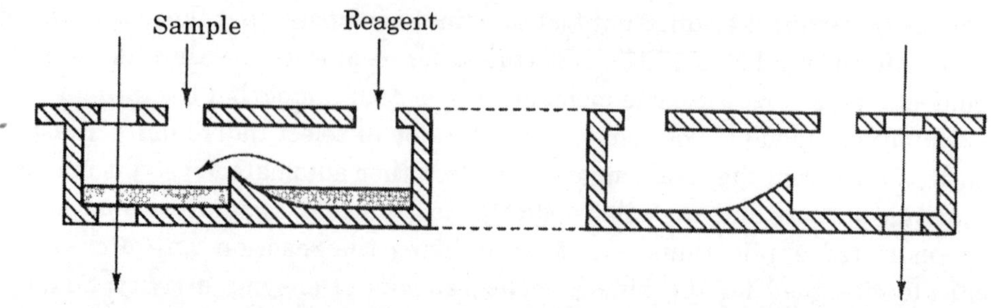

Fig. 1.2. A cross-section of the transfer disc shown in Fig. 1.1.

Small measured volumes of sample and reagent, *e.g.* 5—200 μl, are automatically pipetted into the appropriate wells so that each set contains a different sample but the same reagents. One set is usually prepared as a reagent blank. The wells are designed so that the sample and reagents remain in their separate compartments until centrifugation of the disc commences. During acceleration, mixing occurs and the chemical reaction is initiated. After a suitable time readings are taken while the disc is still spinning. These measurements may be absorbance, nephelometric or turbidimetric depending upon the design of the instrument.

In some instruments the optical cuvettes, one for each set of wells, are incorporated into the perimeter of the transfer disc, while in other systems the fluid is expelled into individual cuvettes inside the centrifuge. Analysis is complete within a few minutes and, if the transfer disc is not disposable, it is automatically washed and dried ready for reloading. A computer is incorporated in the instrument allowing the assay temperature and spinning programme, *i.e.* the speeds and duration of the mixing, and reaction stages, to be preset. It is also used to calculate the concentrations of the test samples by comparison of their photometric readings with those of standards treated identically. Centrifugal analysers are applicable not only to end-point analysis but, because of their continuous monitoring capability, are particularly useful for kinetic assays. Measurements of enzyme activity and enzyme immunoassays are particularly amenable to these types of analyser.

Dry Chemistry Analysers

Advances in the technology associated with the production of immobilized reagents in thin layers has resulted in the development of a range of analytical systems in which the assays

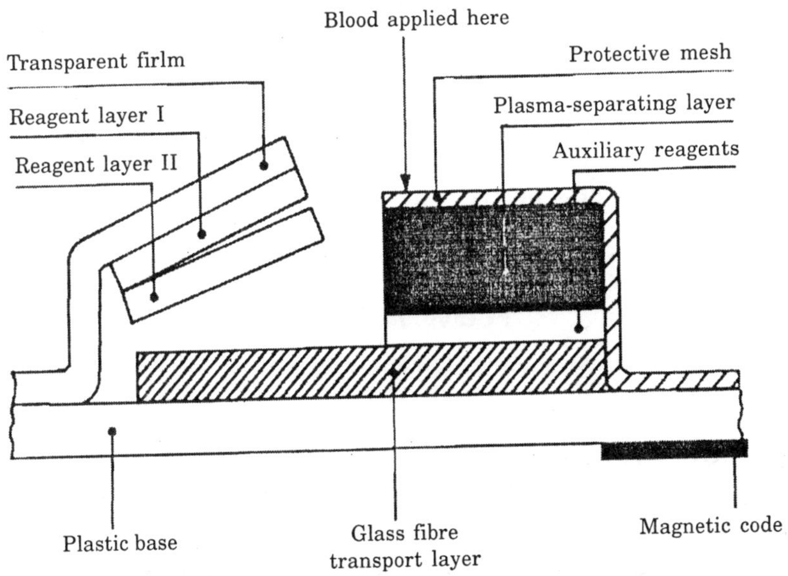

Fig. 1.3. Diagrammatic representation of the dry reagent strips known as reagent carriers. which are used in the Boehringer Reflotron System for the chemical analysis of blood samples. Red blood cells are removed in the separating layer and the plasma passing into the glass fibre transport layer is analysed. The magnetic code carries information about the analytical procedure.

are carried out by the addition of a liquid sample to the reagents in a dry form. This method of analysis requires no reagent preparation and can be conveniently performed without the usual laboratory facilities. A major application is in the analysis of a range of blood serum components for clinical diagnostic purposes.

Assays have been designed to be sensitive and rapid and operation of the instruments is simple. All the analytical parameters are stored within the instruments and these are activated by insertion of the barcoded reagent strip or slide. There is a wide range of assay systems based on colour reactions with detection by reflectance photometry and involving such substances as enzymes and antibodies in addition to the more conventional reagents. The Boehringer Reflotron is an example of a dry reagent strip system whereas the Vitros Chemistry System uses small multilayer slides, some of which incorporate miniature ion-selective electrodes for measuring ions such as sodium and potassium. In the DuPont aca analyser the chemicals are held in pop-out pockets in a small transparent bag where the test reaction takes place.

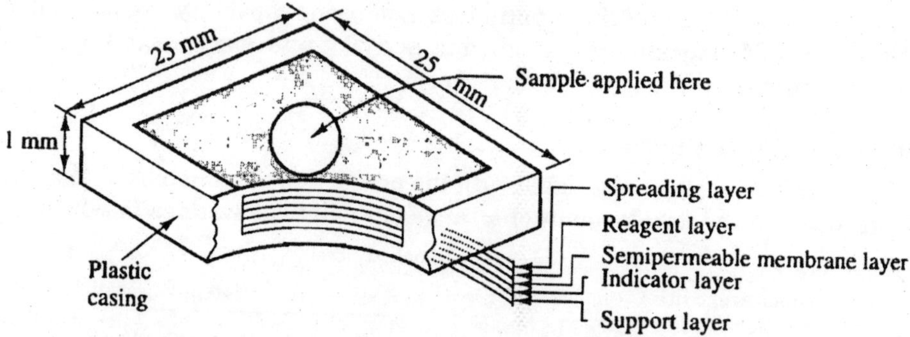

Fig. 1.4. Cross-section through a dry reagent slide for use in the vitros chemistry system., Previously known as the Kodak Ektachem analyser. A range of slides, which vary in the nature, number and composition of the layers, is available for a variety of analytes in blood serum. The sample (approximately 10 /μl) is applied to the spreading layer and reactions take place as it permeates through the various layers. Detection is by reflectance photometry.

FLOW ANALYSIS

In automated flow analysis, samples are analysed as they are pumped in sequence through a series of tubing containing a continuously flowing reagent stream.

Continuous Flow Analysis

In 1957 Professor Skeggs described a method of continuous flow analysis which was developed and marketed by the Technicon Instruments Co. Ltd. under the name of *'Auto Analyser'*. The single- and dual-channel instruments had the flexibility of being able to perform a variety of different tests in laboratories where samples requiring a particular analysis could be batched and analysed-together.

A limited range of instruments is now available from several manufacturers and although they vary in design, showing several modifications from the original model, the concept of continuous flow analysis is common to them all. The sequential multiple analysers (SMAs) were later developments and they were much more complex in design and included a computer.

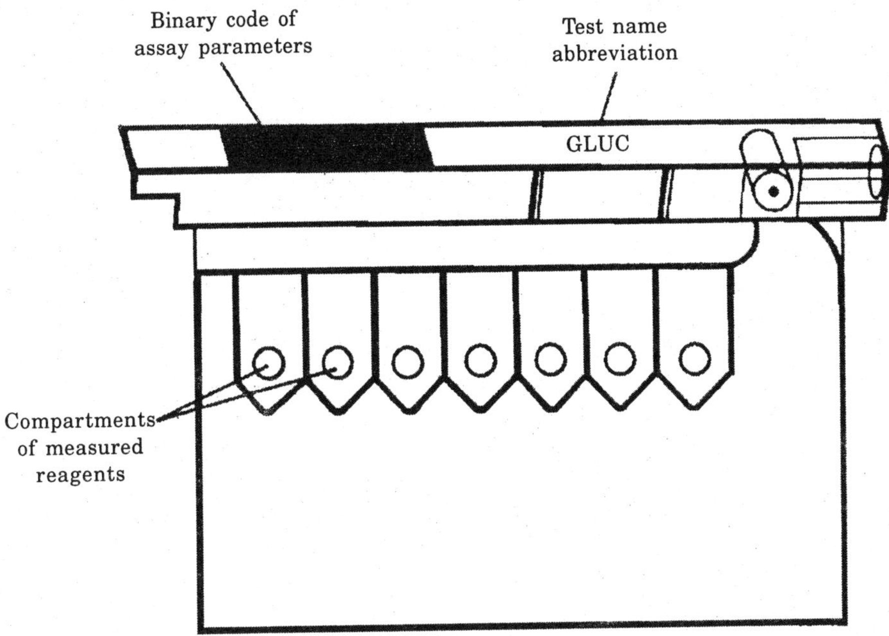

Fig. 1.5. Transparent analytical test pack, (100 mm × 80 mm) for Du Pont aca® analyser containing all the reagents for the assay of glucose. A wide range of test packs is available for use in the automated discrete analyser. Once the pack is loaded into the instrument, sample and diluents are automatically injected. The sealed reagent compartments are then broken open at various stages in the assay and the contents of the pouch are mixed and incubated. The transparent pouch then serves as a cuvette for absorbance measurements.

They could analyse each sample for several constituents simultaneously, the number of channels determining this capacity, *i.e.* 6, 12 or 20. These instruments were developed primarily for hospital clinical chemistry laboratories to allow an overall assessment of the chemical composition of blood samples. They have now been superseded by other types of analysers. Continuous flow analysers can perform all the basic steps involved in a simple quantitative photometric procedure. These include pipetting of the sample and reagents, mixing, heating at specific temperatures, detection and presentation of results.

In addition, sample digestion, phase separation, and the Snoval of inteirfering substances, particularly protein by dialysis, may be included when required. Although colorimetric detection is most frequently employed, fluorimetric, nephelometric, flame photometric, refractive index, radiochemical and electrochemical techniques have also been used, widening the range of analytical applications considerably. Analysis takes place as the samples and reagents are pumped through a system of tubing instead of being retained in separate containers. The stream passes through the tubing to the different units of the analyser, each of which performs a specific analytical function.

As distinct from manual methods and discrete analysers, absolute volumes are not pipetted but the liquids involved are mixed in carefully controlled proportions. These ratios are determined by using delivery tubes of varying internal diameter with a constant-speed pump and it is the different rates of flow (ml min^{-1}) within these tubes that is the important factor. The samples are aspirated in sequence into the reagent stream with a small volume of water separating each one.

Air bubbles are introduced into the flowing stream and in order to maintain a regular liquid-air pattern in the stream, a wetting agent, *e.g.* Brij 35, is usually added to the liquids. The air bubbles have several important functions and are fundamental to this type of analyser. They completely fill the lumen of the tubing and help to preserve the integrity of each individual aliquot of liquid by providing a barrier between them. They reduce contamination between succeeding samples because the pressure of the bubble against the inner wall of the tubing wipes away any droplets from the preceding sample which might contaminate the next. These features permit the use of a shorter water wash between samples and therefore save time.

The bubbles also assist in mixing sample and reagents within each liquid segment. Whenever two air-segmented streams are brought together, mixing within each liquid segment is achieved by passing the stream through a horizontal glass mixing coil, whose circumference should be at least three times greater than the length of the liquid segments. During passage through the coil, each fluid segment is repeatedly inverted allowing the heavier layers to fall through the lighter ones, ensuring complete mixing. Quantitation in continuous flow analysis is based on the direct comparison of the peak heights on a recorder trace of the samples and standards. It is not necessary for the reactions to go to completion, because all the measurements in a particular method are made after the same fixed reaction time, which is determined by the length of analyser tubing and relies on a constant pump speed.

Single- and Dual-Channel Instruments

The design of all single-channel instruments stems from the original Technicon AAI system, which comprises several interconnected units, called modules. The liquid streams flow through the modules, each of which performs a different analytical function. The subsequently developed A AII analysers did not consist of separate modules but included an analytical cartridge which combined several functions in one unit.

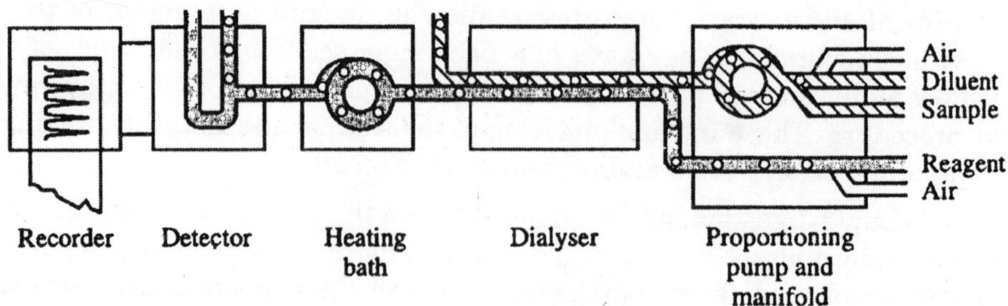

Fig. 1.6. Diagrammatic representation of the function of the continuous flow analyser modules.

Dual-channel instruments are very similar to their single-channel counterparts and differ only in the fact that they analyse each sample for two constituents simultaneously. The sample stream is divided into two fractions, each one flowing independently through its own system of tubing with the appropriate reagents and detectors.

Proportioning Pump and Manifold

The mixing of the sample and reagents in predetermined proportions rather than specific volumes is a fundamental principle of continuous flow analysis and is achieved by using a constant-speed roller bar peristaltic pump and flexible plastic delivery tubes of different internal diameters. These pump tubes are stretched between two plastic end-blocks, which hold them taut between a spring-loaded platen and the pump rollers. The amount of air required to produce the bubbles is similarly introduced using a pump tube with a particular internal diameter. There is a choice of pump (manifold) tubes which deliver between 0.015 and 3.9 ml min^{-1} and are colour coded for easy identification.

Sampler

The sampler unit holds a tray (plate) which can be loaded with up to 40 cups containing the individual samples. The samples are aspirated in turn with water between each one. The length of time for which the sample is aspirated determines the volume of sample that is analysed.

Separation Modules

A separation step is sometimes an essential part of an analytical method and may be as diverse as distillation, filtration, digestion, extraction, phase-separation or dialysis. These can all be performed by continuous flow analysers either by adding a specially designed glass fitting to the manifold or analytical cartridge or by the addition of a separate module to the analyzer.

Many biological samples contain protein and dialysis is often used to remove this protein, which would otherwise affect the analysis. The dialyser unit consists of a thermostatically controlled water bath (37 ± 0.1 °C) which houses a matched pair of spirally grooved plastic (Lucite) plates, which are clamped together with a semipermeable cellophane membrane stretched between them. This in effect forms a long narrow tube which is divided lengthways by the membrane. As the air-segmented sample (donor) and recipient (usually saline or a reagent) streams flow side by side along the semicircular channels, small analyte molecules diffuse across the membrane into the recipient stream.

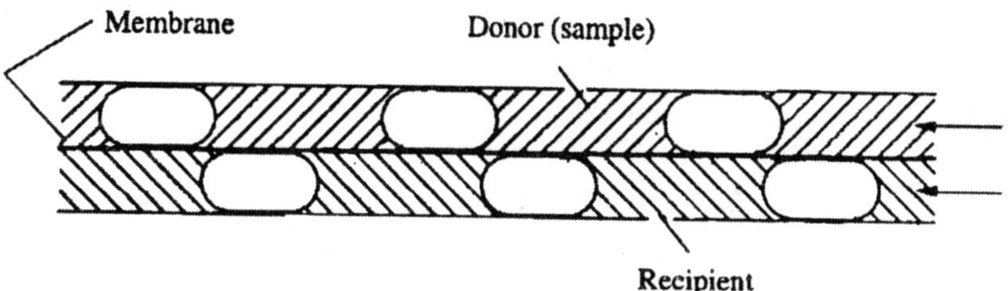

Fig. 1.7. Schematic representation of dialysis, Small analyte molecules from the donor (sample) stream dialyse across the semipermeable membrane into the recipient stream. After emerging from the dialyser, the donor stream is usually pumped to waste while the analysis continues with the recipient stream.

The rate of dialysis is dependent on the concentration of the dialysable molecules if all the other factors which affect dialysis are kept constant. This makes it possible to compare standards and samples directly and although transfer is never complete, it is constant, usually between 5 and 25%.

Flow Diagram

A diagrammatic representation of the components and layout of the system which is required to perform a particular method is known as a flow diagram. This is a convenient and well-recognized way to record information about a particular method, e.g. the sampling rate, the temperature of the heating bath, the type of detection system and the size of pump tubes required for each reagent.

GLUCOSE

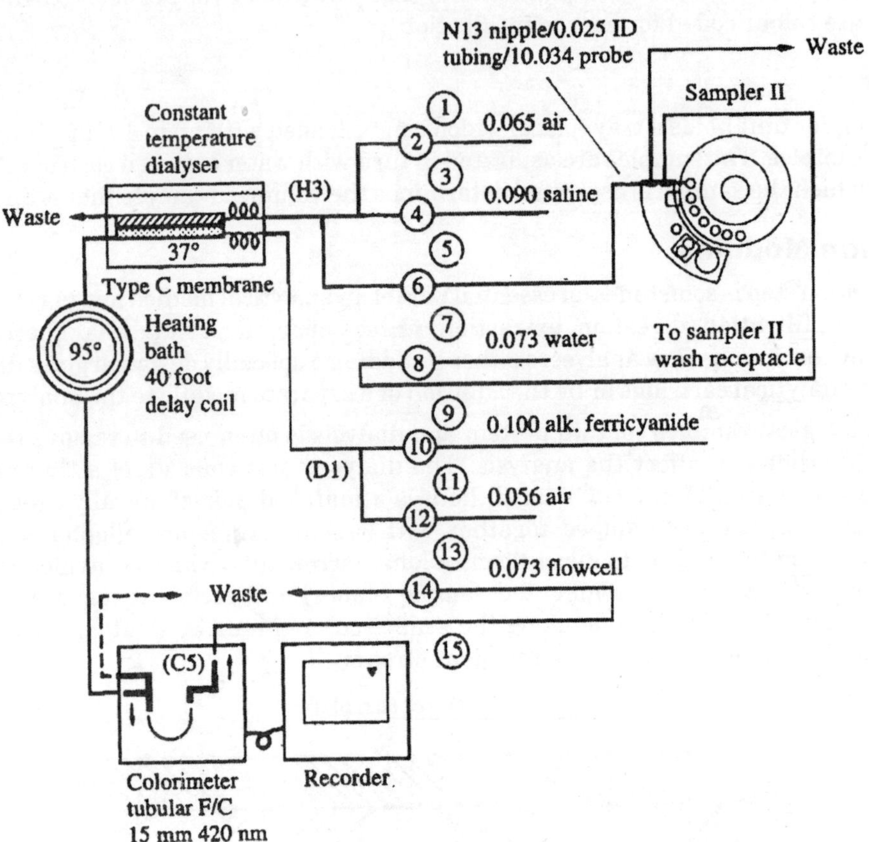

Fig. 1.8. Flow diagram for the measurement of glucose using alkaline ferricyamae.

The pump tubes are specified with respect to their internal diameters and position on the end-blocks. The colour coding of the tubes is:

Position 2	0.065	Blue/Blue
Position 4	0.090	Purple/Black
Position 6	0.020	Orange/Yellow
Positions 8	0.073	Green/Green

Position 10	0.100	Purple/Orange
Position 12	0.056	Yellow/Yellow
Position 14	0.073	Green/Green

The odd-numbered positions on the end-blocks are those on the lower level and are unfilled. H3 and D1 refer to the glass connectors.

Theoretical Aspects of Continuous Flow Analysis

In continuous flow analysis, samples follow one another through the tubing and interaction of adjacent samples, which is called 'carry-over', does occur. Carry-over manifests itself on

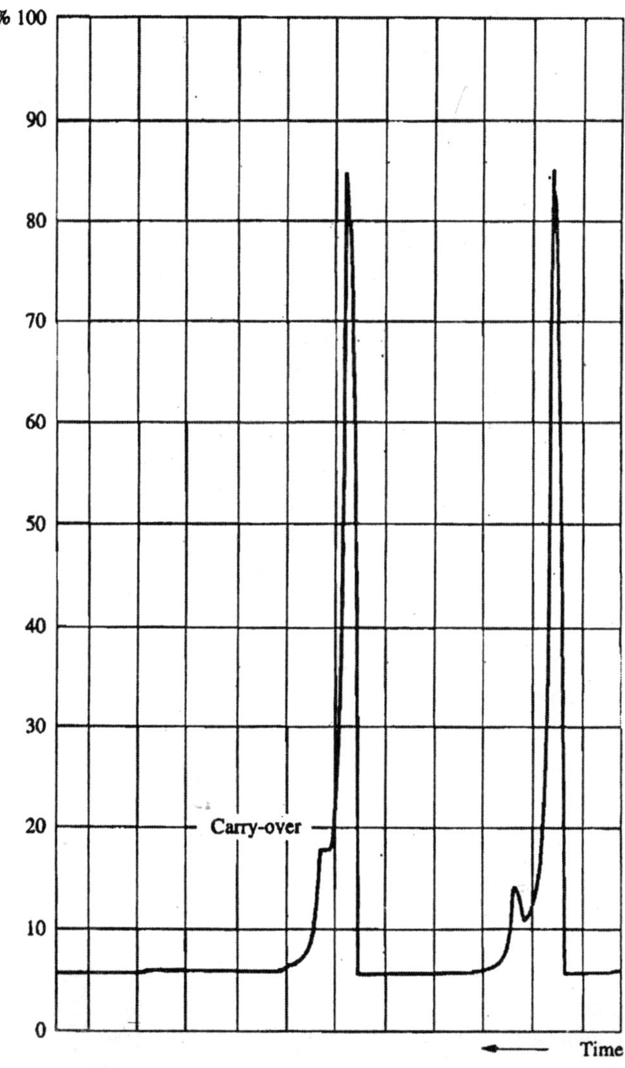

Fig. 1.9. Effect of carry-over. Samples of high and low concentration were analysed with different wash times. Carry-over was more apparent with the shorter wash time.

the recorder trace as peaks that are not completely differentiated. This is most noticeable when a sample of low concentration follows one of high concentration and is either seen as a shoulder on the high peak or obliterated. However, carry-over may also result in the reduction of peak height when a high sample follows a low one, but this is not apparent from the trace and is therefore undetected.

The introduction of water segments between the samples and of air bubbles into the liquid streams significantly reduces carry-over but does not eliminate it. Appropriate sample and water volumes, expressed as aspiration times (seconds), must be determined for each method. This involves aspirating a standard solution of the analyte for .varying times and observing the peak heights. A wash time may then be selected by keeping the sample time constant and varying the wash time. The highest sampling rate that results in the least amount of carry-over is usually sought.

Assessment of Carry-over

An assessment of carry-over can be made by analysing a sample with a high concentration of the analyte (A) in triplicate, giving results a_1, a_2 and a_3, followed by a sample of low concentration (B) in triplicate giving results b_1, b_2 and b_3. The carry-over between a_3 and b_1 is given by the equation:

$$\text{Percentage interaction } (k) = \frac{b_1 - b_3}{a_3 - b_3} \times 100$$

where b_3 is assumed to be the 'true' value for sample B since it is preceded by two samples of equal concentration; the effect of carry-over should be negligible.

Flow Injection Analysis

In this technique, which was developed in the 1970s, microlitre volumes of liquid sample are injected, at intervals, into a continuously flowing carrier stream which is not airsegmented. Various reagent streams are introduced as required and controlled mixing of reagents and sample occurs. The fact that flow injec-tion analysis does not involve air-segmented streams makes it possible to include such separation steps as solvent extraction and gas diffusion. When the merged stream flows through the detector, transient peaks are recorded in proportion to analyte concentration.

Excessive dispersion of the sample zone into the carrier stream is reduced by the use of small sample volumes, narrow bore tubing and short residence times in the system. Rapid analysis and easy starting up procedures make this technique particularly useful for single samples which arrive infrequently and require immediate analysis.

Table 1.1. Typical parameters for flow injection analysis

Sample volume	*40–200 μl*
Tube diameter	0.35–0.9 mm
Flow rate	0.5-2.5 ml min^{-1}
Coil lengths	10-200 cm
Flow cell volume	8-40 μl
Residence time	15-60 s
Sample through-put	60-200 h^{-1}

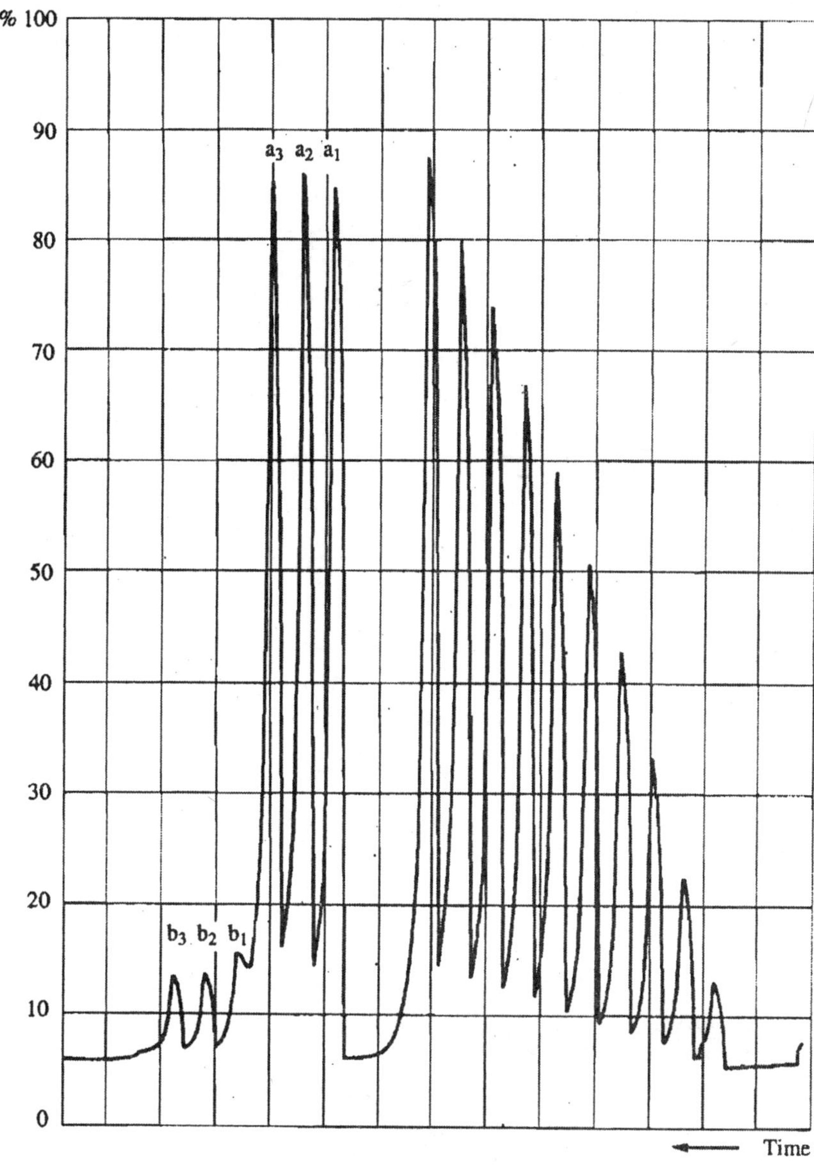

Fig. 1.10. Assessment of carry-over.

Instrumentation

A precision injection device is required to minimize sample dispersion and keep the sample volume and length of sample zone reproducible. This is normally a rotary valve similar to that used for injection in HPLC. Exact timing from sample injection to detection is critical because of rapidly occurring reactions which are monitored before they reach completion. This demands a constant flow rate with low amplitude pulsing, normally achieved by a peristaltic pump. Alternatively, a syringe or gas pressure may be used to propel the fluids.

Correct proportions of sample and reagents are achieved by the use of tubes of varying diameters. Tecator produces a series of complete flow injection analysers. Their manifolds, which provide a means of bringing together the fluid lines and allowing rinsing and chemical reactions to take place in a controlled way, are marketed under the name Chemifolds and a range is available for a variety of applications.

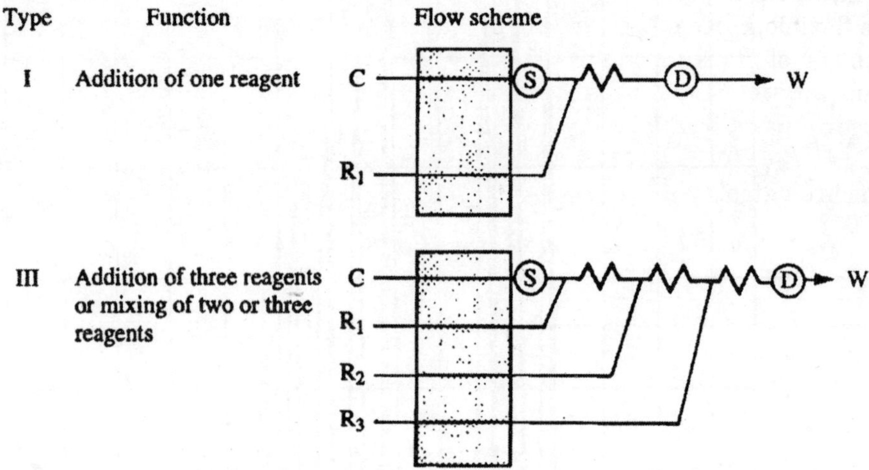

Fig. 1.11. Examples of tecator chemifold types for flow injection analysis. S, sample injection port; C, carrier stream; R_1, R_2, R_3, reagent streams; D, detector; W, waste.

ROBOTICS

A major shortcoming of the automated analysers developed in the 1960s and 1970s was that, to a large extent, sample preparation remained unautomated. Also their use was generally

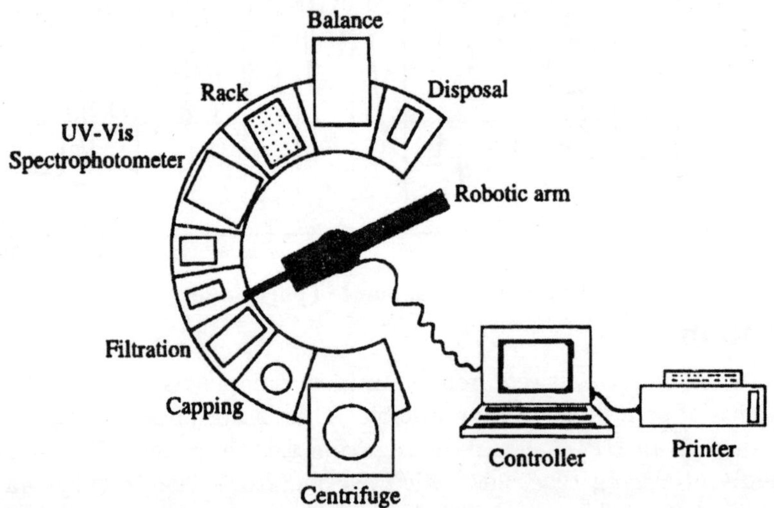

Fig. 1.12. Schematic representation of zymate robotic system (zymark corporation). The robotic arm is programmed to move between the various modules required for a particular analysis.

limited to quantitative photometric procedures on liquid samples. Robotic systems were developed to extend the range of tasks which could be performed. The Zymark Corporation based its robots on the fact that all laboratory procedures, regardless of complexity, can be broken down into a series of simple steps called Laboratory Unit Operations, LUOs.

The various modules or workstations required to perform these tasks can be linked by a robotic arm, which is programmed to transfer samples between them in any sequence. This provides a flexible, custom-built system by which a method can be fully automated irrespective of the number of steps and the sequence required. It combines robotics, computer control and instrumentation and produces a final report. Less complex workstations are also available which perform more limited functions, either individually or in sequences, *e.g.* evaporation of organic solvents, small volume pipetting, reactions of hazardous reagents and auto-injection in column chromatographic techniques.

Table 1.2. Laboratory Unit Operations (LUOs)

Class	*Definition*	*Examples*
Weighing	Quantitative sample mass measurement	Using a balance
Homogenization	Reducing sample particle size and creating a uniform sample	Sonication, grinding
Manipulation	Physical handling of laboratory materials	Moving test-tubes, capping, uncapping
Liquid handling	Physical handling of liquid reagents	Pipetting, transferring, dispensing
Conditioning	Modifying and controlling the sample environment	Incubating, timing, shaking
Measurement	Direct measurement of physical properties	Absorbance, fluorescence, pH
Separation	Coarse mechanical or precision separation	Filtration, extraction, centrifugation
Control	Use of calculation and logical decisions in laboratory procedures	—
Data reduction	Conversion of raw analytical data to useable information	Peak integration, spectrum analysis
Documentation	Creating records and files for retrieval	Notebooks, listings, computers

CHAPTER

2 Analysis of Biological Fluids

Control of the composition of the culture media in which bacterial or mammalian cells are grown is important. Nutritional depletion, accumulation of side products, and pH may change significantly during the incubation period, thereby reducing cell viability as well as interfering with the production of the desired product. This chapter is a summary of techniques and protocols to quantify most of the standard ingredients in complex biological media.

MATERIALS (SELECTED)

Glutamate dehydrogenase from bovine liver [EC 1,4, 1, 3] [Sigma, 49390]

INT, p-iodonitrotetrazolium violet [Sigma, i8377]

Diaphorase [Sigma, D5540]

Aspartate aminotransferase (AST/GOT) [Sigma, 505]

Asparaginase [Sigma, A4887]

Glutaminase [Sigma, G8880]

Phosphatase, alkaline [Sigma, P4978]

BCIP/NBT solution, premixed [Sigma, B6404]

ANALYTES

Ions and Metals

Ammonia (-ium)

Enzymatic Methods

The determination of ammonia is of considerable value in fermentation control as well as clinical diagnosis. If any of the processes involved in ammonia's metabolic disposal is impaired, the level of ammonia, especially ammonium ion, will increase, leading to ammonia intoxication.

In eucaryontic cell fermentation, this induces rapid cell death in humans. It induces complications, including nausea, vomiting, ataxia, convulsions, lethargy, coma, and finally death. Mostly, patients with hepatic intoxication or reduced liver function are affected, as the liver is the site of the conversion of ammonium ion to urea. Therefore, it is particularly important to monitor the concentration of both ammonia and carbon dioxide. Ammonia is determined via reductive amination using glutamate dehydrogenase (GLDH). A decrease of NADPH is measured photometrically at 340 nm.

$$2-\text{Oxoglutarate}+\text{NH}_3+\text{NADPH}\xrightarrow{\text{Glutamate dehydrogenase}}\text{Glutamate}+\text{NADP}^+$$

For this assay relatively minor amounts of glutamate dehydrogenase might be used, as there is only one enzymatic step involved.

Protocol 1 Ammonia

1. Freshly prepare before use NaH_2PO_4 buffer 50 mM, pH 7, with 2-oxoglutarate, 0.11 mM, and NADH, 0.7 mM.
2. Prepare a series of ammonia dilution in expected concentration range.
3. Mix 950 µl of buffer, 40 µl GIDH, and 10 µl of samples or standard ammonia dilution.
4. Shake well and incubate for 10 min.
5. Read the absorbance at 340 nm against a reagent blank.
6. Plot calibration graph from standard series and determine concentration of unknown sample.

Electrochemical Methods

The NH_4^+-selective membrane is based on a mixture of the antibiotics Nonactin and Monactin, which are neutral carriers in a PVC matrix.

Bicarbonate

In fermentation broth as well as in body fluids (*e.g.,* blood, plasma or serum) at about pH 8, CO_2 is present substantially as bicarbonate ion. In animals HCO_3 is carried via the blood to the lungs, which release it into the air. The kidneys filter some of the bicarbonate through the urine, but the lungs release most of it as carbon dioxide each time we exhale. The level of bicarbonate may be too high or too low if the lungs are not functioning properly or in fermentation if carbon dioxide *versus* base supply is not balanced. Abnormal bicarbonate may be seen with certain metabolic problems.

Enzymatic Methods

1. $\text{Phosphoenolpyruvate}+\text{HCO}_3^- \xrightarrow{\textit{Phosphoenolpyruvate carboxylase}} \text{Oxalacetate}+\text{H}_2\text{PO}_4$
2. $\text{Oxalacetate}+\text{NADH}\xrightarrow{\textit{Malatdehydrogenase}}\text{Malate}+\text{NAD}^+$

Protocol 2 Bicarbonate

1. The specimen is first alkalinized to convert all carbon dioxide and carbonic acid to HCO^{3-}
2. Phosphoenolpyruvate and HCO^{3-} reacts with phosphoenolpyruvate-carboxylase (1) (PEPC) and malat dehydrogenase (MDH) to Malate (2).
3. The decrease in absorbance of NADH at 340 nm is proportional to the total carbon dioxide content.

   CO_2 - electrodes

A pCO_2-sensitive electrode determines the gaseous CO_2 after acidification. The rate of pH change of the buffer inside the membrane of the measuring electrode is taken as the measure of total CO_2 in the sample.

Calcium

Calcium is the most abundant mineral in animals, 99% of which is located in the bones and teeth. Calcium is needed to form teeth and bones and is also required for muscle contraction, blood clotting, and signal transmission in nerve cells. Its probably most well-known role is to prevent osteoporosis. Severe deficiencies of both calcium and vitamin D is called "rickets" in children and "osteomalacia" in adults. Most laboratories measure calcium with photometric methods, but atomic absorption spectrometry (AAS) or ion-sensitive electrodes also have been used.

Photometric Methods

Metal-complexing dyes such as O-Cresolphthalein or Arsenzo III form stable complexes with Ca^{2+} in alkaline or acidic solution. O-Cresolphthalein forms a red complex with calcium in alkaline solution with an absorption peak between 570 and 580 nm.

The sample is first acidified to release protein-bound and complexed calcium. Organic base is added to buffer the reaction and to produce an alkaline pH (1). Arsenazo III exhibits a high affinity for calcium, at pH 6, that is much higher than for magnesium. The solution must be thoroughly buffered because binding of Ca^{2+} to Arsenazo III can be influenced by buffer and concentrations (2).

1. Calcium + Cresolphthalein Complexone $\xrightarrow{\textit{Alkaline pH}}$ Calcium-Cresolphthalein Complexone
2. Calcium + Arsenazo III $\xrightarrow{\textit{Acid pH}}$ Calcium-Arsenazo III Complex

Ion-sensitive Electrodes

Calcium ion-sensitive electrodes use liquid membranes containing the calcium-sensitive ionophore dissolved in an organic liquid trapped in a polymeric matrix (PVC).

Atomic Absorption Spectrometry

Total calcium is determined after diluting the sample with lanthanum chloride. The diluted specimen is aspirated into an air-acetylene flame, in which the ground-state calcium ions absorb light from a calcium hollow lamp. The absorption at 422.7 nm is directly proportional to the ground state calcium atoms in the flame.

Chloride

Chloride represents a major extra cellular anion, functions as an electrolyte balance associated with sodium, and preserves the electrolyte neutrality/maintenance of homeostasis. Spectroscopy and ion-sensitive electrodes are the most common methods for determination of chloride.

Spectrophotometric Methods

Chloride ions react with undissociated mercury thiocyanate to form undissociated mercuric chloride and free thiocyanate (1). The thiocanate ions react with Fe^{3+} to form a reddish-brown complex of ferric thiocyanate (2) with absorption maximum at 480 nm. Addition of perchloric acid increases the color intensity.

1. $2Cl^- + Hg(SCN)_2 \longrightarrow HgCl_2 + 2\ SCN^?$

2. $SCN^- + Fe_3^{\ +} \longrightarrow Fe(SCN)_3$

Ion-sensitive Electrodes

Solvent polymeric membranes that incorporate quaternary ammonium salt anion-exchangers (e.g., tri-*n*-octyl propyl ammonium chloride) are used for construction of Cl^- sensitive electrodes.

Iron

At acid pH and in the presence of a reducing agent, transferrin-bound iron is released and reduced from Fe^{3+} to Fe^{2+}. It reacts with a specific chromogen (ferrozine or bathophenathroline), forming a colored complex. The color formed is proportional to the serum total iron concentration. For determination of the serum unsaturated iron-binding capacity (UIBC), a standard amount of ferrous iron is added to serum at alkaline pH and is sequestered by transferrin, filling all available binding sites on the protein.

The remaining unbound ferrous ions are then measured colorimetrically by use of ferrozine or bathophenathroline. The difference between the amount of unbound iron and the total amount of iron added is the UIBC. The serum total iron-binding capacity (TIBC) is the sum of the serum total iron and the unsaturated iron-binding capacity. This is a measurement of the maximum concentration of iron that serum proteins can bind.

Magnesium

Today laboratories measure magnesium with photometric methods, but atomic absorption spectrometry (AAS) and ion-sensitive electrodes also have been used.

Photometric Methods

A number of metallochromic indicators that change color on selectively binding magnesium have been used for measuring it in biological samples. Magnesium forms a colored complex with Calmagite, the most commonly used indicator, in alkaline solution.

The intensity of the color, measured at 520 nm; is proportional to the magnesium concentration measured at 530 to 550 nm. EGTA is added to reduce interference by calcium. Xylidyl blue (Magon) forms a red complex with magnesium under alkaline conditions. This complex is measured at 520 nm. The intensity of the color is proportional to the magnesium concentration in the sample.

Atomic Absorption Spectrometry

Magnesium is determined after diluting the specimen with a lanthanum chloride solution to eliminate interference from anions. Then the sample is aspirated into an air-acetylene flame, in which the ground-state magnesium ions absorb light from a magnesium hollow lamp. The absorption at 285.2 nm is directly proportional to the magnesium atoms in the flame.

Ion-sensitive Electrodes

These instruments use ionophores in a hydrophobic membrane for measuring Mg^{2+} concentration. But current ionophores have insufficient selectivity for Mg^{2+} over Ca^{2+}. Free calcium is simultaneously determined and used to correct free Mg^{2+} levels.

Phosphate

All common methods of phosphate determination are based on the reaction of phosphate ions with ammonium molybdate at acid pH to form a phosphomolybdate complex (1). The colourless phosphomolybdate complex can be measured either directly at 340 nm or after reduction by various reducing agents to produce molybdenum blue. Reduction of this compound produces a blue phosphomolybdenum complex measured at 620 to 700 nm (2).

1. Phosphate + Ammonium Molybdate + H^+ $\longrightarrow$ Ammonium Phosphomolybdate Complex

2. AP-complex + Aminonaphtholsulfonic Acid $\longrightarrow$ blue complex

Sodium and Potassium

Determination of sodium and potassium in body fluids is carried out with flame emission spectroscopy (FES), atomic absorption spectroscopy (AAS), spectrophotometrically, or electrochemically via ion-selective electrodes.

Flame Emission Spectrophotometry

Methods using flame photometry have been developed for both sodium and potassium, but today it is no longer a common laboratory method.

Spectrophotometric Methods

The spectral shift produced when either sodium or potassium binds to a macrocyclic chromphore is detected spectrophotometrically. Macrocyclic ionophores (crown ethers, cryoptands, etc.) are molecules whose atoms are organized to form a specific structure into which metal ions fit and bind with high affinity. For example, ChromoLyte is a sodium-specific ionophore. Valinomycin in conjugation with a pH indicator is used to determine serum potassium concentration.

Ion-sensitive Electrodes

Analyzers contain sodium-sensitive electrodes with glass membranes. Potassium-sensitive electrodes contain liquid ion-exchange membranes that incorporate valinomycin. The measuring system is calibrated by introduction of calibrator solutions containing Na^+ and K^+.

Low-molecular-weight Components

Cholesterol

Enzymatic methods are the most commonly used methods for cholesterol measurements. Cholesteryl esters are hydrolyzed by an esterase to free cholesterol and fatty acids (1). The 3-OH group of cholesterol is then oxidized to a ketone by cholesterol oxidase (2). The resulting hydrogen peroxide is measured in a peroxidase-catalyzed reaction forming a quinoneimine dye (3).

1. Cholesterol Ester $\xrightarrow{\text{Cholesterol esterase}}$ Cholesterol + Fatty Acids

2. Cholesterol + O_2 $\xrightarrow{\text{Colesterol oxidase}}$ Cholest-4-en-3-one + H_2O_2

3. $2H_2O_2$ + 4-Aminoantipyridine + Phenol $\xrightarrow{\text{Peroxidase}}$ Quinone imine Dye + 4 H_2O

Cholesterol, High-density Lipoprotein

High-density lipoprotein (HDL) cholesterol is measured after serum low-density (LDL) and very low-density (VLDL) lipoproteins are selectively precipitated by dextrane sulfate or phosphotungstic acid and removed by centrifugation. In another method, the precipitant is complexed with magnetic particles.

The supernatant contains cholesterol associated with the soluble HDL fraction and is measured by a standard method described as above. In a direct method, an anti-human (β-lipoprotein antibody binds to lipoproteins (LDL, VLDL, and chylomicrons). The antigen-antibody complexes formed block enzyme reactions when an enzymatic cholesterol reagent is added. The blue color complex formed in that reaction is measured photometrically at 600 nm.

Glucose

Almost all commonly used techniques for glucose determination are enzymatic.

Hexokinase Method

ATP phosphorylates glucose in the presence of hexokinase and magnesium (1). The glucose-6-phosphate formed is oxidized by glucose-6-phosphate dehydrogenase in the presence of NAD^+ or $NADP^+$. The amount of NADH generated is directly proportional to the concentration of glucose in the sample and is measured by absorbance at 340 nm (2).

Alternatively, an oxidation/reduction system containing phenazine methosulfate (PMS) and iodonitrotetrazolium (INT) reacts with the NADH produced in the reaction. The reduction of INT forms a colored INT-formazan that is measured at 520 nm (3, 4).

1. Glucose + ATP $\xrightarrow{\text{Hexokinase}}$ Glucose-6-phosphate + ADP

2. Glucose-6-phosphate + NAD^+ $\xrightarrow{G\text{-}6\text{-}PDH}$ 6-Phosphogluconate + NADH + H^+

3. NADH + PMS $\longrightarrow$ NAD^+ + PMSH

4. PMSH + INT $\longrightarrow$ PMS + INT-Formazan

Glucose Oxidase Method

Glucose is oxidized to gluconic acid and hydrogen peroxide by the enzyme glucose oxidase (5). The hydrogen peroxide produced reacts with a chromogenic oxygen acceptor, such as 4-aminoantipyridine (6) or dianisidine (7), and results in a colored compound that can be measured.

5. $\text{Glucose} + 2H_2O + O_2 \xrightarrow{\text{Glucose Oxidase}} \text{Gluconic Acid} + H_2O_2$

6. $2H_2O_2 + \text{4-Aminoantipyridine+p-Hydroxybenzene Sulfonate} \xrightarrow{\text{Peroxidase}} \text{Quione imine Dye} + 4H_2O$

7. $H_2O_2 + \text{o-Dianisidine} \xrightarrow{\text{Peroxidase}} \text{Oxidized o-Diansidine} + H_2O$

Glucose Dehydrogenase Method

β-D-glucose is oxidized to gluconolactone by glucose dehydrogenase (GDH). The amount of NADH generated is proportional to the glucose concentration and can be measured photometrically at 340 nm.

8. $\text{Glucose} + NAD^+ \xrightarrow{\text{Glucose dehydrogenase}} \text{D-Gluconolactone} + NADH + H^+$

Glutamate

L-Glutamic acid plays an important role in metabolism; it is, e.g., the precursor of other amino acids. L-Glutamic acid is part of many fermenter solutions and occurs in large quantities in most proteins.

1. $\text{L-Glutamate} + NAD^+ + H_2O \xrightarrow{\text{Glutamate dehydrogenase}} \text{2-oxoglutarate} + NADH + NH^{4+}$

2. $NADH + H^+ \text{ INT} \xrightarrow{\text{Diaphorase}} NAD^+ + \text{formazan}$

The formation of NADH in the first reaction is not favored; therefore, the produced NADH is oxidized, producing the colored formazan. An alternative flow injection system using immobilized Glutamate Dehydrogenase (GIDH) and aspartate aminotransferase is reported.

Protocol 3 L-Glutamic acid

1. Prepare buffer 350 mM imidazole / HCl pH 8.6, 1 g/1 NAD^+, 0.1 g/1 INT.
2. Prepare standard glutamate concentration series in 50 mM imidazole pH 7 buffer in the expected sample range.
3. Mix 960 μl buffer, 10 μl glutamate solution (standard or sample), 10 μl Diaphorase, and 10 μl GIDH.
4. Measure at 520 nm after 5 min.
5. Plot calibration from standard series and determine concentration of unknown sample.

Remarks and Troubleshooting

The method is specific for L-glutamic acid, and the detection limit is 0.02 mg/l. High concentrations of ammonium ions or reducing substances disturb the reaction, as do very acid or basic samples, which influence the working pH. For high glutamate concentrations, the observed absorbance will be non-linear. In complex matrices it is useful to monitor the assay during the 10 min. of incubation. In the case of saturation, the enzyme or substrate needs to be reduced.

Glutamine

The major carbon and nitrogen sources in cultivation media for mammalian cells are glucose and L-glutamine. After conversion of glutamine to glutamate, a standard glutamate assay is used to quantify glutamine. An appropriate glutaminase exists, but the assay also can be run with the more cost-efficient asparaginase. An alternative flow injection system using the immobilized enzymes Glutamate Dehydrogenase (GIDH), aspartate aminotransferase, and asparaginase is reported.

1. L-Glutamine + $H_2O \xrightarrow{Glutaminase/Asparaginase}$ Glutamate + NH_3
2. L-Glutamate + NAD^+ + $H_2O \xrightarrow{Glutamate\ dehydrogenase}$ 2-oxoglutarate + NADH + NH_4^+
3. NADH + H^+INT + $\xrightarrow{Diaphorase}$ NAD^+ + formazan

Protocol 4 L-Glutamine

1. Prepare buffer 350 mM imidazole / HCl pH 8.7, 1 g/1 NAD^+, 0.1 g/l INT.
2. Prepare buffer 50 mM imidazole / HCl pH 5.0.
3. Prepare standard glutamine concentration series in 50 mM imidazole pH 5 buffer in the expected sample range.
4. Mix 480 ml pH 5 buffer, 10 µl glutamate solution (standard or sample), and 10 µl asparaginase.
5. Incubate at room temperature for 5 min.
6. Add 480 µl pH 8.7 buffer, 10 µl diaphorase, and 10 ml GIDH.
7. Measure at 520 nm after 5 min.
8. Plot calibration graph from standard series and determine concentration of unknown sample.

Remarks and Troubleshooting

The method is specific for L-glutamine; aspartate does not disturb because it is not oxidized by the GIDH. Clearly, the presence of glutamate will interfere with the assay, and the value for glutamate has to be determined with a separate experiment and subtracted from the "total" glutamine value. The detection limit is 0.05 mg/l. High concentrations of ammonium ions or reducing substances disturb the reaction, as do very acid or basic samples, which influence the working pH.

Lactate

Lactate is oxidized to pyruvate and hydrogen peroxide by lactate oxidase (1). In the presence of the H_2O_2 generated, peroxidase catalyzes the oxidation of a chromogen to produce a colored dye with absorption maximum at 540 nm (2). The increase in absorbance at 540 nm is directly proportional to the lactate concentration in the specimen. Lactate is oxidized to pyruvate by lactate dehydrogenase (LD) in the presence of NAD^+. The NADH formed in this reaction is measured spectrophotometrically at 340 nm (3). The equilibrium of the reaction normally lies on the lactate side. At pH 9 to 9.6, an excess of NAD^+, and use of tris-buffer, the equilibrium can be shifted to the right.

1. Lactate + O_2 $\xrightarrow{\text{Lactate oxidase}}$ Pyruvate + H_2O_2
2. 2 H_2O_2 + 4-Aminoantipyridine + 1,7-Dihydroxynaphthalene $\xrightarrow{\text{Peroxidase}}$ Red Dye
3. Lactate + NAD^+ $\xrightarrow{\text{Lactate dehydrogenase}}$ Pyruvate + NADH + H^+

Triglycerides

Triglycerides are measured enzymatically directly in biofluids such as plasma or serum. In all of the methods, the first step is the lipase-catalyzed hydrolysis of triglycerides to glycerol and fatty acids (1). Glycerol is then phosphorylated by glycerokinase (2). In the most used methods, glycerophosphate is oxidized to dihydroxyacetone and H_2O_2 (3), which is measured in a peroxidase-catalyzed reaction forming a quinone imine dye (4). The increase in absorbance, measured at 540 run, is directly proportional to the triglyceride content in the sample.

Free glycerol is assayed using a reagent without lipase. True triglyceride is calculated by subtracting the free glycerol concentration in the sample from the total triglycerides. Alternatively, glyceroposphate can be measured in an NADH-producing reaction and the NADH measured spectrophotometrically at 340 nm (5) or in a diaphorase-catalyzed reaction forming a product measured at 500 nm (6). Another method measures the ADP produced in reactions shown in (7) and (8). The decrease of NADH is measured photometrically at 340 nm.

1. Triglyceride $\xrightarrow{\text{Lipase}}$ Glycerol + Fatty Acids
2. Glycerol + ATP $\xrightarrow{\text{Glycerokinase}}$ Glycerol-1-phosphate + ADP
3. Glycerol-1-phosphate + O_2 $\xrightarrow{\text{Glycerophosphate oxidase}}$ Dihydroxyacetone Phosphate + H_2O_2
4. H_2O_2 + 4-Aminoantipyridine + Phenol- $\xrightarrow{\text{Peroxidase}}$ Quinone imine Dye + 2 H_2O
5. Glycerol-1-Phosphate + NAD + $\xrightarrow{\text{Glycerophosphate dehydrogenase}}$ Dihydroxyacetone Phosphate + NADH + H^+
6. NADH + Tetrazolium Dye $\xrightarrow{\text{Diaphorase}}$ Formazan + NAD^+
7. ADP + Phospho-enol Pyruvate $\xrightarrow{\text{Pyruvate kinase}}$ ATP + Pyruvate
8. Pyruvate + NADH + H^+ $\xrightarrow{\text{Lactate dehydrogenase}}$ Lactate + NAD^+

Urea

Enzymatic Methods

These methods are based on hydrolysis of urea with urease (1) followed by ammonium quantitation with glutamate dehydrogenase (GLDH) (2). A decrease of NADH concentration from the GLDH-reaction is measured photometrically at 340 nm in either an equilibrium or kinetic mode.

1. Urea + $2H_2O$ $\xrightarrow{\text{Urease}}$ 2 NH_4^+ + CO_3^{2-}
2. NH_4^+ + 2 – Oxoglutarate + NADH + H^+ $\xrightarrow{\text{GLDH}}$ Glutamate + H_2O + NAD^+

Chemical Methods

Most chemical methods for urea are based on the condensation of urea with diacetyl monoxime to form pink diazine (Fearon reaction), which absorbs strongly at 515 to 540 nm. Ferric ions and thiosemicarbazide are added to the system to enhance the color.

$$\text{Urea} + \text{Diacetyl} \xrightarrow{H^+/heat} \text{Diazine} + 2H_2O$$

Uric Acid

Uric acid reacts with uricase at pH 9.4 to form allantoin. The absorbance of the reaction is measured before and after the addition of uricase. Since uric acid absorbs at 292 nm and the end products do not react; the decrease in absorbance is proportional to the uric acid concentration.

Many enzymatic assays for uric acid in serum involve a peroxidase system coupled with oxygen acceptors to produce a chromogen. The most common oxygen acceptor is a 4-aminophenazone together with a substituted phenol.

$$\text{Uric Acid} + 2H_2O + O_2 \xrightarrow{Uricase} \text{Allantoin} + CO_2 + H_2O_2$$

Proteins

Total Protein
Biuret Method

This method is based on the presence of peptide bonds in all proteins. The copper ions in the Biuret reagent react with peptide bonds of serum proteins in alkaline medium to form a purple color with an absorbance maximum at 540 nm. The intensity of the color is proportional to the protein concentration.

Lowry Method

A pyrogallol red-molybdate complex binds basic amino acid groups (Tyr, Trp) of protein molecules, which results in a shift in the reagent absorbance. The increase in absorbance at 600 nm is directly proportional to protein concentration in the sample. The reagent is intended for the quantitative determination of total protein in various fluids. A modification of the Lowry micro method uses Folin & Ciocalteau's phenol reagent to intensify the color of the purple-blue complex.

The absorbance, which is read at 550 to 750 nm, is used to determine results from a standard curve. Another modification of the Lowry method utilizes SDS for dissolving lipoproteins and offers excellent recovery of membrane proteins. An optional deoxycholate-trichloroacetic acid (TCA) precipitation technique permits removal of interfering substances such as tris, sucrose, EDTA, etc. The absorbance, read at 500 to 800 nm, is used to determine protein concentration.

Bradford Method

Coomassie brilliant blue G reacts with proteins in an alcohol-acid medium to form a blue-colored protein dye complex. The absorbance is measured at 595 nm and is proportional to the protein concentration.

Protocol 5 Standard protein quantification

1. Prepare dye reagent by diluting 1 part Dye Concentrate with 4 parts distilled DDI water.
2. Filter through Whatman #1 filter (or equivalent) to remove particles.
3. The diluted reagent may be used for approximately 2 weeks when kept at room temperature.
4. Prepare three to five dilutions of a protein standard, which is representative of the protein solution to be tested.
5. The linear range of the assay for BSA is 0.2 to 0.9 mg/ml, whereas with IgG the linear range is 0.2 to 1.5 mg/ml.
6. Transfer 100 µl of each standard and sample solution into a clean, dry test tube.
7. Add 5.0 ml of diluted dye reagent to each tube and vortex.
8. Incubate at room temperature for at least 5 min.
9. The absorbance will increase over time. Thus, samples should incubate at room temperature for no more than 1 h.
10. Measure absorbance at 595 nm.

Protocol 6 Protein in microtiter plates

1. Prepare dye reagent as cited above.
2. The linear range of this microtiter plate assay is 0.05 mg/ml to approximately 0.5 mg/ml.
3. Protein solutions are normally assayed in duplicate or triplicate.
4. Transfer 10 µl of each standard and sample solution into separate microtiter plate wells.
5. Add 200 µl of diluted dye reagent to each well.
6. Mix the sample and reagent thoroughly using a microplate mixer.
7. Alternatively, use a multi-channel pipette to dispense the reagent. Depress the plunger repeatedly to mix the sample and reagent in the well. Replace with clean tips and add reagent to the next set of wells.
8. Incubate at room temperature for at least 5 min.
9. Absorbance will increase over time; samples should incubate at room temperature for no more than 1 h.
10. Measure absorbance at 595 nm.

Occasionally, a protein is assayed with exceptionally low color response to the Bradford reagent (*e.g.*, gelatine). Nevertheless, quantitation of the protein is possible when the sample-to-dye ratio is changed. By using the sample-to-dye ratio of 800 µl sample + 200 µl dye reagent concentrate, a usable standard curve with reduced sensitivity is obtained. If the absorbance is very low, the dye concentrate may be old. If it is over one year old, replace with a new bottle of reagent. Sometimes the sample may contain a substance that interferes with the reaction, such as detergents or sodium hydroxide. For low-molecular-weight proteins or peptides (< 5000 Daltons), this assay is not well suited.

Albumin

Albumin is the cheapest, well-defined single protein and represents the most abundant protein in blood (35 to 58 g/l, i.e., 55% to 65% of the total protein content). It has diverse functions, including the maintenance of the colloid osmotic pressure of the blood. Moreover, it is responsible for transportation of various ions, hormones, and drugs. Various chemical and immunological methods are available for the analysis of serum albumin. Mostly, methods are applied that are based on interaction with dyes, such as bromcresol green (BCG) and bromcresol purple (BCP). Bromcresol green binds quantitatively with human serum albumin at pH 4.2 to form an intensive blue-green complex that is determined spectrophotometrically at 628 nm. At pH 4.2, albumin acts as a cation to bind the anionic dye.

Bromcresol purple forms a stable complex with albumin at pH 5.2. The absorbance maximum is measured at 600 nm. The intensity of the color produced is directly proportional to the albumin concentration in the sample.

Protocol 7 Albumin

1. Transfer 3 ml BCG reagent into a spectrophotometer cell.
2. Add 20 µl of serum sample or standard.
3. Mix well and incubate for 2 min.
4. Read the absorbance at 640 nm against a reagent blank.
5. Plot calibration graph from standard series and determine concentration of unknown sample.

Hemoglobin

Hemoglobin reacts with Drabkin's reagent, which contains potassium ferricyanide, potassium cyanide, and sodium bicarbonate. The Fe^{2+} of hemoglobin is oxidized to Fe^{3+} of methemoglobin by ferricyanide (1). Methemoglobin is converted into stable cyanomethemoglobin (2) by potassium cyanide (KCN). The absorbance of this derivative at 540 nm is proportional to the hemoglobin content.

For measurements of hemoglobin in complex fluids, 3, 3′, 5, 5′-tetramethyl benzidine (TMB) is used as chromogen. Hemoglobin catalyzes the oxidation of TMB by hydrogen peroxide. The color formed in that reaction is proportional to the hemoglobin concentration in the test sample.

1. $HBFe^{2+} + Fe^{3+}(CN)_6^{3-} \longrightarrow HBFe^{3+} + Fe^{2+}(CN)_6^{4-}$
2. $HBFe^{3+} + CN \longrightarrow HBFe^{3+}CN$

Enzymes

Alanine Aminotransferase

1. $\text{L-Alanine} + \text{2-Oxoglutarate} \xrightarrow{\textit{Alanine aminotransferase}} \text{Pyruvate} + \text{Glutamate}$
2. $\text{Pyruvate} + \text{NADH} \xrightarrow{\textit{Lactate dehydrogenase}} \text{Lactate} + \text{NAD}^+$
3. $\text{Pyruvate} + \text{2,4-Dinitrophenylhydrazine} \longrightarrow \text{Dinitrophenylhydrazone}$

Protocol 8 Alanine aminotransferase (ALT)

1. Alanine aminotransferase (transaminase) catalyzes the reaction from L-alanine and 2-oxoglutarate, forming glutamate and pyruvate (1).
2. The pyruvate produced is reduced to lactate by lactate dehydrogenase (LDH) with simultaneous oxidation of NADH to NAD^+ that can be measured at 340 nm (2).
3. The decrease in NADH concentration is directly proportional to the ALT-activity.
4. The pyruvate that formed can react with 2,4-dinitrophenylhydrazine.
5. The resulting hydrazone of pyruvate is highly colored and its absorbance at 490-520 nm is proportional to ALT-activity (3).

Alkaline Phosphatase

Alkaline phosphatase is a broadly used enzyme that is applied for an easy, optically detectable interaction analysis. It catalyzes the hydrolysis of p-nitrophenyl phosphate (colorless) forming phosphate and free p-nitrophenol (yellow) under alkaline conditions. The rate of increase in absorbance at 405 nm that is due to the formation of p-nitrophenol is proportional to the AP activity.

$$\text{p-Nitrophenyl phosphate} \xrightarrow{\textit{Alkaline phosphatase}} \text{p-Nitrophenol} + P_i$$

Protocol 9 Alkaline phosphatase

1. Mix one part each of BCIP (5-bromo-4-chloro-3-indolyl-phosphate) concentrate and NBT (nitroblue tetrazolium) concentrate with 10 parts of Tris buffer solution in a glass container prior to use (*i.e.,* 1 ml BCIP + 1 ml NBT + 10 ml Tris / HCl pH 8.2 Buffer). Alternatively, a premixed BCIP/NBT solution can be obtained from Sigma.
2. Prepare standard alkaline phosphatase concentrations in expected sample range.
3. Add 10 µl of samples or standard alkaline phosphatase dilution.
4. Mix well and incubate for 2 min.
5. Read the absorbance at 405 nm against a reagent blank.
6. Plot calibration curve from standard series and determine concentration of unknown sample.

Aspartate Aminotransferase

1. $\text{L-Aspartate} + \text{2-Oxoglutarate} \xrightarrow{\textit{Aspartate aminotransferase}} \text{Oxalacetate} + \text{Glutamate}$
2. $\text{Oxalacetate} + \text{NADH} \xrightarrow{\textit{Malate dehydrogenase}} \text{Malate} + NAD^+$
3. $\text{Oxalacetate} + \text{2, 4-Dinitrophenylhydrazine} \longrightarrow \text{Dinitrophenyl-hydrazone}$

Protocol 10 Aspartate aminotransferase (AST)

1. Aspartate aminotransferase (transaminase) catalyzes the reaction from L-aspartate and 2-oxoglutarate, forming glutamate and Oxalacetate (1).
2. The produced oxalacetate is reduced to malate by malate dehydrogenase (MDH).

3. Simultaneously, NADH is oxidized to NAD^+.
4. The reaction is measured at 340 nm (2).
5. The decrease in NADH concentration is directly proportional to the AST activity.
6. The pyruvate that is formed can react with 2,4-dinitrophenylhydrazine.
7. The resulting hydrazone of pyruvate is highly colored and its absorbance at 490-520 nm is proportional to AST activity (3).

Creatine Kinase

Creatine phosphate is converted to creatine with a phosphorylation of ADP to ATP by creatine kinase (CK) (1). The ATP produced is necessary for the hexokinase-catalyzed reaction of glucose to glucose-6-phosphate, which is converted to glucose-6-phosphate by glucose-6-phosphate dehydrogenase (G-6-PDH) (2). The rate of increase in absorbance at 340 nm that is due to the formation of NADH is directly proportional to the creatine kinase activity (3).

1. ADP + Phospho-creatine $\xrightarrow{\text{Creatine kinase}}$ ATP + Creatine
2. ATP + Glucose $\xrightarrow{\text{Hexokinase}}$ ADP + Glucose-6-Phosphate
3. Glucose-6-Phosphate + NAD^+ $\xrightarrow{\text{G-6-PDH}}$ 6-Phosphogluconate + NADH

Creatine Kinase Isoenzymes

CK isoenzymes are separated by electrophoresis, ion-exchange chromatography, or immunological methods.

Electrophoretic Methods

The most commonly used technique is based on electrophoresis on agarose gels or cellulose acetate. The isoenzyme bands are visualized via incubation in a medium that promotes the following reactions: -

1. ADP + Creatine Phosphate $\xrightarrow{\text{Creatine kinase}}$ Creatine + ATP
2. ATP + Glucose $\xrightarrow{\text{Hexokinase}}$ ADP + Glucose-6-Phosphate
3. Glucose-6-Phosphate + NAD^+ $\xrightarrow{\text{Glucose 6-phosphate dehydrogenase}}$ 6-Phosphogluconate + NADH
4. NADH + Phenazine Methosulfatc (oxid.) $\longrightarrow$ NAD^+ + PMS (red.)
5. PMS (red.) + Tetrazolium Dye $\longrightarrow$ PMS (oxid.) + Formazan

The highly colored, insoluble formazan (NET, MTT, or INT) localizes in electrophoresis zones of activity. The electrophoretic patterns may be visualized by photo-scanners.

Immunological Methods

Immunological methods for measuring the CK iso-enzymes require specific antisera against M and B subunits. Immune-inhibition techniques are simpler and faster than immune-precipitation.

Serum is incubated with a CK-MB reagent containing antibody specific to the CK-M subunit, which completely inhibits the CK-M monomer. The activity of the CK-B that is not inhibited by the antibody is measured by the following reaction sequence:

1. ADP + Creatine Phosphate $\xrightarrow{\textit{Creatine kinase}}$ Creatine + ATP
2. ATP + Glucose $\xrightarrow{\textit{Hexokinase}}$ ADP + Glucose-6-Phosphate
3. Glucose-6-Phosphate + NAD^+ $\xrightarrow{\textit{Glucose-6-phosphate dehydrogenase}}$ 6-Phosphogluconate + NADH

The rate of change in absorbance, that is measured at 340 nm, is directly proportional to the CK activity.

γ-Glutamyl Transferase

γ-Glutamyl transferase (γ-GT) catalyzes the transfer of glutamyl from γ-glutamyl-3-carboxy-4-nitroanilide to glycylglycine at pH 8.2. The increase in absorbance at 405 nm that is due to the p-nitroaniline formed in the reaction is measured spectrophotometrically.

γ-Glutamyl-3-Carboxy-4-Nitroanilide + Glycylglycine $\xrightarrow{\gamma\textit{-Glutamyl-Transferase}}$

γ-Glutamylglycylglycine + 5-Amino-2-Nitrobenzoate

Lactate Dehydrogenase

Lactate dehydrogenase (LDH) catalyzes the conversion of 1-lactate to pyruvate at alkaline pH 9.4. The rate of increase in absorbance at 340 nm that is due the formation of NADH is directly proportional to the LDH activity. In the presence of 2,4-dinitrophenylhydrazine, the pyruvate produces a highly colored phenyl hydrazone. The absorbance of that color formed is directly proportional to the LDH activity.

Lactate + NAD^+ $\xrightarrow{\textit{Lactate dehydrogenase}}$ Pyruvate + NADH + H^+

Lipase

Titrimetric, turbidimetric, spectrophotometric, and fluorometric methods, and methods based on immunoassays, are commercially available for the measurement of Upases in biological samples. The most popular are spectrophotometric and titrimetric assays.

Spectrophotometric Method

Pancreatic lipase catalyzes the reaction from 1,2-diacylglycerol to 2-monoacyl-glycerol and fatty acids at pH 8.7 (1). In additional reaction steps (2)-(4), hydrogen peroxide is produced, which is reduced by peroxidase together with oxidation of a redox dye (5). The rate of increase in absorbance at 550 nm that is due to the formation of quinone diimine dye is directly proportional to the lipase activity in the specimen.

1. 1,2-Diacylglycerol + H_2O $\xrightarrow{\textit{Pancreatic lipase}}$ 2-Monoacylglycerol + Fatty Acid
2. 2-Monoacylglycerol + H_2O $\xrightarrow{\textit{Monoglyceride lipase}}$ Glycerol + Fatty Acid
3. Glycerol + ATP $\xrightarrow{\textit{Glycerol kinase}}$ Glycerol-3-Phosphate + ADP

4. Glycerol-3-Phosphate + $O_2 \xrightarrow{Glycerol-3-Phosphate\ oxidase}$ Dihydroxyacetone Phosphate + H_2O_2

5. 2 H_2O_2 + 4-Aminoantipyridine + TOOS* $\xrightarrow{Peroxidase}$ Quinone Diimine Dye + 2 H_2O

   *N-Ethyl-N-(2-hydroxy-3-sulfopropyl)-m-toluidine

Titrimetric Method

Lipase catalyzes the hydrolysis of fatty acids from an emulsion of olive oil triglycerides. The quantity of fatty acids formed is measured by titration with dilute sodium hydroxide solution. The quantity of alkali required to reach the indicator (thymolphthalein) endpoint is proportional to lipase activity.

Olive Oil Triglycerides $\xrightarrow{Lipase}$ Fatty Acids + Diglycerides

Immunoassay

There are some ELISA-based assays for lipase measurement based on the sandwich technique. Specific anti-lipase antibodies, conjugated to peroxidase (PO) or alkaline phosphatase (AP), bind to various lipases from cell, serum, or microorganisms. The measured PO/AP activity is proportional to the amount of lipase bound.

STANDARD VALUES OF SELECTED PARAMETERS

Standard values of selected parameters in **blood plasma** and serum.

Substance	*Value*	*Conc. unit*
Alanine aminotrasferase	*m* 10-40	U/l
	f 7-35	
Albumin	35-55	g/l
Alkaline phosphatase	*m* 38-94*	u/l
	f 28-111*	
Ammonia-N (whole blood)	52-144	μ mol/l
Asparte aminotransferase	8-20	U/l
Bicarbonate	21-25	mmol/l
Bilirubin, direct	<6.8	μmol/l
Bilirubin, total	5-18.8	μmol/l
Calcium	2.2-2.7	mmol/l
Chloride	94-112	mmol/l
Cholesterol	3.35-6.70*	mmol/l
Creatine	*m* 11-22	μmol/l
	f 13.3-24.3	
Creatinine	*m* 23-61	μmol/l
	f 23-92	
Creatine kinase	*m* 62-106	U/l
	F 44-88	

Substance	*Value*	*Conc. unit*
C-reactive protein	0-9	mg/l
γ-glutamyl transferas)	*m* 2-30	U/l
	f 1-24	
Glucose	3.31-5.57*	mmol/l
Hemoglobin	*m* 131-174*	g/l
	f 118-157*	
Iron	*m* 1-25	μmol/l
	f 16-25	
Lactate	4-1.8	mmol/l
Lactate dehydrogenase	100-190*	U/l
Lipase	30-190	U/l
Magnesium	0.6-0.9	mmol/l
Phosphate	0.81-1.55	mmol/l
Potassium	4.1-5.6	mmol/l
Protein, Total	66-86	g/l
Sodium	135-149	mmol/l
Triglycerides	0.95-2.7*	μmol/l
Urea nitrogen	3.3-6.7	mmol/l
Uric acid	*m* 150-400	μmol/l
	f 120-375	

*age dependent m = male f = female

CHAPTER

3

Bioseparation of Membrane

Membranes have gained an important place in chemical technology and are used in a broad range of applications ranging from pharmaceuticals to water treatment. Industrial applications are divided into six main subgroups: reverse osmosis, nanofiltration, ultrafiltration, microfiltration, gas separation, pervaporation and electrodialysis. The key property that is exploited is the ability of a membrane to control the permeation rate of a chemical species through the membrane. In controlled drug delivery, the goal is to moderate the permeation rate of a drug from a reservoir to the body. In separation applications, the goal is to allow one component of a mixture to permeate the membrane freely, while hindering permeation of other components.

TYPES OF MEMBRANE

This chapter is limited to synthetic membranes, excluding all biological structures, but the topic is still large enough to include a wide variety of membranes that differ in chemical and physical composition and in the way they operate. In essence, a membrane is nothing more than a discrete, thin interface that moderates the permeation of chemical species in contact with it. This interface may be molecularly homogeneous, i.e., completely uniform in composition and structure; or the interface may be chemically or physically heterogeneous, e.g., containing holes or pores of finite dimensions or consisting of some form of layered structure. A normal filter meets this definition of a membrane, but, by convention, the term 'filter' is usually limited to structures that separate particulate suspensions larger than 1-10 μm. The principal types of membrane are shown schematically in Fig. 3.1 and are described briefly in the following pages.

Isotropic Membranes

Microporous Membranes

A microporous membrane is very similar in structure and function to a conventional filter. It has a grid, highly voided structure with randomly distributed, interconnected pores. However, these pores differ from those in a conventional filter by being extremely small, of the order of 0.01-10 (μm in diameter. All particles larger than the largest pores are completely rejected by the membrane. According to the pore size distribution of the membrane, the particles smaller than the largest pores and the particles larger than the smallest pores are partly rejected.

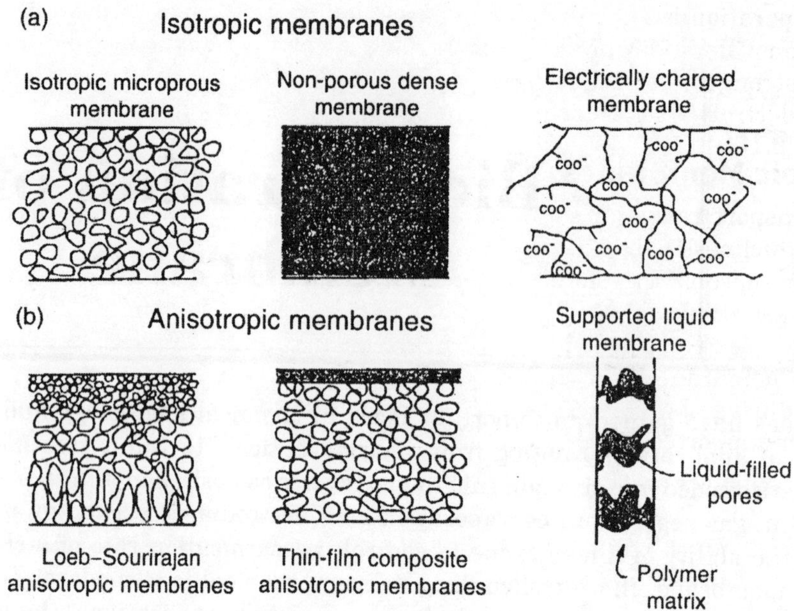

Fig. 3.1. Schematic diagrams of the principal types of membrane.

Particles much smaller than the smallest pores are absolutely passed through the membrane. Thus, separation of solutes by microporous membranes is mainly a function of molecular size and pore size distribution. In general, only molecules that differ considerably in size can be separated effectively by microporous membranes, e.g., in ultrafiltration and microfiltration.

Non-porous, Dense Membranes

Non-porous, dense membranes consist of a dense film through which permeats are transported by diffusion under the driving force of a pressure, concentration or electrical potential gradient. The separation of various components of a mixture is related directly to their relative transport rates within the membrane, which are determined by their diffusivity and solubility in the membrane material. Thus, non-porous, dense membranes can separate permeats of similar size if their concentration in the membrane material (i.e. their solubility) differs significantly. Most gas separation, pervaporation and reverse osmosis membranes use dense membranes to perform the separation. Usually these membranes have an anisotropic structure to improve the flux.

Electrically Charged Membranes

Electrically charged membranes are dense or microporous. Most commonly these membranes are very fine microporous, with the pore walls carrying fixed positively or negatively charged ions. A membrane with fixed positively charged ions is referred to as an anion-membrane because it binds anions in the surrounding fluid. Similarly, a membrane containing fixed negatively charged ions is called a cation-exchange membrane.

Separation with charged membranes is achieved mainly by exclusion of ions of the same charge as the fixed ions of the membrane structure, and to a much lesser extent by the pore size. The separation is affected by the charge and concentration of the ions in solution. For example, monovalent ions are excluded less effectively than divalent ions, and in solutions of high ionic strength, selectivity decreases. Electrically charged membranes are used for processing electrolyte solutions in electrodialysis.

Anisotropic Membranes

The transport rate of a species through a membrane is inversely proportional to the membrane thickness. High transport rates are desirable in membrane separation processes for economic reasons; therefore, the membrane should be as thin as possible. Conventional film fabrication technology limits manufacture of mechanically strong, defect-free films to about 20 μm thickness. The development of novel membrane fabrication techniques to produce anisotropic membrane structures has been one of the major breakthroughs of membrane technology during the past 30 years. Anisotropic membranes consist of an extremely thin surface layer supported on a much thicker, porous structure.

The surface layer and its struc-ture may be formed in a single operation or separately. In composite membranes, the layers are usually made from different polymers. The separation properties of permeation rates of the membrane are determined exclusively by the surface layer; the substructure functions as a mechanical support. The advantages of the higher fluxes provided by anisotropic membranes are so great that almost all commercial processes use such membranes.

Ceramic, Metal and Liquid Membranes

The discussion so far implies that membrane materials are organic polymers, and in fact most membranes used commercially are polymer-based. However, in recent years, interest in membranes made of less conventional materials has increased. Ceramic membranes, a special class of microporous membranes, are being used in ultrafiltration and microfiltration applications for which solvent resistance and thermal stability arȩ required. Dense, metal membranes, particularly palladium membranes, are being considered for the separation of hydrogen from gas mixtures, and supported liquid films are being developed for carrier-facilitated transport processes.

Table 3.1. Classification of membrane separation processes for liquid systems

Process	*Driving force*	*Separation size range*	*Example of materials separated*
Microfiltration	Pressure gradient	0.1-10 μm	Small particles, large colloids, microbial cells
Ultrafiltration	Pressure gradient	<0.1 μm-5 nm	Emulsions, colloids, macromolecules, proteins
Reverse osmosis	Pressure gradient	< 5 nm	Dissolved salts, small organics
Electrodialysis	Electric field gradient	< 5 nm	Dissolved salts
Dialysis	Concentration gradient	< 5 nm	Treatment of renal failure

MEMBRANE PROCESSES

Industrial membrane processes may be classified according to the size range of materials that they are to separate and the driving force used in separation. There is always a degree of arbitrariness about such classifications, and the distinctions that are typically drawn. Table 3.1 presents classification of membrane separation processes for liquid systems. The four developed industrial membrane separation processes are microfiltration (MF), ultrafiltration (UF), reverse osmosis (RO) and electrodialysis. These processes are well-established large-scale industrial processes. The range of application of the three pressure driven membrane water separation processes, reverse osmosis, ultrafiltration and microfiltration is illustrated in Fig. 3.2.

Microfiltration membranes filter colloidal particles and bacteria from 0.1 to 10 μm in diameter. Ultrafiltration membranes can be used to filter dissolved macromolecules, such as proteins, from solutions. The mechanism of separation by reverse osmosis membranes is quite different. In reverse osmosis membranes, the membrane pores are so small, from 3 to 5 angstroms (1 angstrom = 10^{-10} m) in diameter that they are within the range of thermal motion of the polymer chains that form the membrane. The accepted mechanism of transport through these membranes is called the solution-diffusion model. According to this model, solutes permeate the membrane by dissolving in the membrane material and diffusing down to concentration gradient.

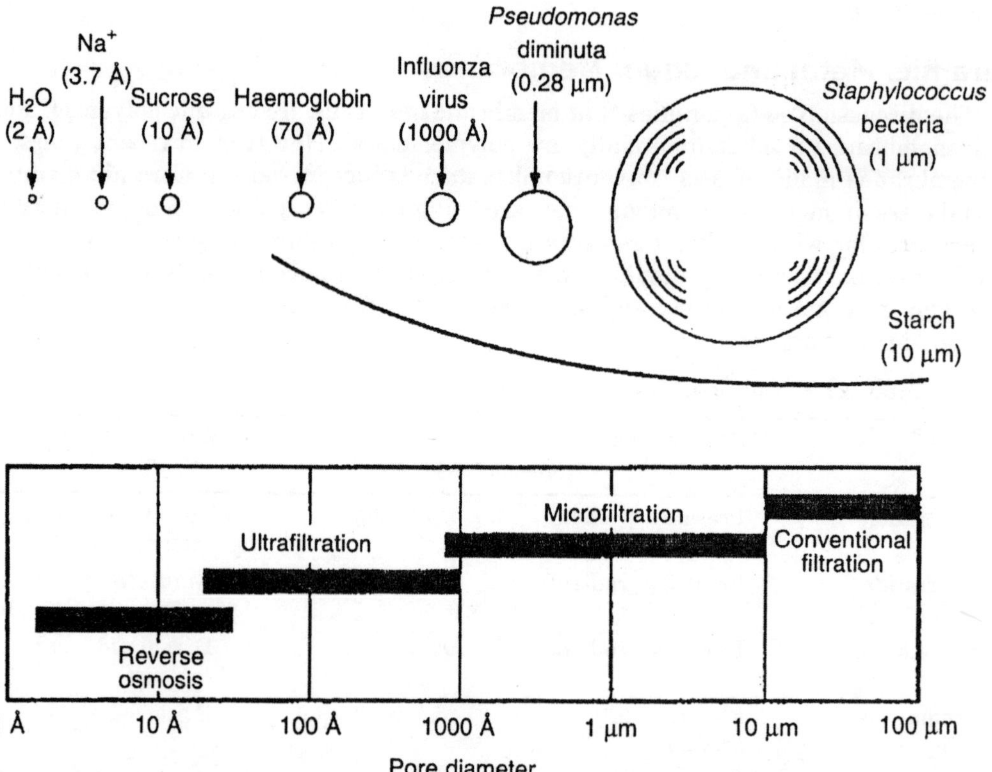

Fig. 3.2. Reverse osmosis, ultrafiltration, microfiltration and conventional filtration with distinct pore size.

Separation occurs because of the difference in solubilities and mobilities of different solutes in the membrane. The principal application of reverse osmosis is the desalination of brackish groundwater or seawater. Fig. 3.2 show the reverse osmosis, ultrafiltration, microfiltration and conventional filtration processes, which are related but differ principally in the average pore diameter of the membrane filter. Reverse osmosis membranes are so dense that discrete pores do not exist; transport occurs by statistically distributed free volume areas. The relative size of different solutes removed by each class of membrane is illustrated in the schematic. The fourth fully developed membrane process is electrodialysis, in which charged membranes are used to separate ions from aqueous solutions under the driving force of an electrical potential difference.

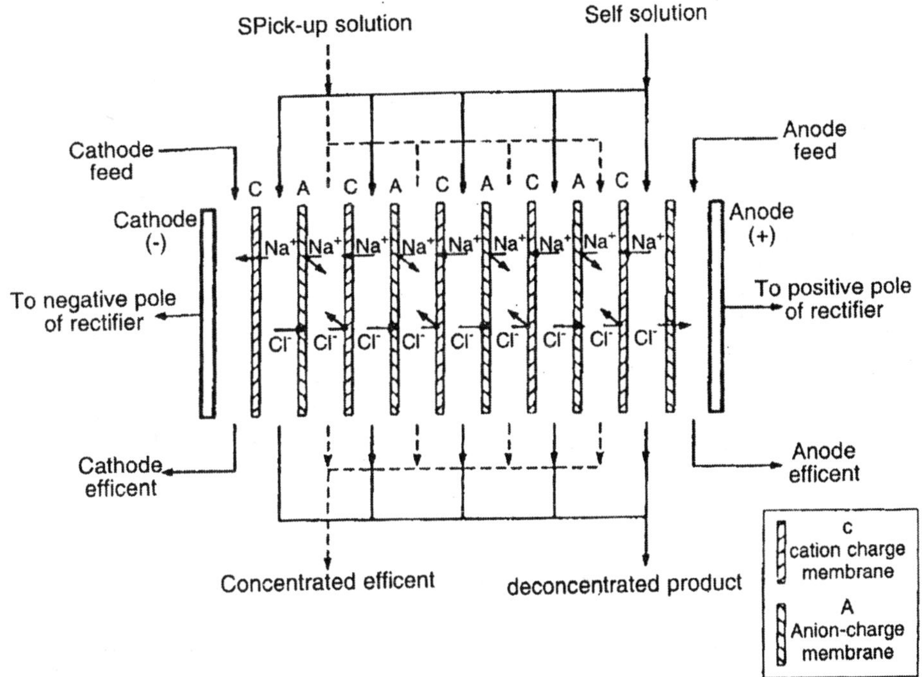

Fig. 3.3. Schematic diagram of an electrodialysis.

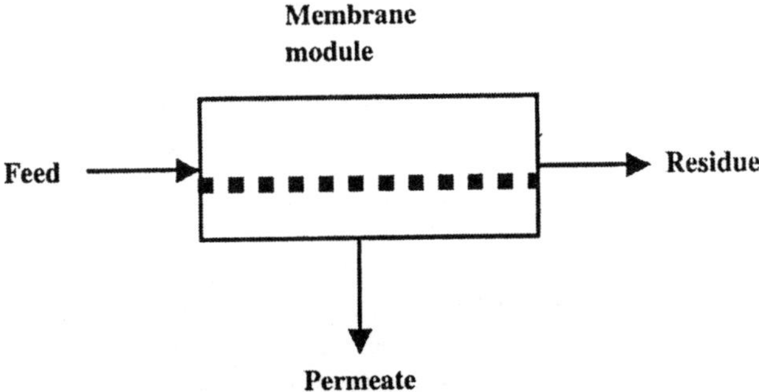

Fig. 3.4. Schematic diagram of the basic membrane gas separation process.

The process utilises an electrodialysis stack, built on the filter press principle and containing several hundred individual cells, each formed by a pair of anion and cation exchange membranes. The principal application of electrodialysis is the desalting of brackish groundwater. However, industrial use of the process in the food industry, e.g., to deionise cheese whey, is growing, as is its use in pollution control applications. In gas separation with membranes, a gas mixture at an elevated pressure is passed across the surface of a membrane that is selectively permeable to one component of the mixture. Major current applications of gas separation membranes include the separation of hydrogen from nitrogen, argon and methane in ammonia plants; the production of nitrogen from air; and the separation of carbon dioxide from methane in natural gas operations.

Membrane gas separation is an area of considerable research interest and the number of applications is expanding rapidly. Pervaporation is a relatively new process thatlias elements in common with reverse osmosis and gas separation: In pervaporation, a liquid mixture contacts one side of a membrane, and the driving force for the process is low vapour pressure on the permeate side of the membrane generated by cooling and condensing the permeate vapour.

The attraction of pervaporation is that the separation obtained is proportional to the rate of permeation of the components of the liquid mixture through the selective membrane. Therefore, pervaporation offers the possibility of separating closely boiling mixtures or azeotropes that are Difficult to separate by distillation or other means. A schematic of a simple pervaporation process using a condenser to generate the permeate vacuum is shown in Fig. 3.5. Currently, the main industrial application of pervaporation is in the dehydration of organic-solvents, in particular, the dehydration of 90-95% ethanol solutions, a difficult separation problem because of the ethanol-water azeotrope at 95% ethanol. Pervaporation membranes can produce more than 99% ethanol from a 90% ethanol feed solution. Pervaporation processes are also being developed for the removal of dissolved organics from water and for the separation of organic mixtures.

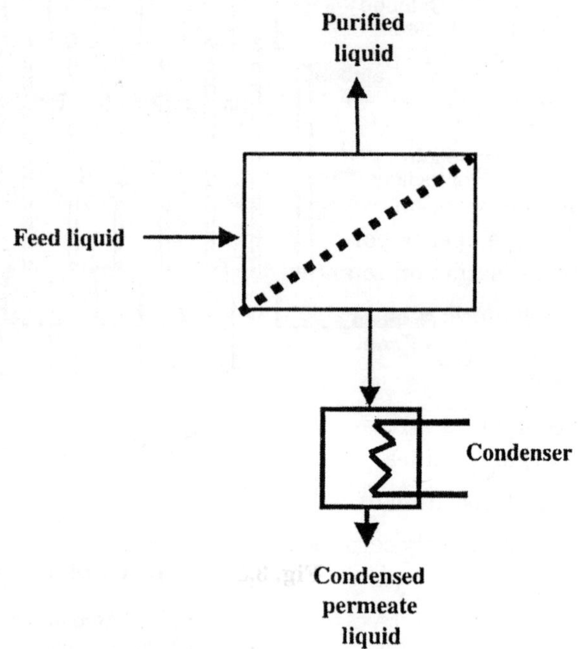

Fig. 3.5. Schematic diagram of the basic pervaporation process.

SYNTHETIC MEMBRANES

Membranes used for the pressure driven separation processes, microfiltration (MF), ultrafiltration (UF) and reverse osmosis (RO), as well as those used for dialysis, are most commonly made of polymeric materials. Initially most such membranes were cellulosic in nature. These are now being replaced by polyamide, polysulphone, polycarbonate and several other advanced polymers. These synthetic polymers have improved chemical stability and better resistance to microbial degradation. Membranes have most commonly been produced by a form of phase inversion known as immersion precipitation. This process has four main steps:

1. The polymer is dissolved in a solvent to 10-30% by weight;
2. The resulting solution is cast on a suitable support as a film of thickness of about 100 mm;
3. The film is quenched by immersion in a non-solvent bath, typically water or an aqueous solution;
4. The resulting membrane is annealed by heating.

The third step gives a polymer-rich phase forming the membrane, and a polymer-depleted phase forming the pores. The ultimate membrane structure results as a combination of phase separation and mass transfer, variation of the production conditions giving membranes with different separation characteristics. Most MF membranes have a systematic pore structure, and they can have porosity as high as 80%. Figure 3.6 shows an atomic force microscope image of polycarbonate MF membrane (0.2 mm mean pore *size).* UF and RO membranes have an asymmetric structure comprising a l-2 mm thick top layer of finest pore size sup-ported by a approximately 100 mm thick more openly porous matrix. Such an asymmetric structure is essential if reasonable membrane permeation rates are to be obtained.

Another important type of polymeric membrane is the thin-film composite membrane. This consists of an extremely thin layer, typically about 1 μm, of finest pore structure deposited on a more openly porous matrix. The main layer is formed by phase inversion or interfacial polymerisation on to an existing microporous structure.

Polymeric membranes are most commonly produced in the form of flat sheets, but they are also widely produced as tubes of diameter 10-25 mm and in the form of hollow fibres of diameter 0.1-2 mm.

A significant recent advance has been the development of MF and UF membranes composed of inorganic oxides. These are currently produced by two main techniques:

1. Deposition of colloidal metal oxide on to a supporting material such as carbon;
2. As purely ceramic materials by high temperature sintering of spray-dried oxide microspheres.

Other innovative production techniques lead to the formation of membranes with very regular pore structures. The main advantages of inorganic membranes compared with the polymeric types are their higher temperature stability, allowing steam sterilisation in biotechnological and food applications, increased resistance to fouling, and a narrower pore size distribution. The physical characterisation of membrane structure is important if the correct membrane is to be selected for a given application. The pore structure of microfiltration membranes is relatively easy to characterise, SEM and AFM being the most convenient method and allowing three-dimensional structure of the membrane to be determined. Other techniques such as the bubble point, mercury intrusion or permeability methods use measurements of the permeability of membranes to fluids. Both the maximum pore size and the pore size distribution may be determined.

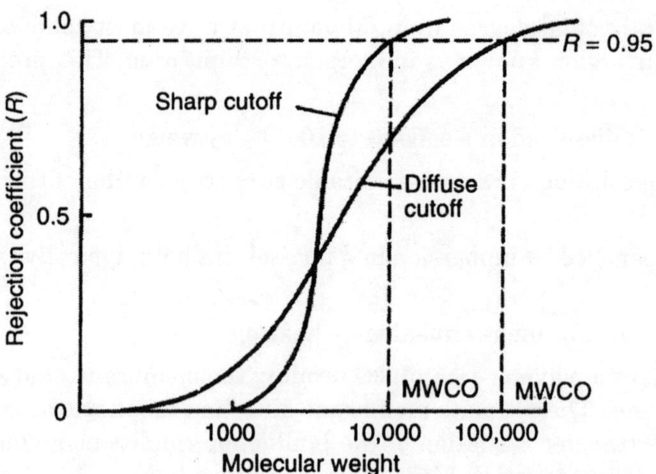

Fig. 3.6. Dependence of rejection coefficient on molecular weight for ultrafiltration membranes.

A parameter often quoted in manufacturer's literature is the nominal molecular weight cut-off (MWCO) of a membrane. This is based on studies of how solute molecules are rejected by membranes. A solute will pass through a membrane if it is sufficiently small to pass through a pore, if it does not significantly interact with the membrane and if it does not interact with other (larger) solutes. It is possible to define a solute rejection coefficient R by:

$$R = 1 - (C_p / C_f) \qquad ...(3.1)$$

where C_f is the concentration of solute in the feed stream and C_p is the concentration of solute in the permeate. For a given UF membrane with a distribution of pore sizes there is a relation between R and the solute molecular weight.

The nominal molecular weight cut-off is normally defined as the molecular weight of a solute for which $R = 0.95$. Values of MWCO typically lie in the range 2000-100,000, with values of the order of 10,000 being most common. Fig. 3.10 shows an AFM scan of 30,000 MWCO membrane.

GENERAL MEMBRANE EQUATION

It is not possible at present to provide an equation, or set of equations, that allows the prediction from first principles of the membrane permeation rate and solute rejection for a given real separation. Research attempting such prediction for model systems is underway, but the physical properties of real systems, both the membrane and the solute, are too complex for such analysis. An analogous situation exists for conventional filtration processes. The general membrane equation is an attempt to state the factors that may be important in determining the membrane permeation rate for pressure driven processes. This takes the form:

$$J = \frac{|\Delta P| - |\Delta \Pi|}{(R_m + R_c + R'_f)\mu} \qquad ...(3.2)$$

where J is the membrane permeation rate (flux expressed as volumetric ate per unit area), $|\Delta P|$ is the pressure difference applied across the membrane (transmembrane pressure), $\Delta \Pi$

is the difference in osmotic pressure across the membrane, R_m is the resistance of the membrane, R_c is the resistance of layers deposited on the membrane (filter cake, gel solution) and $R'f$ is the 'resistance' of the film layer. If the membrane is only exposed to pure solvent, say water, then reduces to $J = | dP | R_m \mu$.

Knowledge of such water fluxes is useful for characterising new membranes and for assessing the effectiveness of membrane cleaning procedures. In the processing of solutes, shows that the transmembrane pressure must exceed osmotic pressure for flow to occur. It is generally assumed that the osmotic pressure of most retained solutes is likely to be negligible in the formation of a gel when the concentration of macromolecules at the membrane surface exceeds their solubility giving rise to a precipitation, or due to materials in the process feed that adsorb on the membrane surface producing an additional barrier to solvent flow. The separation of a solute by a membrane gives rise to an increased concentration of that solute at the membrane surface, an effect known as *'concentration polarisation'*. This may be described in terms of an additional *"resistance,"* R'_f. The limitation is that the resistances are not readily calculable. However, it is within the framework of this equation that the factors influencing membrane permeation rate will be discussed in the following section.

CROSS-FLOW MICROFILTRATION

The solid-liquid separation of slurries containing particles below 10 μm is difficult by conventional filtration techniques. A conventional approach would be to use a slurry thickener in which the formation of a filter cake is restricted and the product is discharged continuously as concentrated slurry. Such filters use filter cloths as the filtration medium and are limited to concentrating particles above 5 μm in size.

Dead end membrane microfiltration, in which the particle-containing fluid is pumped directly through a polymeric membrane is used for the industrial clarification and sterilisation of liquids. Such process allows the removal of particles down to 0.1 μm or less, but is only suitable for feeds containing very low concentrations of particles as otherwise the membrane becomes too rapidly clogged.

The concept of cross-flow microfiltration represents a cross-section through a rectangular or tubular membrane module. The particle-containing fluid to be filtered is pumped at a velocity in the range 1-8 m/s parallel to the face of the membrane and with a pressure difference of 0.1-0.5 MN/m^2 (MPa) across the membrane.

The liquid permeates through the membrane and the feed emerges in a more concentrated form at the exit of the module. All of the membrane processes are listed in Table 3.2. Membrane processes are operated with such a cross-flow of the process feed.

The advantages of cross-flow filtration over conventional filtration are:

1. A higher overall liquid removal rate is achieved by prevention of the formation of an extensive filter cake.
2. The process feed remains in the form of mobile slurry suitable for further processing.
3. The solids content of the product slurry may be varied over a wide range.
4. It may be possible to fractionate particles of different sizes.

A flow diagram of a simple cross-flow system is shown in Fig 3.12. This is the system likely to be used for batch processing or development rigs; it is in essence a basic pump recirculation loop. The process feed is concentrated by pumping it from the tank and across the membrane in the module at an appropriate velocity. The partly concentrated retentate is recycled into the tank for further processing while the permeate is stored or discarded as required. In cross-flow filtration applications, product washing is frequently necessary and is achieved by a process known as diafiltration in which wash water is added to the tank at a rate equal to the permeation rate. In practice, the membrane permeation rate falls with time owing to membrane fouling; that is, blocking of the membrane surface and pores by particulate materials.

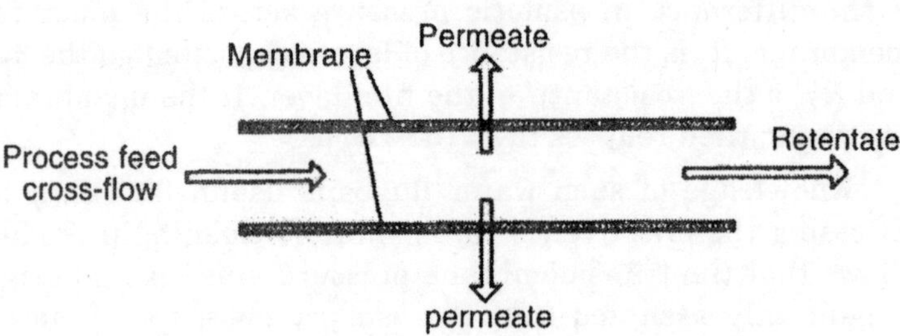

Fig. 3.7. The concept of cross-flow filtration.

The rate of fouling depends on the nature of the materials being processed, the nature of the membrane, the cross-flow velocity and the applied pressure. For example, increasing the cross-flow velocity results in a decreased rate of fouling. Backflushing the membrane using permeate is often used to control fouling. Further means of control-ling membrane foulingare discussed later. ldeally, cross-flow microfiltration would be the pressure-driven removal of the process liquid through a porous medium without the deposition of particulate material. The flux decrease occurring during cross-flow microfiltration shows that this is not the case. If the decrease is due to particle deposition resulting from incomplete

Table 3.2 Module designs most commonly used in major separation processes

Application	*Module type*
Reverse osmosis: seawater	Both hollow-fibres and spiral-wound modules
Reverse osmosis: industrial and brackish water	Spiral-wound modules used almost exclusively; fine fibres too susceptible to scaling and fouling
Ultrafiltration	Tubular, capillary and spiral-wound modules all used. Tubular generally limited to highly fouling feeds (automotive paint), spiral-wound to clean feeds (ultrapure water).
Gas separation	Hollow-fibre for high-volume applications with low-flux, low-selectivity membranes in which concentration polarisation is easily controlled (nitrogen from air).
	Spiral-wound when fluxes are higher, feed gases more contaminated, and concentration polarisation a problem (natural gas separations, vapour permeation).
Pervaporation	Most pervaporation systems are small so plate-and-frame systems were used in the first systems. Spiral-wound and capillary modules are being introduced.

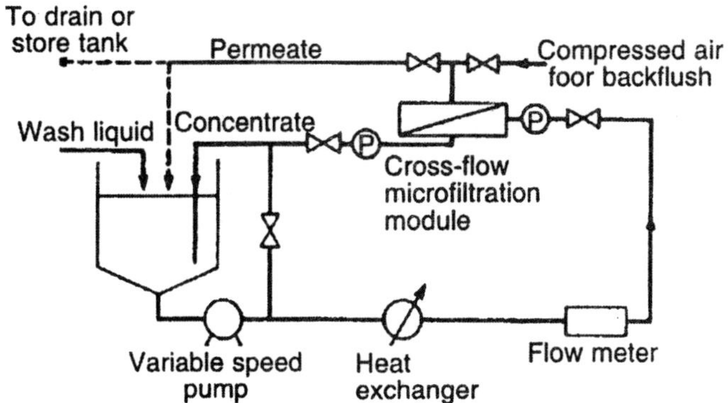

Fig. 3.8. Flow diagram for a simple cross-flow system

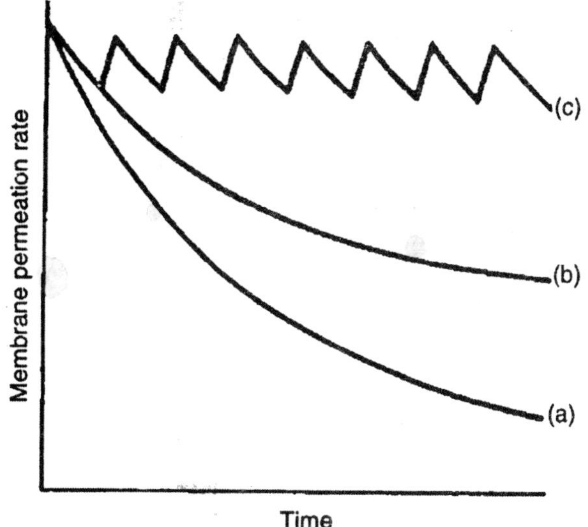

Fig. 3.9. Time dependence of membrane permeation rate during cross-flow filtration: (a) low cross-flow velocity, (b) increased cross-flow velocity, (c) back flusing at the bottom of each 'saw-tooth'.

removal by the cross-flow liquid, then may be written as:

$$J = \frac{|\Delta P|}{(R_m + R_c)\mu} \qquad \text{...(3.3)}$$

where R_c now represents the resistance of the cake, which if all filtered particles remain in the cake, may be written as:

$$R_c = \frac{rVC_b}{A_m} = \frac{rV_s}{A_m} \qquad \text{...(3.4)}$$

where r is the specific resistance of the deposit, V is, the total volume filtered, V_s the *volume of particles* deposited, C_b the bulk concentration of particles in the feed (particle volume/feed volume) and A_m the membrane area. The specific resistance can be related to the properties for spherical particles by the Carman relation as:

$$r = 180\frac{1-e}{e^3}\frac{1}{d_s^2} \qquad ...(3.5)$$

where e is the void volume of the cake and d_s the mean particle diameter.

Combining (3.3) and (3.4) gives:

$$J = \frac{1}{A_m}\frac{dV}{dt} = \frac{|\Delta P|}{(R_m + rVC_b / A_m)\mu} \qquad ...(3.6)$$

Solution of (3.6) for V at constant pressure gives:

$$\frac{t}{V} = \frac{R_m\mu}{|\Delta P| A_m} + \frac{C_b r\mu V}{2\Delta P A_m^2} \qquad ...(3.7)$$

yielding a straight line on plotting t/V versus V.

The early stages of cross-flow microfiltration often follow such a pattern. However, the growth of the cake is limited by the cross-flow of the process liquid. There are several ways of accounting for the control of cake growth. A useful method is to rewrite as:

$$J = \frac{1}{A_m}\frac{dV}{dt} = \frac{|\Delta P|}{(R_m + R_{sd} - R_{sr})\mu} \qquad ...(3.8)$$

where R_{sd} is the resistance that would be caused by deposition of all particles and R_{sr} is the resistance removed by cross-flow. Assuming the removal of solute by cross-flow to be con-stant and equal to the convective particle transport at steady-state (= $J_{ss}C_b$), then,

$$\frac{1}{A_m}\frac{dV}{dt} = \frac{|\Delta P|}{\left(R_m + \left(V/A_m - J_{ss}t\right)rC_b\right)\mu} \qquad ...(3.9)$$

where J_{ss} can be obtained experimentally or from the film-model.

In several cases, a steady rate of filtration in never achieved. In such cases it is possible to describe the time dependence of filtration by introducing an efficiency factor β representing the fraction of filtered particles remaining in the filter cake rather than being swept along by the bulk flow. Equation then becomes

$$R_c = \frac{\beta rVC_b}{A_m} \qquad ...(3.10)$$

where $0 < \beta < 1$. The layers deposited on the membrane during cross-flow microfiltration are sometimes considered to constitute dynamically formed membranes with their own rejection and permeation characteristics.

Film and gel-polarisation models are developed for ultrafiltration. These models are also widely applied to cross-flow microfiltration.

ULTRAFILTRATION

Ultrafiltration is one of the most widely used of the pressure-driven membrane separation processes. The solute retained or rejected by ultrafiltration membranes are those with molecular weights of 1000 or greater, depending mostly on the MWCO of the membrane chosen. The process liquid, dissolved salts and low molecular weight organic molecules (molecular weight 500-1000) generally pass through the membrane. The pressure difference applied across the membrane is usually in the range 0.1-0.7MN/m^2 (MPa) and membrane permeation rates are typically 0.01-0.2 m^3/m^{2}h.

In industry, ultrafiltration is always operated in the cross-flow mode. The separation of process liquid and solute that takes place at the membrane during ultrafiltration gives rise to an increase in solute concentration close to the membrane surface. This is termed 'concentration polarisation' and takes place within the boundary film generated by the applied cross-flow. With a greater concentration at the membrane, there will be a tendency for solute to diffuse back into the bulk feed according Pick's law. At steady-state, the rate of back-diffusion will be equal to the rate of removal of solute at the membrane, minus the rate of solute leakage through the membrane:

$$J\left(C - C_{\mathrm{p}}\right) = -D\frac{dC}{dy} \qquad \text{...(3.11)}$$

Here solute concentration C and C_{p} (in permeate) are expressed as mass fractions, D is the diffusion coefficient of the solute and y is the distance from the membrane. Rearranging and integrating from $C = C_f$ when $y = l$ the thickness of the film, to $C = C_{\mathrm{w}}$, the concentration of solute at the membrane wall, when $y = 0$, gives:

$$-\int_{C_{\mathrm{w}}}^{C_{\mathrm{f}}} \frac{dC}{C - C_p} = \frac{J}{D}\int_0^1 dy \qquad \text{...(3.12)}$$

$$\frac{C_{\mathrm{w}} - C_{\mathrm{p}}}{C_{\mathrm{f}} - C_{\mathrm{p}}} = \exp\left(\frac{Jl}{D}\right) \qquad \text{...(3.13)}$$

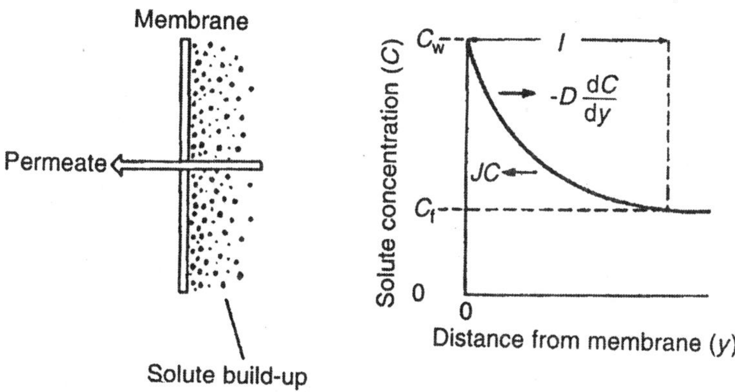

Fig. 3.10. Concentration polarization at a membrane surface.

If it is further assumed that the membrane completely rejects the solute, that is, $R = 1$ and $C_{\mathrm{p}} = 0$, then:

$$\frac{C_{\mathrm{w}}}{C_{\mathrm{f}}} = \exp\left(\frac{Jl}{D}\right) \qquad \text{...(3.14)}$$

where the ratio $C_{\mathrm{w}}/C_{\mathrm{f}}$ is known as the polarisation modulus. Note that it has been assumed that l is independent of J and that D is constant over the whole range of C at the interface.

The film thickness is usually incorporated in an overall mass transfer coefficient h_D, where $h_D = D/l$, giving:

$$J = h_D \text{ In}\left(\frac{C_w}{C_f}\right) \qquad ...(3.15)$$

The mass transfer coefficient is usually obtained from correlations for flow in non-porous ducts.

There is an initial pressure-dependence region followed by a pressure-independent region. The convergence of plots of the membrane permeation rate against C_f is a test of.

The basic features of the flux-pressure profiles are:

1. At low $|\Delta P|$ the slope is similar to that for pure solvent flow;
2. As $|\Delta P|$ increases, the slope declines and approaches zero at high pressure.

REVERSE OSMOSIS

A classic demonstration of osmosis is to stretch a parchment membrane over the mouth of a tube, fill the tube with a sugar solution, and then hold it in a beaker of water. The level of solution in the tube rises gradually until it reaches a steady level. The static head developed would be equivalent to the osmotic pressure of the solution if the parchment were a perfect semipermeable membrane, such a membrane having the property of allowing the solvent to pass through but preventing the solute from passing through. The pure solvent has a higher chemical potential than the solvent in the solution and so diffuses through until the difference is cancelled out by the pressure head.

If an additional pressure is applied to the liquid column on the solution side of the membrane then it is possible to force water back through the membrane. This pressure driven transport of water from solution through membrane is known as 'reverse osmosis'. Note that it is not quite the reverse of osmosis because, for all real membranes, there is always a certain tran and this is not reversed. This phenomenon of reverse osmosis has been extensively developed as an industrial process for the concentration of low molecular weight solutes and especially for the desalination, or more generally demineralisation, of water.

Typically, a membrane that rejects 93% of Na^+ or Cl^- will reject 98% of Ca^{2+} or SO_4^{2-} when rejections are measured on solutions of a single salt. With mixtures of salts in solution, the rejection of a single ion is influenced by its relative proportion in the mixture. Thus for 0.1 kg/m^3 Cl^- in the presence of 1 kg/m^3 SO_4^{2-} there would be only 50-70% rejection compared with 93% for solutions of a single salt. The rejection of organic molecules depends on mole-cular weight. The molecular weights less than 100 are usually not rejected, those with molecular weights of about 150 have about the same rejection as NaCl, and those with molecular weights greater than 300 are effectively entirely rejected.

MEMBRANE MODULES

Industrial membrane plants often require hundreds of thousands of square metres of membrane to perform the separation required on a useful scale. Before a membrane separation can be used industrially, therefore, methods of economically and efficiently packaging large areas of membrane are required. These packages are called membrane modules. The areas

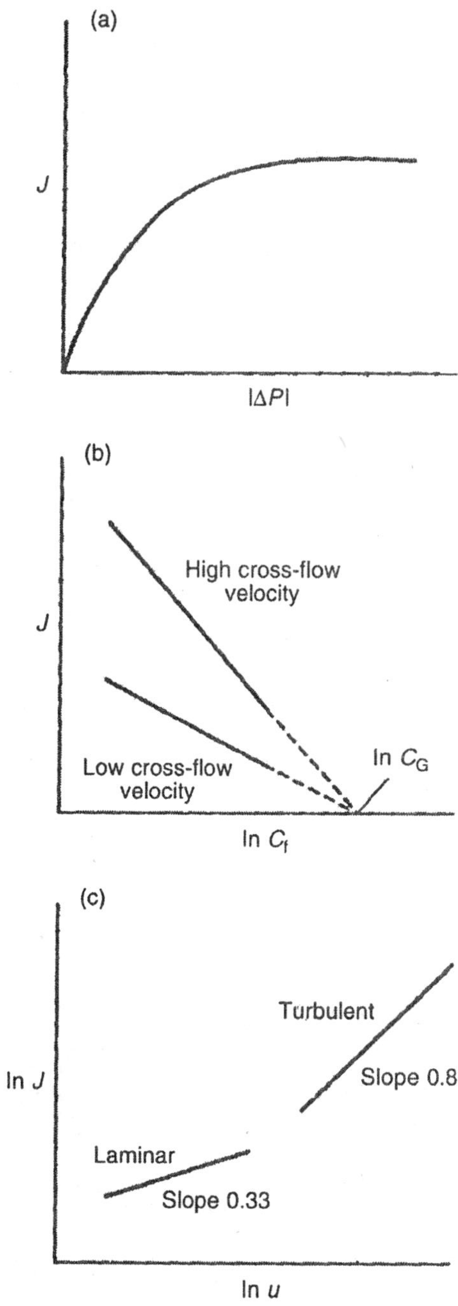

3.11. Dependence of membrane permeation rate J on (a) applied pressure difference, (b) feed solute corlcefrtnrtron C_f and (c) cross-flow velocity (u) for ultrafiltration.

of membrane contained in these basic modules are in the range 1-20 m^2. The modules may be connected together in series or in parallel to form a plant of the required performance.

The four most common types of membrane module are tubular, spiral, wound and hollow fibre. Despite the importance of membrane module technology, many researchers are astonishingly uninformed about module design issues.

In part this is because module technology has been developed within companies, and many developments are only found in patents, which are ignored by many academics. An overview of the principal module types are given below, followed by a summary of the factors governing selection of particular types for different membrane processes. Cost is always important, but perhaps the most important issues are membrane fouling and concentration polarisation. This is particularly true for reverse osmosis and ultrafiltration systems, but concentration polarisation issues also affect the design of gas separation and pervaporation modules.

Tabular Modules

Tubular modules are widely used where it is advantageous to have a turbulent flow regime; for example, in the concentration of high solids content feeds. The membrane is cast on the inside of a porous support tube which is often housed in a perforated stainless steel pipe. Individual modules contain a cluster of tubes in series held within a stainless steel permeate shroud. The tubes are generally 10-25 mm in diameter and 1-6 m in length. The feed is pumped through the tubes at Reynolds numbers greater than 10,000. Tubular modules are easily cleaned and a good deal of operating data exist for them. Their main disadvantages are the relatively low membrane surface area contained in a module of given overall dimensions and their high volumetric hold-up.

Flat-Sheet Modules

Flat-sheet modules are similar in some ways to conventional filter presses. This consists of a series of annular membrane discs of outer diameter 0.3 m placed on either side of polysulphone support plates which also provide channels through which permeate can be withdrawn. The sandwiches of membrane and support plate are separated from one another by spacer plates which have central and peripheral holes, through which the feed liquor is directed over the surface of the membranes, the flow is laminar. A single module contains 19 m^2 of membrane area. Permeate is collected from each membrane pair so that damaged membranes can be easily identified, though replacement of membranes requires dismantling the whole stack.

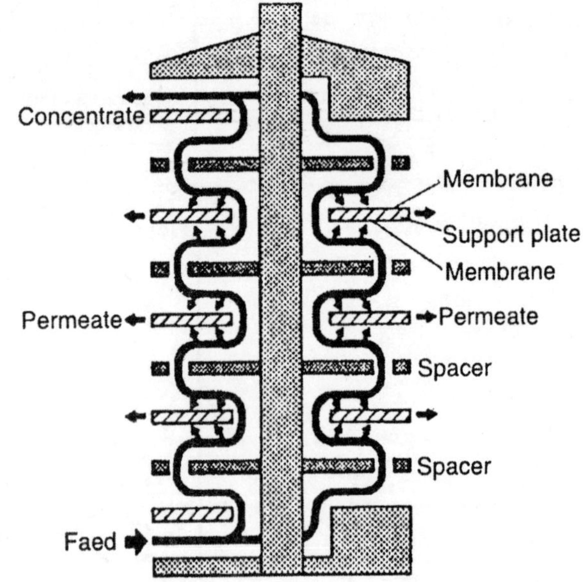

Fig. 3.12. Schematic diagram of a flat-sheet module.

Spiral-Wound Modules

Spiral-wound modules consist of several flat membranes separated by turbulence-promoting mesh separators and formed into a Swiss roll. The edges of the membranes are sealed to each other and to a central perforated tube. This produces a cylindrical module which can be installed within a pressure tube. The process feed enters at one end of the pressure tube and encounters a number of narrow, parallel feed channels formed between adjacent sheets of membrane.

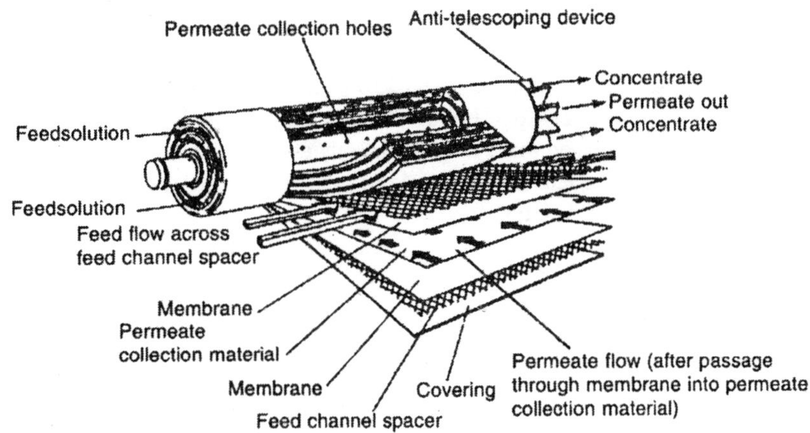

Fig. 3.13. Schematic diagram of spiral-wound module.

Permeate spirals toward the perforated central tube for collection. A standard size spiral-wound module has a diameter of about 0.1 m, a length of about 0.9 m and contains about 5 m^2 of membrane area. Up to six such modules may be installed in series in a single pressure tube. These modules make better use of space than tubular or flat sheet types, but they are rather prone to fouling and difficult to clean.

Hollow-Fibre Modules

Hollow fibre modules consist of bundles of fine fibres, 0.1-2.0 mm in diameter, sealed in a tube. For reverse osmosis desalination applications, the feed flow is usually around the outside of the unsupported fibres with permeation radially inward, as the fibers cannot withstand high pressure differences in the opposite direction. This gives very compact units capable of high pressure operation. However, the flow channels are less than 0.1 mm wide and are therefore readily fouled yet difficultto clean.

The flow is usually reversed for biotechnological appli-cations so that the feed passes down the centre of the fibres giving better controlled laminar flow and easier cleaning. However, this limits the operating pressure to less than 0.2 MN/m^2 (MPa), that is, to microfiltration and ultrafiltration applications. A single ultrafiltration module will typically contain up to 3000 fibres and be 1 m long. Reverse osmosis modules will contain larger numbers of finer fibres. This is a very effective means of incorporating a large membrane surface area in a small volume. Membrane modules can be configured in various ways to produce a plant of the required separation capability.

A simple batch recirculation system has already been described in cross-flow filtration. Such an arrangement is most suitable for small-scale batch operation, but larger scale plants will operate as 'feed and bleed' or 'continuous single pass' operation.

(*a*) Feed and Bleed

The start up is similar to a batch system in that the retentate is initially totally recycled. When the final required solute concentration is reached within the loop, a fraction of the loop is continuously bled off. Feed into the loop is controlled at a rate equal to the permeate plus concentrate flow rates. The main advantage is that the final connection is then continuously available as feed is pumped into the loop. The main disadvantage is that the loop is operating continuously at a concentration equivalent to the final concentration in the batch system and the flux is therefore lower than the average flux in the batch mode, with a correspondingly higher membrane area required. Large-scale plants usually use multiple stages operated in series to overcome the low-flux disadvantage of the feed and bleed operation and yet to maintain its continuous nature.

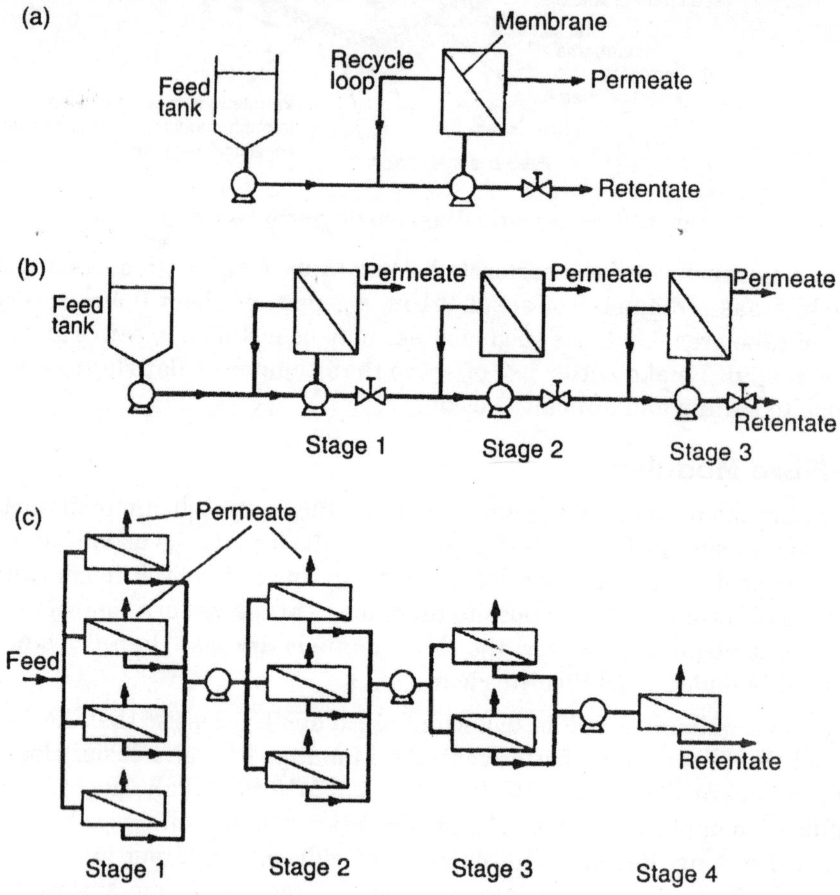

Fig. 3.14. Schematic flow diagrams of (*a*) single-stage 'feed and bleed', (*b*) multiple-stage 'feed and bleed, and (*c*) continuous single-pass membrane plants.

Only the final stage is operating at the highest concentration and lowest flux, while the other stages are operating at lower concentrations with higher flux. Thus, the total membrane area is less than that required for a single stage operation. Usually a minimum of three stages is required, and seven to ten stages are quite common. The residence time, volume hold-up and tankage required are much less than for the same duty in batch operation. Feed and bleed systems also require less frequent sterilisation than batch processes in biotechnological applications, allowing longer effective operating times.

(*b*) Continuous Single Pass

In such a system the concentration of the feed stream increases gradually along the length of several stages of membrane modules arranged in series. The feed only reaches its final concentration at the last stage. There is no recycle and the system has a low residence time. However, such systems must either be applied on a very large scale or have only a low velocities to control concentration polarisation. The smallest possible single-pass system will have a single module in the final stage, with a typical feed flow rate of about 0.1 m^3/min. Such systems are used in large-scale reverse osmosis desalination plants but are unlikely to be used in biotechnological applications.

MODULE SELECTION

The choice of the most suitable membrane module type for a particular membrane separation must balance several factors. The principal module design parameters that enter into the decision are summarised in Table 3.3.

Table 3.3. Parameters for membrane module design

Parameter	*Hollow fine Fibres*	*Capillary fibres*	*Spiral-wound*	*Plate-and-frame*	*Tubular*
Manufacturing cost ($/$m^2$)	2-10	5-50	5-50	50-200	50-200
Concentration polarization couling control	Poor	Good	Moderate	Good	Very good
Permeate-side pressure drop	High	Moderate	Moderate	Low	Low
Suitability for high-pressure operation	Yes	No	Yes	Marginal	Marginal
Limitation to specific types of membrane material	Yes	Yes	No	No	No

Cost, always important, is difficult to quantify because the actual selling price of the same module design varies widely, depending on the application. Generally, high-pressure modules are more expensive than low-pressure or vacuum modules. Hollow fibre modules are significantly cheaper per square metre of membrane than spiral-wound or plate-and-frame modules but can only be economically produced for very-high-volume applications that justify the expense of developing and building the spinning and mcjdiile fabrication equipment.

An estimate of module manufacturing cost is given in Table 3.2; the selling price is typically two to five times higher. Two other major factors determining module selection are concentration polarisation control and resistance to fouling. Concentration polarisation control is a particularly important issue in liquid separations such as reverse osmosis and ultrafiltration. Hollow-fine-fibre modules are notoriously prone to fouling and concentration polarisation and can be used in reverse osmosis applications only when extensive, costly feed solution pretreatment removes all particulates. These fibres cannot be used in ultrafiltration applications at all. Another factor is the ease with which various membrane materials can be fabricated into a particular module design.

Almost all membranes can be formed into plate-and-frame, spiral-wound and tubular modules, but many membrane materials cannot be fabricated into hollow fine fibres or capillary fibres. Finally, the suitability of the module design for high-pressure operation and the relative magnitude of pressure drops on the feed and permeate sides of the membrane can be important factors. The types of module generally used in some of the major membrane processes are listed in Table 3.2.

Example

An ultrafiltration plant is required to treat 50 m^3/day of protein-containing waste stream. The waste contains 0.05% by weight of protein which has to be concentrated to 2% to allow recycling to the main process stream. The tubular membranes to be used are available as 30 m^2 modules. Pilot-plant studies show that the flux J through these membranes is given by:

$$J = 0.02 \ln\left(\frac{30}{C_f}\right)$$

where C_f is the concentration of protein in kg/m^3. However, owing to fouling the flux never exceeds 0.04 m/h.

Estimate the minimum number of membrane modules required for the operation of this process (a) as a single feed and bleed stage and (b) as two feed and bleed stages in series. Assume operation for 20 hours each day.

Solution

(a) As a single feed and bleed stage:

- Let Q_0 be the volumetric flow rate of feed, Q_2 the volumetric flow rate of concentrate, C_0 the solute concentration in the feed, C_2 the solute concentration in the concentrate, F the volumetric flow rate of membrane permeate, and A the required membrane area.
- We assume there is no loss of solute through the membrane.
- First we check the concentration (C_1) at which the flux becomes fouling-limited:

$$0.04 = 0.02 \ln\left(\frac{30}{C_1}\right) \Rightarrow C_1 \approx 4 \text{ kg/m}^3$$

That is, below this concentration the membrane flux is 0.04 m/h.

This does not pose a constraint for the single stage, as the concentration of solute (C_2) will be that of the final concentrate (20 kg/m^3).

Conservation of solute gives:

$$Q_0C_0 = Q_2C_2, \qquad (a)$$

A fluid balance gives:

$$Q_0 = F + Q_2 \qquad (b)$$

Combining these equations and substituting known values:

$$2.438 = A0.02 \ln\left(\frac{20}{30}\right) \Rightarrow A = 301 \text{ m}^2$$

Thus, 10 modules will almost meet the specification for a single-stage process.

As two feed and bleed stage in series:

In additition to the symbols previously defined, let Q_1 be the volumetric flow rate of retentate at the intermediate point, C_1 be the concentration of solute in the retentate at this point, F_1 and F_2 be the volumetric flow rate of membrane permeate in the first and second stages respectively, and A_1 and A_2 the required membrane area in these respective stages.

Conservation of solute now gives:

$$Q_0C_0 = Q_1C_1 = Q_2C_2 \quad (c)$$

A fluid balance on stage 1 gives:

$$Q_0 = Q_1 + F_1 \quad (d)$$

A fluid balance on stage 2 gives:

$$Q_1 = Q_2 + F_2 \quad (e)$$

Substituting given values in equations (*d*) and (*e*) :

$$2.5 = \frac{1.25}{C_1} + 0.02A_1 \ln\left(\frac{30}{C_1}\right) \quad (f)$$

$$\frac{1.25}{C_1} = 0.0625 + 0.00811\,A_1 \quad (g)$$

The procedure now is to use trial and e ror to es ima e the value of C_1 that gives the optimum values of A_1 and A_2.

- If C_1 = 5 kg/m^3, the A_1 = 63 m^2 and A_2 = 23 m^2. That is, an arrangement of three modules—one module is required.
- If C_1 = 4 kg/m^3, the A_1 = 55 m^2 and A_2 = 31 m^2. That is, an arrangement of two module is one module is almost sufficient.
- If C_1 = 4.5 kg/m^3, the A_1 = 59 m^2 and A_2 = 27 m^2. That is, an arrangement of two modules – one module meets the requirement. This arrangement is that requiring the minimum number of modules.

MEMBRANE FOULING

A limitation to the more widespread use of membrane separation processes is membrane fouling, as would be expected in the industrial application of such finely porous materials. Fouling results in a continuous decline in membrane permeation rate, an increased rejection of low molecular weight solutes and eventually blocking of flow channels. On start-up of a process, a reduction in membrane permeation rate to 30-10% of the pure water permeation rate after a few minutes of operation is common for ultrafiltration. Such a rapid decrease may be even more extreme for microfiltration. This is often followed by a more gradual decrease throughout processing.

Fouling is partly due to blocking or reduction in effective diameter of membrane pores, and partly due to the formation of a slowly thickening layer on the membrane surface. The extent of membrane fouling depends on the nature of the membrane used and on the properties of the process feed. The first means of control is therefore careful choice of membrane type. Secondly, a module design that provides suitable hydrodynamic conditions for the particular application should be chosen. Process feed pretreatment is also important. In biological applications pretreatment might include prefiltration, pasteurisation to destroy bacteria, or adjustment of pH or ionic strength to prevent protein precipitation. When membrane fouling has occurred, back-flushing of the membrane may substantially restore the permeation rate.

However, this is seldom totally effective so that chemical cleaning is eventually required. This involves interruption of the separation process, and possibly quite substantial time losses due to the extensive nature of cleaning required. Thus, a typical cleaning procedure would involve: flushing with filtered water at 35-50 °C to displace residual retentate; recirculation or back-flushing with cleaning agent, possibly at elevated temperature; and rinsing with water to remove sterilising solution. More recent approaches to the control of membrane fouling include the use of more sophisticated hydrodynamic control affected by pulsated feed flows or non-planar membrane surfaces, and the application of further perturbations at the membrane surface such as continuous or pulsated electric fields.

NOMENCLATURE

C_f	Concentration of solute in feed stream, mg-l^{-1}
C_p	Concentration of solute in permeate, mg.l^{-1}
C	Solute concentration, mg. l^{-1}
J	Membrane permeation rate, mg.l^{-1}.h^{-1}
ΔP	Pressure difference applied across the membrane
ΔII	Difference in osmotic pressure across the membrane
R_m	Resistance of the membrane
R_c	Resistance of layers deposited on the membrane
R_f	Resistance of the film layer
R_c	Resistance of the cake
E	Void volume of the cake
d_s	Mean particle diameter
r	Specific resistance of the deposit
V	Total volume filtered
V_s	Volume of particles deposited
C_b	Bulk concentration of particles in the feed
A_m	Membrane area
R_{sd}	Resistance of all particles deposition
R_{sr}	Resistance removed by cross-flow
J_{ss}	Flux obtained experimentally or from the film-model
β	Fraction of filtered particles remained in the filter cake

MEMBRANE ON CERAMIC SUPPORT

Membrane separations represent a new type of unit operation, which, ultimately, is expected to be replaced by significant proportions of conventional separation processes. The advantage of membrane separation lies in its relatively low energy requirements. The reason for low energy consumption is that, unlike conventional processes such as distillation, extraction and crystallisation, it generally does not feature phase transition. The rapid development of membranes in wastewater treatment encourages the development and fabrication of an inorganic membrane. The results expand knowledge and produce various types of ceramic membrane. The ceramic membrane has a great potential and market. It represents a distinct class of inorganic membrane.

In particular, metallic coated membranes have many industrial applications. The potential of ceramic membranes in separation, filtration and catalytic reactions has favoured research on synthesis, characterisation and property improvement of inorganic membranes because of their unique features compared with other types of membrane. Much attention has focused on inorganic membranes, which are superior to organic ones in thermal, chemical and mechanical stability and resistance to microbial degradation. Alumina membranes have recently received considerable attention for their use in a wide range of applications.

Alumina is the most cost effective and widely used material in the family of engineering ceramics. The raw materials from which this high-performance technical grade ceramic is made are readily available and reasonably priced, resulting in good value for the cost in fabricated alumina shapes, with an excellent combination of properties and an attractive price. For commercial membrane applications, α-Al_2O_3 and ~γ-Al_2O_3 are the most common membranes. α-Al_2O_3 membranes are well known for their thermal and hydrolheimal stabilities beyond 1000 °C. An inorganic membrane can be prepared by various methods such as sol-gel, phase separation and leaching. The sol-gel process is considered the most practical method among those used to prepare inorganic membrane. Sol-gel processing is a simple technology in principle but requires considerable effort to become of practical use.

The advantage of this technology is the ability to produce high-purity "γ-alumina membranes at low temperatures (400-600 °C). The pore size distribution is very narrow and additional elements can be added. However, only γ-alumina membranes have the disadvantage of a poor chemical and thermal stability. Below pH 4, they dissolve to form Al^{3+} ions; above pH 12, the membrane will form $Al(OH)^{-4}$. γ-Al_2O_3 has a highly detective spinal structure and can be considered as an AlO_xH_y species, which dissolves more slowly, in low and high pH of the aqueous phase as strong hydroxides behave. Meanwhile, at higher temperatures (600-900 °C), pore growth occurs. Above 850 °C, the transformation of 7-alumina upon heating: $\gamma \rightarrow \theta \rightarrow \alpha$-alumina.

The stable α-Al_2O_3 phase forms above 1000 °C by nucleation and growth. In the transition of alumina derived from boehmite gels, the pores are as large as the grains. These pores are unable to stop grain boundary migration when α-Al_2O_3 forms above 1100 °C. Hao *et al.* stated that the phase transformation from γ-Al_2O_3 to θ-Al_2O_3 occurred at 900 °C for pure alumina membranes. With further heat treatment, 1200 °C and higher, a complete transformation of the θ-Al_2O_3 to the high temperature α-Al_2O_3 phase is observed. Foreign additives can affect the phase transformation as well as the pore structure at high temperatures.

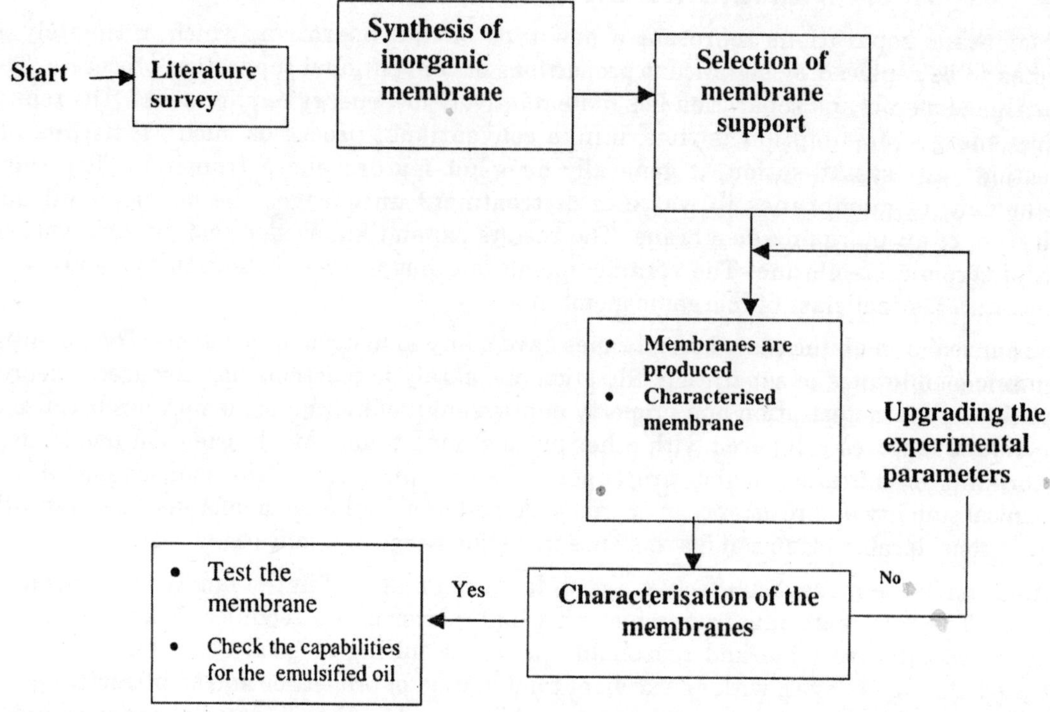

Fig. 3.15. Overall scheme and stages involved for fabrication of inorganic membrane.

To inhibit dissolution of alumina species or pore growth, the particles are coated with an inert component such as zirconia, titania and lanthana. It has been reported that the effect of zirconia addition in alumina retarded the grain growth behaviour of the alumina membrane. It was found that the ZrO_2 prevents abnormal growth of Al_2O_3 during heat treatment. The effect of zirconia on the phase transformation could be explained by the fact the presence of the zirconia on the γ-alumina surface may reduce the possibility of the nucleation of α-alumina, thus raising the phase transformation temperature. In this case study, a zirconia-alumina membrane has been developed using the sol-gel technique with and without support.

The porous ceramic was prepared to fabricate the membrane support. A thin film of aluminum and zirconium were formed on the porous ceramic support. Unsupported membrane was also prepared. The unsupported membrane was not strong enough to hold a high-pressure gradient; it was very fragile and not useful for independent use. The supported membrane on a highly porous ceramic was quite strong, but the coating covered the solid surface. The ceramic support was strong enough to hold any mechanical forces.

Apart from the ceramic support, poly-vinyl alcohol (PVA) was prepared to be used as a binder in the preparation of zirconia-alumina membranes to avoid the formation of cracks on the membrane surface.

The uniform coating and crack-free membranes were examined by scanning electron microscopy (SEM) and light microscop (LM). The main part of the research activities was the preparation and development of an inorganic membrane with and without support (zirconia-coated on γ-alumina). The membrane surface was characterised by SEM and LM. In preparing the membrane, a clear sol was obtained by the addition of acid into the aluminum sec-butoxide sol to peptise the sol and obtain a stable colloid solution.

Aluminum monohydroxides formed by the hydrolysis of aluminum alkoxides, which are peptisable to a clear sol. Peptisation was performed by the addition of acid and heat treatment for a sufficient time. It was found that stable sols cannot be obtained when the concentration of the peptisation acid is too low. The critical range for inorganic acids such as nitric, hydrochloric and perchloric acids is 0.03-0.1 mole/mole of hydroxide. In this study, nitric acid was used as the peptising agent. The resulting sols are poured into Petri dishes and dried in an oven at a controlled drying rate to obtain a gel layer. The molar ratio of zirconia salt to alumina was a controlling variable in obtaining crack-free zirconia-coated γ-alumina membranes. Finally, the dried gel is sintered and heat treated at 500 °C for 3 hours. Support materials were prepared by blending the fine power as stated in the later part of this case study. PVA was used as the binder. The organic macromolecule was used to create sufficient pores in the material when burned out after a solid strong support was formed.

Drying was performed in a high-temperature furnace. The shape and thickness of the support were based on the mass of the material and the way it was moulded. Once the membrane was successfully produced, it was analysed for characterisation and scanning. The sol-gel technique was successfully used to obtain a crack-free unsupported membrane, which was expected to have pore size of 1-2 nm. The development of the crack-free membrane may not have the same strength without strong, solid support. The next stage of this work was to characterise the fabricated membrane. The objectives of this study were to develop a zirconia-coated γ-alumina membrane with inorganic porous support by the sol-gel method and to characterise the surface morphology of the membrane and ceramic support.

Materials and Methods

The fabricated membrane consists of inorganic ceramic material with a fine coating. The support materials were prepared by tape casting or slip casting. Slips were prepared by dispersing the alumina powder in water with the dispersing agent. The slips were homogenized by ball milling for 12–18 h or by ultrasonification for 20 min. The starch was added during vigorous stirring. The addition of starch to the sip created pores in the material when burned out after forming. Then the binder was added while gently stirring the suspension. PVA has been used for a long time as the organic binder of choice in many ceramics applications. The high binding strength, ease of plasticization, water solubility and cost effectiveness of this polymer have made it a reliable performer.

Drying was provided by application of heat from underneath the tape and by a controlled passage of air over the surface of the tape. Square supports were cut from the tapes and several tapes laminated together in a press at 60 MPa at room temperature to build up desired thickness of about 1 mm. Finally, to evaluate the membranes, analysis such as X-ray diffraction (XRD), SEM, TEM and light scattering were performed at the School of Mineral and Material Engineering, Unversiti Sains Malaysia. The last part of the work, testing the produced membrane to remove emulsifier oil from domestic waste water, was accomplished on a limited budget. An experimental rig and membrane module were required. Also the need for experimental data for the application of the supported membrane may show the real success of this project.

Preparation of PVA Solution

PVA was used as a temporary binder owing to its water solubility, excellent binding strength and clean burning characteristics. To prepare the PVA solution, 4 g of PVA were added to 100 ml distillated water. The mixture was heated and stirred vigorously until all the PVA was dissolved in the water. This took about half an hour. Peptisation was done by addition of 5 ml 1M HNO_3 to the solution. Finally, the solution was refluxed for 4 hours. The PVA solution was used in the preparation of the zirconia-alumina sol–gel solution. The preparation of the PVA solution can be summarised as follows:

(*i*) 4 g PVA were added into a 250 ml beaker;

(*ii*) 100 ml of double distilled water was stirred vigorously until all the PVA was dissolved;

(*iii*) The solution was boiled until all the PVA was dissolved. This step took at least 0.5-1 h;

(*iv*) 5 ml of 1M HNO_3 was added into the solution;

(*v*) the solution was reflexed for 4 h.

Preparation of Zirconia-coated Alumina Membrane

The supported and unsupported membranes were produced by a sol–gel dipping technique. The sol-gel solution was made by using aluminium tri-sec-butoxide and zirconium (IV) oxide. Compared with the pure alumina membrane, the zirconia-coated alumina membrane had high chemical resistances, which allowed steam sterilisation and cleaning procedures in the pH range of 0–14. It possessed good pure-heating permeability and high membrane flux in separation and filtration. It had high thermal stability, which was an attractive criterion for catalytic membrane reactors used at high temperatures. The molar ratio of Al^3+:H+:Zr:H_2O was 1:0.07:0.15:100, whereas the volumetric ratio on the basis of 100 ml of H_2O is Al^3+:H+:Zr:H_2O^+ 1:3.88 ml: 1.0275 g: 100 ml. The 100 ml of double-distilled water was heated on a hot plate between 80 and 85 °C to ensure the formation of γ-AlOOH.

Aluminium tri-sec-butoxide (14.25 ml) was added to the double-distillated water. The solution was vigorously stirred with a magnetic stirrer until a homogeneous mixture was obtained. HNO_3 (1M) (3.88 ml) was added to the sol for peptisation, and the mixture was again stirred for an additional 15 minutes. After that, 1.0275 g of ZrO_3 was added into the mixture. The sol was kept at boiling condition in the open flask for 1 h. The PVA solution was added to the mixture according to a PVA: sol ratio of 1:20, and stirred in the mixture continuously. The product was refluxed under continuous stirring at 90 °C for 16–20 h to ensure complete mixing and hydrolysis. The sol was cooled down slowly and left for a few hours.

The above steps were repeated with different amounts of ZrO_3. The sol was dried under ambient conditions until gelation and viscous gel was obtained. The sol was transferred to the support by the dip-coating method to prepare a thin layer of gel on the support. The membrane thickness firstly increased linearly with the dipping time, until it may reach a limit. It was noted that a thicker and non-uniform gel layer was more liable to crack than a thinner one. A uniform gel layer was obtained by heating and calcination. After dip coating the sol on the surface of the support, the membrane was heated in the furnace. The furnace temperature started at 30 °C and was raised to 300 °C at a fate of 0.5 °C/min. The furnace temperature was kept at 300 °C for 0.5 h for relaxation of the gel and to avoid the stress exceeding the elastic strength of the gel. The temperature was raised from 300 °C to 700 °C at a rate of 0.5 °C/min. Sintering of the inorganic membrane was done at 700 °C for 5 h. Cooling was conducted at a rate of 1 °C/min, to ambient temperature, 30 °C.

Preparation of Porous Ceramic Support

The preparation of porous ceramic support is summarised in the following sections.

Raw Material

(*i*) Feldspar potash

(*ii*) Ball clay

(*iii*) Kaolin

(*iv*) Silica power

(*v*) Sodium hexametaphosphate flake

(*vi*) Distilled water

(*vii*) PVA 3% solution

Preparation of the PVA Solution

(*i*) 7.5 g of PVA was gradually added to 250 ml to double-deionised water at ambient temperature.

(*ii*) Stir the solution until all the PVA distributed equally.

(*iii*) Boil the solution until all PVA is dissolved and stir the solution carefully. The step should take at least 40 minutes.

(*iv*) After that, 10 ml of 1M HNO_3 is added to the solution.

(*v*) Reflux the solution for 4 hours.

Sieve all the feldspar potash, ball clay, kaolin and silica powder by mesh, size 90 μm

All the powders were mixed according to mass ratio given below:

(*i*) 25% feldspar potash

(*ii*) 25% ball clay

(*iii*) 25% kaolin

(*iv*) 25% silica powder

Mix 20 g of mixed powder and 25 ml PVA solution to form the concentrated slurry support. A different composition of mixed powder and PVA solution was used.

Dip the sponge in the mixture. After that put the sponge inside the furnace with the starting temperature at 30 °C, raise the temperature at 1 °C/min up to 1200 °C. Maintain the temperature for 2 h. After heating, decrease the furnace temperature by 1 °C/min back to 30 °C.

Some experience is needed to perform the last step efficiently. From our previous work, the strength of the ceramic support will increase with a decrease in the amount of dispersion solution. The advantage of sol–gel technology is the ability to produce a highly pure γ-alumina and zirconia membrane at medium temperatures, about 700 °C, with a uniform pore size distribution in a thin film.

However, the membrane is sensitive to heat treatment, resulting in cracking on the film layer. A successful crack-free product was produced, but it needed special care and time for suitable heat curing. Only γ-alumina membrane have the disadvantage of poor chemical and thermal stability.

RESULTS AND DISCUSSION

The vast increase in the application of membranes has expanded our knowledge of fabrication of various types of membrane, such as organic and inorganic membranes. The inorganic membrane is frequently called a ceramic membrane. To fulfil the need of the market, ceramic membranes represent a distinct class of inorganic membrane. There are a few important parameters involved in ceramic membrane materials, in terms of porous structure, chemical composition and shape of the filter in use. In this research, zirconia-coated γ-alumina membranes have been developed using the sol–gel technique. Finally, analytical equipment was used for characterisation, such as XRD, SEM, TEM, LM and light scattering.

These were available either in the School of Chemical Engineering or other departments and research centres in the Universiti Sains Malaysia. However, owing to limited access to the high-end analytical equipment to analyse the membrane, the surface morphology of the membrane and the porous ceramic support was only characterised with SEM and LM.

More than 50 samples of successful and unsuccessful membranes for demonstration were fabricated. The porous ceramic supports were fabricated in our research lab. The non-uniformity resulted from the synthetic foam used as a base to absorb the ceramic solution before vaporising any water from the inorganic mixture. Uniform porous media as a solid support for the membrane was obtained.

There were opportunities to fabricate a crack-free ceramic membrane coated with γ-alumina. The supported zirconia-alumina membrane on the ceramic support shows an irregular surface. The non-uniform surface of ceramic support causes the irregular surface on the top layer of the membrane. Some of the membrane sol was trapped in the porous ceramic support during coating, and caused the irregularity of the membrane surface.

The zirconia membrane was obtained in a unique manner. Light micrographs of the zirconia-alumina membrane coated on the ceramic support have been shown. The non-uniformity and crater-filled surface of the ceramic support was covered by the zirconia-alumina membrane layer. Zirconia was mounted by very thin or nanolayers on the ceramic membrane.

A combination of alumina and zirconia was used as a strong nano-film on the ceramic membrane. Observation by SEM shows that the zirconia-alumina membrane layer was properly adhered and could stand on the top of the porous ceramic support.

We have successfully developed a new inorganic ceramic membrane coated with zirconium and alumina. A thin film of alumina and zirconina unsupported membrane was also fabricated. The successful method developed was the sol-gel technique.

CHAPTER 4

Methods of Isolation

One of the major difficulties in the analysis of biological samples is the presence of substances that may be confused with the one under investigation. This means that it is often necessary either to isolate the test substance or to remove the interfering substance before the analysis can proceed. Even the presence of substances that are grossly different from the substance under investigation can often affect the quality of the analytical results. For this reason many methods include a preliminary separation step. Proteins are frequent interfering substances in biological samples and these may be removed by techniques such as dialysis, ultrafiltration and gel permeation chromatography. Alternatives include precipitating the protein with high concentrations of salts, organic acids or solvents or by heating.

Table 4.1. Protein precipitants

Precipitating agent	*Final concentration in mixture*
Anionic	
Perchloric acid	1.0 mol l^{-1}
Picric acid	10 g l^{-1}
Salicylsulphonic acid	25 g l^{-1}
Trichloroacetic acid	10 g l^{-1}
Tungstic acid	Sodium tungstate 10 g l^{-1} sulphuric acid 0.05 mol l^{-1}
Cationic	
Zinc hydroxide	Zinc sulphate 10 g l^{-1} sodium hydroxide 0.05 mol l^{-1}
Copper tungstate	Cupric sulphate 5.0 g l^{-1} sodium tungstate 2.5 g l^{-1}
Neutral	
Sodium sulphate (and other salts)	250 g l^{-1}
Organic solvents	
Acetone	63% (v/v)
Chloroform	25% (v/v)
Ethanol	70% (v/)
Methanol	75% (v/v)

The precipitated protein can be subsequently removed by either filtra tion or centrifugation. Because many small non-polar molecules are solubilized by being bound to proteins, they will also be removed with the protein and in order to quantitate these substances it is necessary to break these complexes, often by extraction into an organic solvent, before removing the precipitated protein. Although the removal of interfering substances is an important application of separation methods, this chapter is more concerned with the use of these procedures to isolate, identify or quantify a particular substance or group of substances in the presence of other very similar substances. The quantitation of one amino acid in the presence of other amino acids is only one example of such an analytical problem.

PRINCIPLES OF SEPARATION TECHNIQUES

All separation procedures depend primarily upon some physical characteristic of the compounds. It is relatively easy to separate substances that have significantly different physical characteristics by simple techniques, such as solvent extraction. However, if the various compounds are similar to each other then it is essential that any slight differences between them are exploited in order to achieve separation. It is possible to exaggerate any differences by repeating a manipulation many times, for instance, in sequential solvent extractions.

Paper chromatography is an example of a separation procedure in which a large number of partition equilibria occur between an organic solvent and the water in the cellulose. The polarity of a molecule is a characteristic that significantly affects many of its properties, in particular its solubility and volatility. Molecules are said to be polar if they have a dipole. This is defined as an unequal distribution of electrons within the molecule, caused by differences in the electronegativity of the atoms involved. Water is a polar substance because oxygen shows greater electronegativity than hydrogen and tends to develop a slight negative charge, leaving the hydrogen atoms with a slight positive charge.

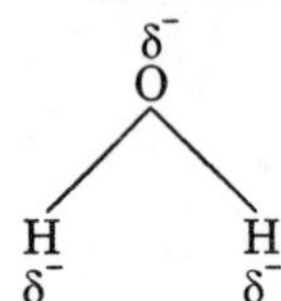

Owing to this partial ionic or polar nature, water molecules tend to associate with one another to form a lattice structure stabilized by these ionic interactions. Other molecules that also show this partial ionic nature will fit more readily into the lattice structure than non-polar molecules and will dissolve in water. This attraction between two polar molecules has other effects besides affecting the solubility of the substance. If one of the substances is a solid, the binding of the other polar substances is known as adsorption and the strength of the adsorptive bonds reflects the ionic interactions between the two molecules.

Separation techniques, such as gas–liquid chromatography, liquid chromatography and adsorption chromatography, all depend to a large extent on polar interactions, and slight differences in polarity can be used to facilitate separation. The possible formation of a dipole is a feature of covalent bonding but it is obvious that an ionic bond results in a definite unequal distribution of electrons within a molecule and such molecules (or ions) are extremely polar. However, the fact that they carry a definite charge enables additional separation techniques to be applied.

The rate of migration in an electric field (electrophoresis) and the affinity for ions of opposite charge (ion-exchange chromatography) are extremely valuable techniques in the separation of ionic species. Molecules that vary significantly in their size can be separated by ultra-filtration or dialysis, while molecules that are only slightly different in size can often be separated by gel permeation chromatography.

Ultracentrifugal techniques, while apparently separating on the basis of size, are strictly speaking more influenced by the mass and density of the molecule and to a lesser extent by its shape. Many molecules show a definite affinity for another molecule based on a shape relationship, e.g. enzymes and their substrates, antibodies and their antigens, and such relationships form the basis of extremely useful and specific methods of separation known generally as affinity chromatography. The classification of separation techniques is concise and easy to remember but it is also simplistic because it appears to imply that only one factor is involved in each technique. In practice, the effectiveness of any method is a composite of many factors, the one indicated in the table usually being the most significant.

Table 4.2. Classification of separation techniques

Molecular characteristic	*Physcial property*	*Separation technique*
Polarity	Volatility	Gas–liquid chromatography
	Solubility	Liquid–liquid chromatography
	Adsorptivity	liquid–solid chromatography
Ionic	Charge	Ion–exchange chromatography
		Electrophoresis
Size	Diffusion	Gel permeation chromatography
(mass)		Dialysis
	Sedimentation	Ultracentrifugation
Shape	Ligand binding	Affinity chromatography

Some of the developments in separation procedures exploit this range of factors involved in any separation technique by using conditions or reagents designed to minimize one or maximize another. As a consequence, the techniques and instrumentation of separation methods are constantly changing but the fundamental principles remain the same and need to be understood in order to appreciate the usefulness and limitations of any particular technique.

General Methods of Separation

The movement of the analyte is an essential feature of separation techniques and it is possible to define in general terms the forces that cause such movement. If a force is applied to a molecule, its movement will be impeded by a retarding force of some sort. This may be as simple as the frictional effect of moving past the solvent molecules or it may be the effect of adsorption to a solid phase. In many methods the strength of the force used is not important but the variations in the resulting net force for different molecules provide the basis for the separation.

In some cases, however, the intensity of the force applied is important and in ultracentrifugal techniques not only can separation be achieved but various physical constants for the molecule can also be determined, *e.g.* relative molecular mass or diffusion coefficient. There are basically three ways in which separation methods can be performed. They are most easily described in relation to chromatography but they are also relevant to other methods such as electrophoresis and ultracentrifugation.

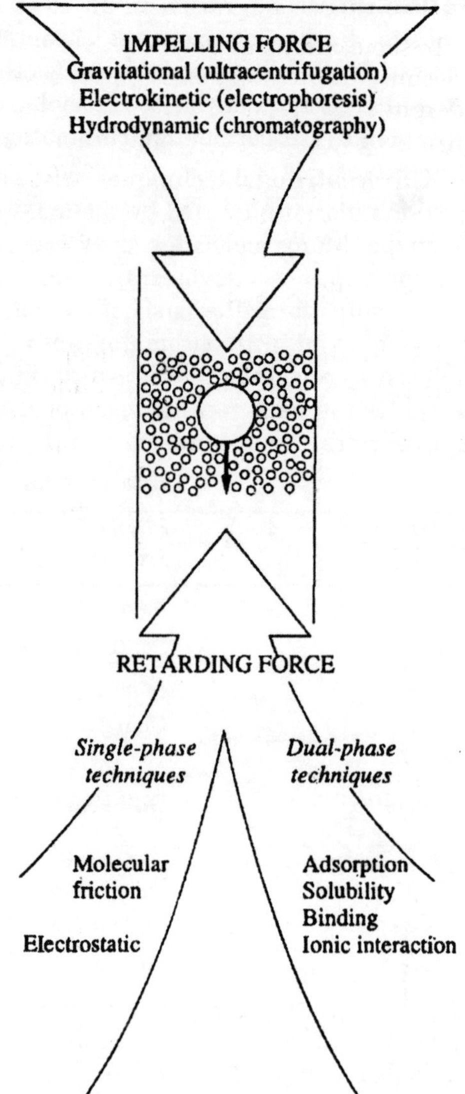

Fig. 4.1. A representation of some of the factors involved in separation techniques.

Zonal Method

In zonal methods the sample is applied as a band or zone and separation takes place in an appropriate solvent system, resulting in the separation of the individual components of the sample into separate bands. The separation process can take place entirely under uniform conditions (normal development) but it may be desirable to alter the conditions during the procedure in order to change either the impelling or retarding forces on a predetermined basis. Such gradient developments may involve gradual changes in the conditions, such as pH or temperature. In zonal separation, the individual components of the sample are separated from each other and it is often possible to use the method in a preparative manner to obtain pure preparations of each component as in some forms of chromatography, electrophoresis and ultracentrifugation.

Frontal Methods

Frontal analysis does not involve the use of separate solvent systems but separation takes place within the liquid sample usually as a result of one component moving faster than another.

The result is that a series of concentration boundaries develop, each one being due to one component of the mixture. While only a small fraction of the fastest component can actually be obtained in a pure form, the movement of each boundary is characteristic of the component and it is feasible to study a component despite never having a pure preparation. This technique is used in ultracentrifugation for the determination of relative molecular mass and sedimentation coefficient.

Displacement Methods

In some chromatographic techniques the components of a mixture show different affinities for a solid phase and can only be displaced from that phase by other molecules that have a greater affinity for it. They are zonal techniques because the sample is applied as a band or zone but separation is achieved using a solution of the displacing solute rather than the pure solvent. It is the solute that is important rather than the solvent. These techniques may be used in a normal development mode (*i.e.* without changing the composition of the eluting solution) or in a gradient mode in which the composition of the solution is progressively changed. While the individual components of the mixture may be separated from each other, they will all be contaminated to some extent with the eluting solute. Techniques such as ion-exchange chromatography and various types of adsorption and affinity chromatographies are examples of displacement methods.

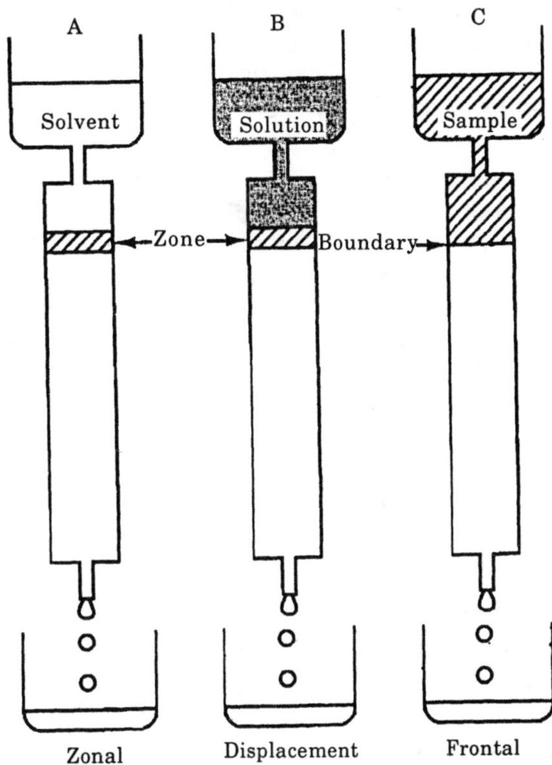

Fig. 4.2. Three major methods in chromatography. The commonest form of chromatography involves the introduction of a small volume of sample onto a column and is known as zonal chromatography. Movement down the column is effected by the mobile phase, which may be simply a solvent (A) allowing partition of the test molecules between the stationary and mobile phases. Alternatively, the mobile phase may be a solvent containing solute molecules (B), which actively displace test molecules from the stationary phase. A less frequently used method known as frontal separation (C) does not involve a separate mobile phase but a large volume of the sample is allowed to pass through the column and as the various components separate, concentration fronts develop and their movement can be monitored.

General Methods of Detection

Some detection procedures involve the loss of the sample: for example, in the flame ionization detectors used in gas chromatography the sample is burnt, while in liquid chromatography the addition of a colour reagent will also result in the loss of the sample. For such methods to be used in a preparative manner, it is necessary to split the effluent stream from the column, testing one part in the normal manner and collecting the remainder until the identity or position of each component is known.

A fraction collector is normally used for this purpose. In thin-layer and electrophoretic techniques, destructive locating methods such as colour reagents and stains can be used in a preparative manner by either masking part of the chromatogram or cutting it into slices and only staining part of it, the remainder being kept for further use. A range of non-destructive physic methods are available to monitor the effluent from a column. Many compounds absorb radiation in either the visible or ultraviolet region of the spectrum, while some are fluorescent. In the absence of specific monitoring methods, changes in the composition of a solution can often be detected potentiometrically because of changes in resistance, degree of ionization, etc.

Changes in the concentration of a solution also tend to alter various general characteristics, *e.g.* refractive index. If the compound shows no convenient intrinsic feature, it may be possible to develop a detectable characteristic by chemical modification and, particularly in thin-layer chromatography and electrophoresis, many locating reagents which are specific for various groups of compounds are available. The inclusion of a fluorescent dye into thin-layer plates can be used to detect substances that quench its fluorescence and so result in dark zones when the chromatogram is examined under ultraviolet radiation. Autoradiography can also be used in thin-layer chromatography and electrophoresis when samples are radio-labelled.

METHODS BASED ON POLARITY

Fig. 4.3 outlines the similarities in those methods in which the polarity of the various test substances is an important consideration. All of these methods are essentially chromatographic, a word coined originally by Tswett in 1906 which now implies the separation of the components of a mixture by a system involving two phases, one of which is stationary and the other mobile.

The range of applications of the different chromatographic methods depends mainly upon the number of alternative phases available. Thin-layer chromatography is restricted to very few stationary phases but has a wide range of mobile phases available. In gas-liquid chromatography (GLC) there are a large number of stationary phases available, while in high performance liquid chromatography (HPLC) there are a wide range of both stationary and mobile phases.

Liquid-solid Chromatography (Adsorption)

When a solution of a polar compound is in contact with a finely divided solid such as charcoal or silica, fairly extensive adsorption takes place on the surface of the solid. The majority of adsorbents are polar, either acidic (*e.g.* silica) or basic (*e.g.* alumina), and a large surface area is necessary for a significant degree of adsorption to take place. Charcoal is a non-polar adsorbent that will bind large or non-polar molecules from an aqueous solution, but its effects are not very predictable. However, several synthetic non-polar adsorbents have been developed, known as XAD resins, which are synthetic polymers, often polystyrene based. They are used mainly as preparative media for extracting substances from samples which, after washing the resin, can be eluted from it with a polarorganic solvent: The adsorptive effects of the polar adsorbents are often due to the presence of hydroxyl groups and the formation of hydrogen bonds with the solute molecules. The strength of these bonds and hence the degree of adsorption increases as the polarity of the solute molecule increases.

In order to separate the solute from the adsorbent it is necessary to use a solvent in which the solute will dissolve and which also has the ability to displace the solute from the adsorbent. Solvents that are too polar will overwhelm the adsorptive effects and result in the simultaneous elution of all the components of a mixture. While only a small number of adsorbents are commonly used, which are all relatively non-specific the versatility of adsorption chromatography is due to the range of solvent mixtures which can be used.

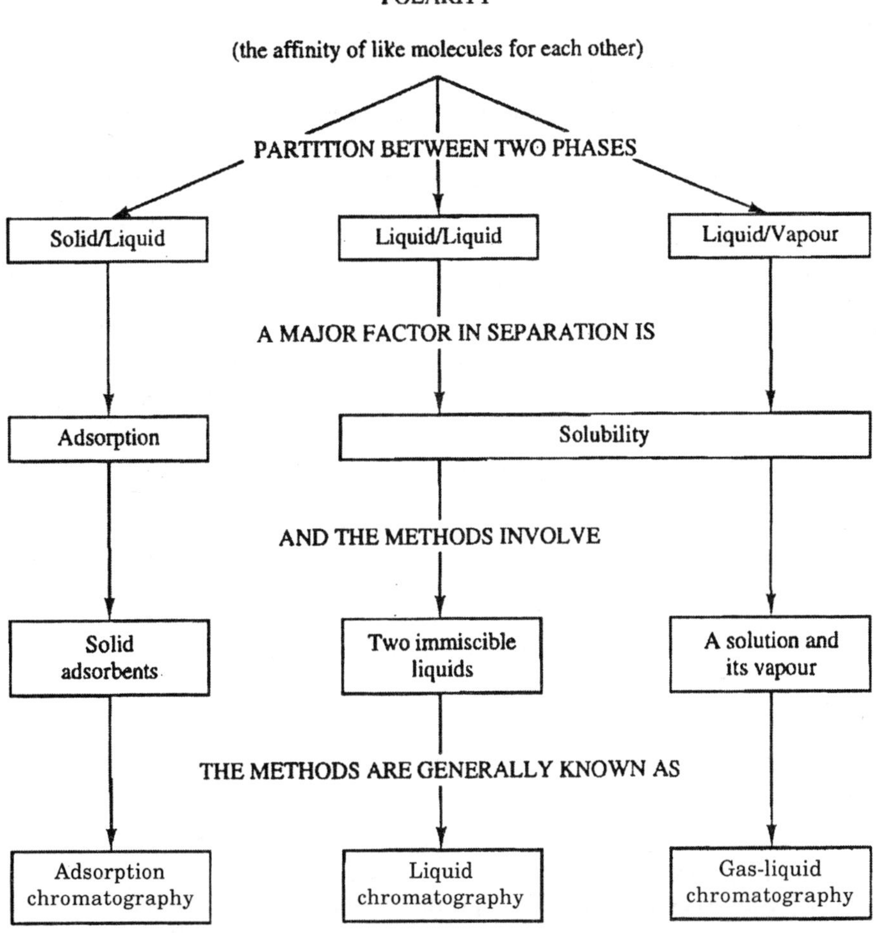

Fig. 4.3. Separation methods in which the polarity of the test molecule plays an important part.

Table 4.3. Polarity of selected solutes and solvents

Solute	*Adsorption energy*		*Solvent*	*Solvent strength*
Hydrocarbons	0.07	Increasing polarity ↓	Hexane	0.01
Halogen derivatives	1.74		Benzene	0.32
Aldehydes	4.97		Chloroform	0.40
Esters	5.27		Acetone	0.55
Alcohols	6.50		Pyridine	0.71
Acids/bases	7.60		Methanol	0.95

Adsorption energy is a measure of the affinity of a solute for an adsorbent and it varies from one adsorbent to another. The above data are relevant to silica as the adsorbent. Solvent strength is a measure of the affinity of a solvent for the adsorbent. Complete lists are known as eluotropic series and will include mixtures of solvents as well as the pure solvents above.

Methods of Adsorption Chromatography

Separations may be achieved using columns with zonal development, a tech-nique that is very useful for preparative and qualitative purposes and in HPLC offers a high level of selectivity. However, thin-layer techniques are particularly popular because of their convenience and speed and they are particularly useful in qualitative analysis.

Table 4.4. Some examples of adsorbents and possible applications

Adsorbent	*Strength*	*Application*
Silicic acid (silica gel)	Strong	Steroids, amino acids, lipids
Charcoal	Strong	Peptides, carbohydrates
Aluminium oxide	Strong	Steroids, esters, alkaloids
Magnesium carbonate	Medium	Porphyrins
Calcium phosphate	Medium	Proteins, polynucleotides
Cellulose	Weak	Proteins

Thin-layer plates consist of a layer of the adsorbent spread over an inert, flat support, usually glass, aluminium or rigid plastic film. The layer is usually stabilised by incorporating a binding agent, such as plaster of Paris (10%) in the mixture. Plates are prepared by shaking a mixture of the adsorbent and the binding agent with an appropriate volume of water. A uniform layer of this slurry is spread over clean plates and allowed to set. The plates are then dried in an oven and stored in a desiccator.

Variations in layer uniformity result in differences in separation patterns, and commercially prepared plates are normally used because of their quality and convenience.

Samples are applied as discrete spots or as streaks with capillary or micropipettes as volumes of between 1 and 20 μl. Damage to the surface must be avoided in order to achieve a regular pattern of the separated components. Appropriate reference mixtures must be applied to the same plate as the samples to assist in identifying the test components. After the samples have been applied and dried, the plate is placed in a tank that contains the solvent (0.5–1 cm deep).

It is particularly important when using volatile solvents that the atmosphere of the tank is saturated with solvent vapour. This prevents both evaporation from the surface of the plate and alteration to the composition of the solvent due to the loss of volatile components. When the solvent has moved to the top the plate (10–20 cm) or, preferably, to a line that has been scored across the plate about 1 cm from the top edge the plate is removed and dried quickly.

The separated bands can be visualized by using a suitable reagent or by viewing under ultraviolet light and their positions marked. Test spots can be identified by comparing their migration with that of reference samples, together with additional evidence, *e.g.* using a specific colour reagent. The R_F value (rate of flow) is a measure of the movement of a compound compared with the movement of the solvent:

$$R_F = \frac{\text{distance compound has moved from the origin}}{\text{distance solvent has moved from the origin}}$$

Ideally, this value should be the same for a given compound using the same solvent regardless of the distance moved by the solvent. However, it is diffi-cult to standardize chromatographic conditions, and variations that may occur in the structure and thickness

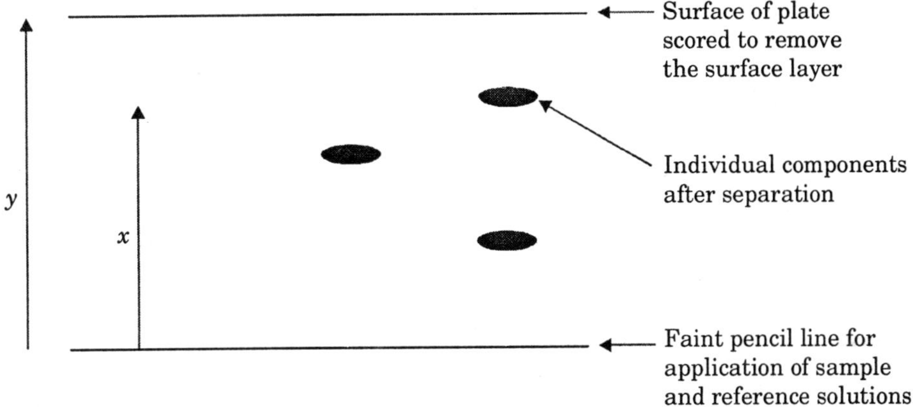

Fig. 4.4. Thin-layer chromatography. Rate of flow (R_F) for the analyte indicated is *x/y*. The R_F values for the other components can be calculated in a similar manner, the value *y* being the same in each case.

of the layer, the amount of water remaining and the effect of the binding agent all contribute to the non-reproducibility of R_F values. Demonstration of identical R_F values is not really conclusive proof of the test identity.

The evidence may be strengthened by demonstrating identical R_F values for the test and reference compounds in a range of solvents. In addition, a mixture of the sample and the reference compound can be run and if the two substances are identical, a single spot should result after chromatography. This latter technique is known as co-chromatography.

Liquid-liquid Chromatography

The partition of a solute between two immiscible liquid phases provides the basis for simple solvent extraction techniques. The polarity of both solute and solvent are important factors in determining the solubility of the solute, and polar solutes will dissolve more readily in polar solvents than in non-polar solvents. Liquid-liquid chromatography in its simplest form involves two solvents that are immiscible.

However, many recently developed media consist of a liquid (the stationary phase) that is firmly bound to a solid supporting medium. As a result, it is possible to use a second solvent (the mobile phase) which under normal conditions would be miscible with the first solvent. The second solvent is permitted to move in one direction across the stationary phase to facilitate the separation process. The presence of a supporting medium introduces some problems in the system and, in theory, it should be completely inert and stable, showing no interaction with the solutes in the sample. However, this is not always the case and sometimes it affects the partitioning process, resulting in impaired separation.

Paper Chromatography

The simplest form of liquid-liquid chromatography is paper chromatography (or commercially prepared cellulose thin-layer plates) in which the water bound to the cellulose acts as the stationary phase (polar) and the mobile phase (less polar) can be a mixture of various organic solvents, butanol, chloroform, etc. The efficiency of any chromatographic technique depends upon the number of sequential separations or equilibria that take place, which in the case of paper cromatography are due to the large number of 'compartments' of cellulose-bound water.

The test solutes are carried up the paper dissolved in the mobile phase and encounter successive compartments of water. At each one, rapid partition between the two phases occurs leaving the mobile phase to carry up the residual solute to the next water compartment and another parti-tioning effect. The solute, which is dissolved in the water and hence not carried up the paper, is now presented with fresh solvent rising up the paper and again is redistributed between the two phases. The extremely large number of partitioning phenomena greatly exaggerates the differences in the relative solubilities of the solutes and results in the effective separation of the components.

Those solutes that show greater solubility in the stationary phase will move less rapidly up the paper than those solutes that show a greater solubility in the mobile phase. In a typical chromatographic system the samples are spotted onto a strip of filter paper, dried and then placed in a tank containing a suitable solvent, which is allowed to run up the paper by capillary action. For long chromatographic runs (exceeding 20 cm), a descending technique may be used in which the paper is hung from the reservoir, the solvent flowing down the paper. Cellulose is itself polar in nature and can cause some adsorption, which may result in the 'tailing' of zones.

However, this adsorptive effect may contribute to the separation process in some instances and the use of a polar mobile phase can enhance this effect further, e.g. the separation of amino acids using an aqueous solution of ammonia as the mobile phase. The combination of partition and adsorption generally influences separations on cellulose thin-layer plates, which have superseded paper chromatography in most instances and offer increased speed and resolution.

High Performance Liquid Chromatography

Column chromatographic techniques were originally used as preparative tools but over the years major advances have taken place. HPLC is now a highly developed technique and a wide range of stationary phases are available. These enable partition adsorption, gel permeation, affinity and ion-exchange chromatography to be performed. For efficient separations it is essential to have very small and regular-shaped support media, a supply of mobile phase pumped at a pressure that is adequate to give a suitable constant flow rate through the column and a convenient and efficient detection system.

For all types of stationary phases the apparatus required consists of five basic components each available with varying degrees of sophistication. The solvent delivery system includes a pump that delivers the appropriate solvent from a reservoir at a preselected rate. The use of fine particle columns, and the high flow rates demanded, require relatively high pressures of solvent, sometimes up to 35 MN m^{-2} (5000 lb in^{-2}) although more usually about

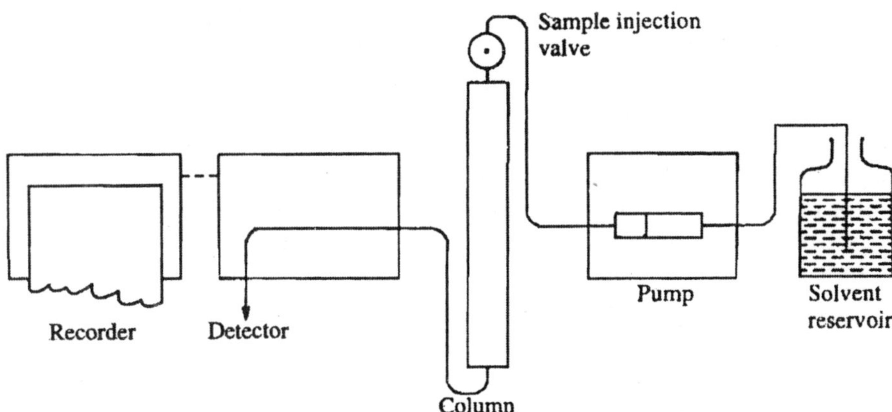

Fig. 4.5. A representation of the components of an HPLC system.

10 MN m^{-2} (1500 lb in^{-2}). In addition to being able to produce these pressures and in order to obtain a steady baseline signal from the detector, it is necessary to eliminate gross fluctuations in pressure which often result from simple pumps.

To achieve this 'pulse-free' pumping it is necessary either to incorporate pulse dampening systems or to use more sophisticated dual-piston pumps. The use of a single solvent is known as an isocratic system (with equal power), but it is often desirable to have the ability to alter the composition of the solvent mixture during the chromatographic run, a process known as gradient elution. The gradient may be as simple as a stepwise change from one previously prepared solvent mixture to another at any given time in the separation.

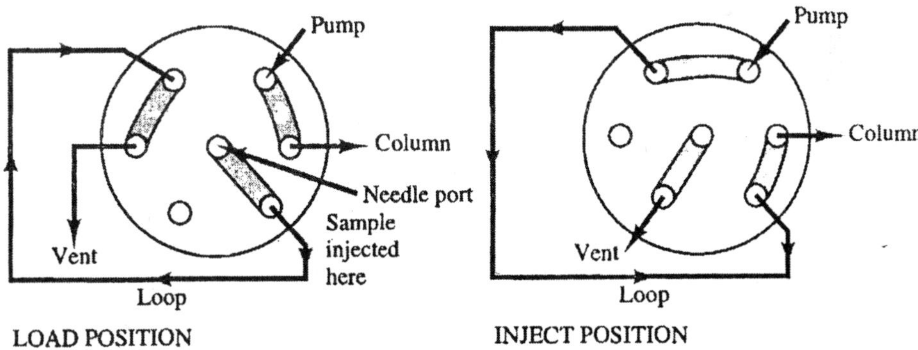

Fig. 4.6. Diagram of a six-port sample injection valve. In the load position the loop can be filled with the sample using a simple syringe. When the valve is turned to the inject position, the sample is washed into the column.

However, more frequently and more satisfactorily, the production of the gradient is microprocessor controlled. The fact that the solvent is delivered under pressure demands a suitable *sample injection device,* usually in the form of a sampling valve, which will give good precision. The typical six-port valve includes a loop of specified volume (10 μl to 2 ml), which is filled using a normal low pressure syringe before the sample is introduced into the column by switching the solvent flow through this loop. A *continuously monitoring detector* of high sensitivity is required and those that measure absorption in the ultraviolet are probably the most popular. These may operate at fixed wavelengths selected by interference filters but the variable wavelength instruments with monochromators are more useful.

Wavelengths in the range of 190–350 nm are frequently used and this obviously means that a mobile phase must not absorb at those wavelengths. Conventional spectroscopic detectors can only monitor at a single selected wavelength but the development of diode array detection permits the continuous, simultaneous monitoring of the column effluent over any selected wavelength range in the ultraviolet and visible region of the spectrum.

A diode array consists of a large number of microdiodes (sometimes up to 500) arranged so that the spectrum from a holographic grating is focused on them. Each diode then will record variations in the intensity of radiation from a particular section of the spectrum. The data from all the diodes are handled by a computer and can be presented as a traditional chromatogram but with the advantage that an absorption spectrum can be produced for any point on the chromatogram or the data can be manipulated in any appropriate manner by the computer.

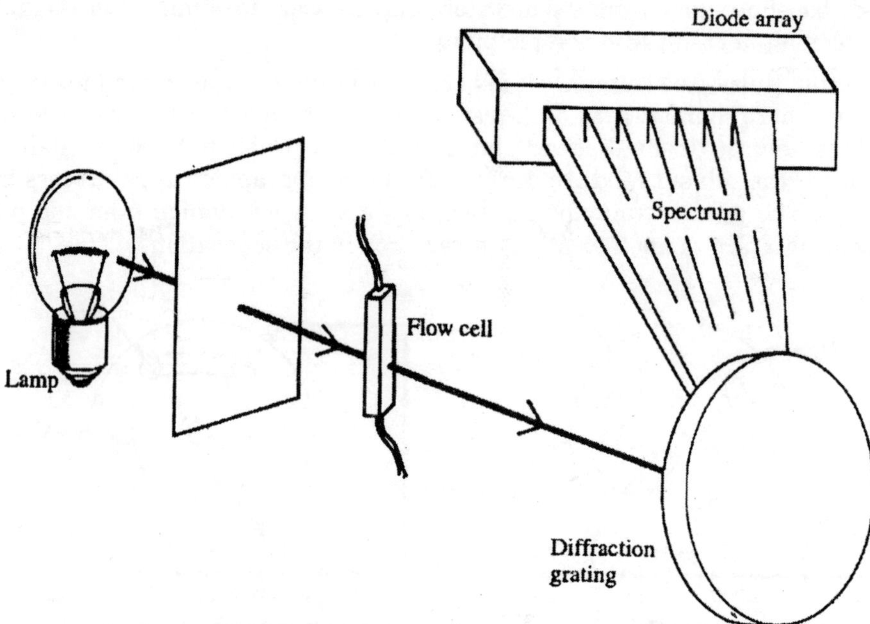

Fig. 4.7. *Diode array detector.* Light from the lamp passes through the flow cell and to holographic reflectance grating and the resulting spectrum is focused on the diode array. Detectors frequently have a spectral range of 190-800 nm and can offer bandwidths as low as 1.0 nm.

Fluorescence detectors can also be used and while their sensitivity may be greater, they are less widely applicable owing to the smaller number of fluorescent compounds. Differential refractometers will detect changes in the refractive index of the solvent due to the presence of solutes and, while they are less sensitive than the other detectors and often cannot be used effectively with gradient elution techniques, they are capable of detecting the presence of any solute. Electrochemical detection systems are becoming more popular.

Although most electrochemical methods have been investigated, *e.g.* potentiometric, conductometric, coulometric, etc., the most frequently used are amperometric. In these detectors a working and a reference electrode are held at a selected potential difference determined by the discharge potential of the analyte. The current resulting from an oxidation/reduction reaction of a small proportion of the test molecules at the working electrode is measured and, under certain conditions, is proportional to their molar concentration. Careful selection of the working voltage can permit the selective determination of one analyte in the presence of another requiring a higher working voltage. Some detectors offer increased selectivity by incorporating multiple working electrodes, each one set at a different working voltage.

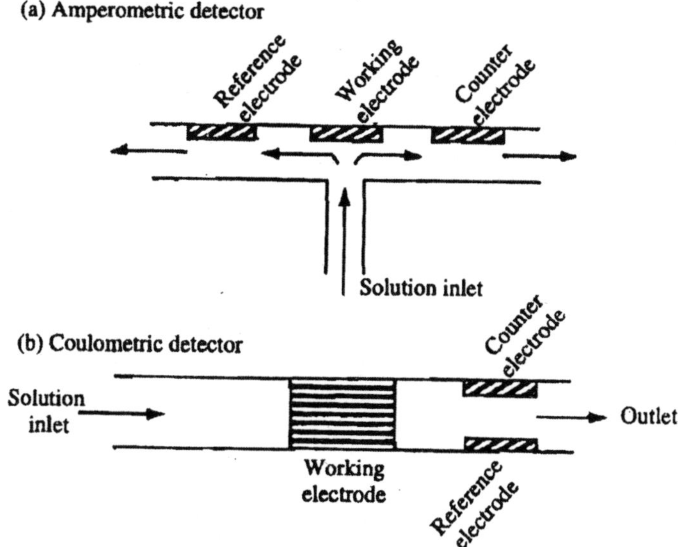

Fig. 4.8. Amperometric detectors *(a)* measure the current that flows between the working electrode, usually a glassy carbon electrode, and a reference electrode, at a fixed voltage, usually close to the discharge potential for the compound. *Coulometric detectors* (b) are less common and are designed with a porous carbon flow cell so that all the analyte reacts in the cell, the amount of current consumed during the process being proportional to the amount of the substance.

When electrochemical detection systems are used, it is essential that the mobile phases used are electrically conductive but only result in a low background current at the voltage selected.

Cell volumes are of the order of 1 µl and the control of flow rate, pH and temperature is critical for reliable detection. The *flow cell* for spectroscopic detection has to be made of material that transmits radiation of the wavelength required and it is important that the volume of the cell is small enough to give good resolution between two sample components that are close together. Generally speaking the volume of a flow cell should be approximately one-tenth that of the peak volume, which can be calculated from the solvent flow rate and the time base of the peak. Problems occur when bubbles of air become trapped in the flow cell and it is important that it should be designed to clear bubbles if they occur. However, this occurrence may be substantially reduced if the solvents have been previously 'degassed', a process which may be required to improve the quality of separation.

The simplest method of degassing is to shake the solvent vigorously in a large flask under a vacuum. Alternative methods include ultrasonication and bubbing nitrogen through the solvent. *Columns* are most frequently made of internally polished stainless steel tubing of approximately 4 mm internal diameter and 25 cm in length, although for preparative work the dimensions may be increased. It is important that the space not occupied by the supporting medium (the dead volume) is kept to an absolute minimum to prevent excessive dilution of the sample by the mobile phase. Microbore columns with internal diameters ranging from 0.5 to 2 mm may be used to reduce considerably consumption of solvent and to improve resolution.

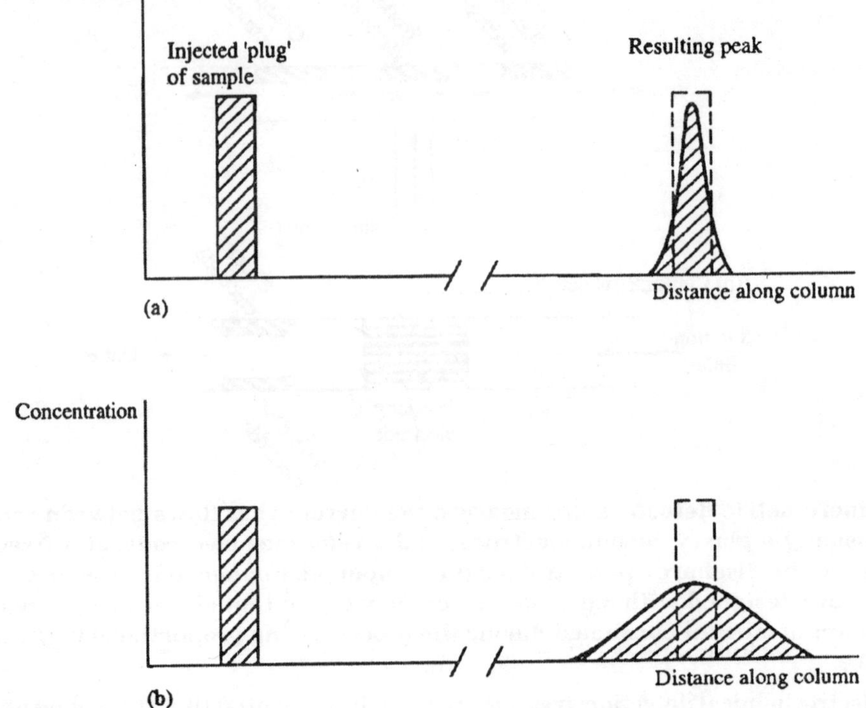

Fig. 4.9. Band broadening. The sharp boundaries of the injected sample are blurred due to diffusion as it is carried down the column (*a*). If there is a large amount of dead space, the diffusion effect becomes excessive, resulting in a very broad peak which is not very useful analytically (*b*).

However, specially designed detection systems are needed as well as special pumps designed to deliver volumes in the order of 10 μl per minute. Such systems have an advantage in that they can be coupled directly to a mass spectrometer. The dimensions of the *packing material* are important in achieving efficient separations and the flow pattern around particles causes mixing of solute and solvent and increases the dilution effects, contributing to the phenomenon known as *'band broadening'* Pellicular packing materials are solid cores of approximately 40 μm in diameter with a porous surface layer approximately 2 μm thick and are most useful when high pressures or extremes of solvent polarity are required. Microparticulate packings consist of totally porous particles usually only 5–10 μm in diameter and result in substantially better performance than the pellicular materials. Columns of pellicular materials may be prepared by dry packing techniques whereas the microparticulate materials require the pumping of a slurry in an appropriate solvent under high pressures in order to achieve uniform packing.

Assessment of column efficiency

Many factors contribute to the quality of the separation of the individual components in a sample and it is necessary to have a defined means of assessing and comparing the efficiency of the process under different conditions.

The ability of a chromatographic process to separate or resolve two similar compounds (Fig. 4.10) is measured as the resolution index (R_S)

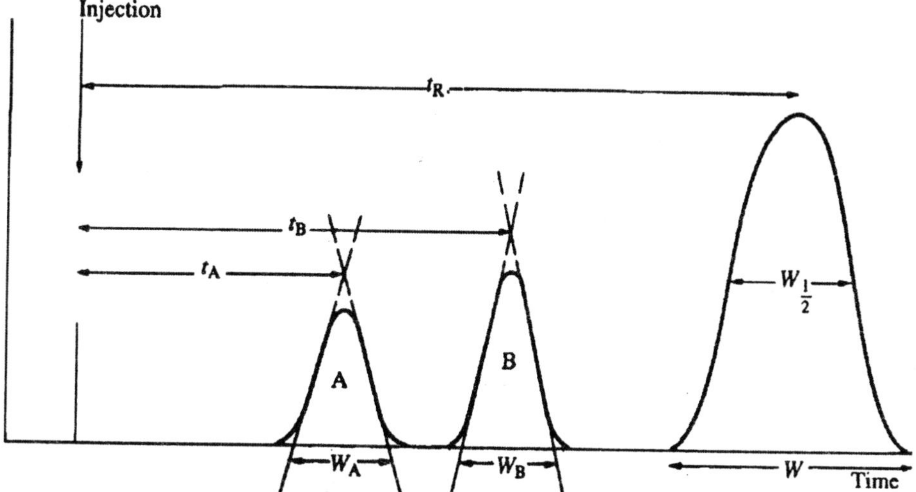

Fig. 4.10. Assessment of column efficiency.

$$\text{Resolution index } (R_s) = \frac{2(t_B - t_A)}{W_A + W_B}$$

$$\text{Number of theoretical plates } (N) = 5.54\left(\frac{t_R}{W_{1/2}}\right)^2$$

where $W_{1/2}$ is the peak width at half height.

Regardless of any quantitative considerations, the presence of two separate peaks in the chromatogram indicates the extent of the resolution and may be quantified using the following equation:

$$\text{Resolution Indes } (R_S) = \frac{\text{twice the distance between the two peaks}}{\text{sum of the base width of the two peaks}}$$

$$= \frac{2(t_B - t_A)}{W_A + W_B}$$

The greater the value for R_s, the better the resolution of the two compounds. However, large values of R_s indicate a significant time difference between the two peaks and R_s values of about 1.5 are ideal. The injection of the sample will theoretically result in a square-sided zone which will be broadened by mixing with the solvent to produce a trace which approximates to a Gaussian distribution about the mean.

Assuming that the base of the original sample zone is small, the extent of this peak broadening may be expressed as the variance (σ^2) and the base of the curve *(W)* will equal 4σ. The concept of a theoretical plate is based on the number of equilibria that may have taken place during the separation process and this number is related to the number of times the effective volume of a column is greater than the peak volume.

The variance (σ^2) for the peak is a measure of the broadening of the injection volume while the square of the retention time (t_R^2) is a measure of the effective column volume for that compound. Hence the number of theoretical plates *(N)* may be calculated from the following equation:

$$N = \frac{t_R^2}{\sigma^2} = \left(\frac{t_R}{\sigma}\right)^2$$

Using information from the dimensions of the peak (Fig. 4.10) this equation can be converted to various forms, *e.g.*:

$$N = 16\left(\frac{t_R}{W}\right)^2$$

Because it is often difficult to measure the base width *(W)* accurately, it is probably better to use the peak width at half peak height, which is known to equal 2.355σ (Fig. 4.10) for a Gaussian distribution. Hence:

$$N = \left(\frac{2.355 \times t_R}{\text{peak width at half peak height}}\right)^2$$

$$= \left(\frac{\text{retention distance}}{\text{width at half peak height}}\right)^2 \times 5.54$$

The *theoretical plate number* (*N*) is inversely related to the amount of zone broadening occurring in a column. The greater the value for *N*, the more efficient is the column but differences of less than 25% are not very significant.

The *height equivalent to a theoretical plate* (HETP) may be calculated from the value for *N* and the column length;

$$\text{HETP} = \frac{\text{length of the column}}{N}$$

While the value for *N* is useful for comparing the relative efficiencies of different columns, the HETP is useful in assessing the varying efficiency of the same column under different conditions. The value decreases as the efficiency of the col-umn increases, a characteristic that is generally better for small particle supports and less viscous fluids. In order to assess column efficiency independently of the particle size, the reduced plate height *(h)* is sometimes used.

$$h = \frac{\text{HETP}}{\text{particle diameter } (\mu\text{m})}$$

Fig. 4.11 illustrates the effect of varying the flow rate of the mobile phase on the efficiency of the separation process and provides a standard method of determining the optimum flow rate for a specific column and mobile phase system.

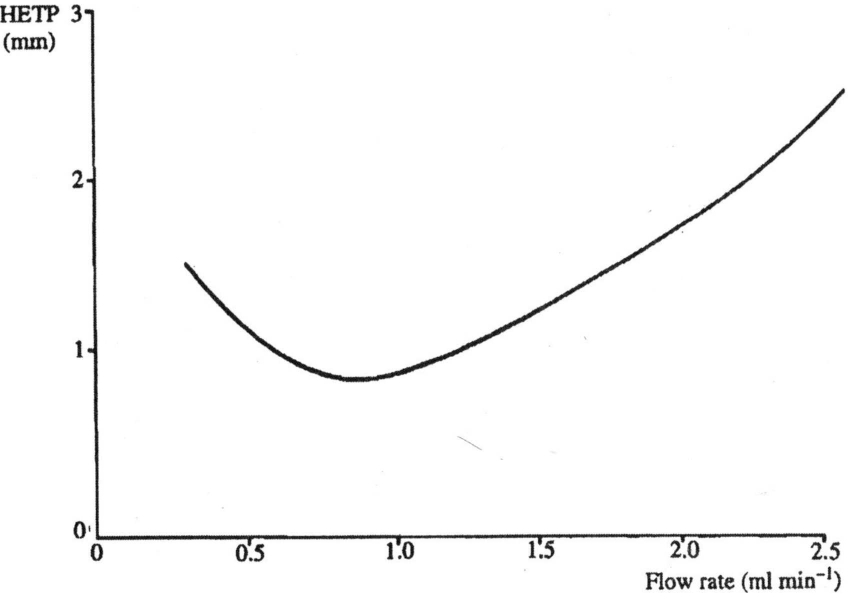

Fig. 4.11. The effect of solvent flow rate on column efficiency. The HETP of the column for the test compound is determined at different solvent flow rates. The flow rate resulting in the lowest value for HETP gives the most efficient separation.

Qualitative Analysis

Retention distance (or time) is normally used to aid the identification of a component of a mixture, provided that a known sample of the component has been subjected to separation under identical conditions. Because of the variations that can occur in the retention time due to technical factors, e.g. fluctuations in flow rate, condition of the column, the relative *retention* or *selectivity factor* (a) is sometimes used. This expresses the test retention time as a ratio ol the retention time of another component or reference compound when both are injected as a mixture:

$$\text{Relative retention} = \frac{\text{test retention time}}{\text{reference retention time}}$$

For instance, the relative retention time for phenobarbitone to barbitone in procedure 4.1 is calculated as follows:

$$\text{Relative retention} = \frac{4.4}{3.4} = 1.3$$

Quantitative Analysis

Although the response of the detector is usually proportional to the concentration of the test substance, this relationship can vary and it is essential that the response to a series of standard solutions is measured and a calibration factor or curve dgtermined. The relationship between the concentration of the solute and the peak produced in the chromatogram is, strictly speaking, only valid for peak area measurements, but in most instances it is more convenient to measure peak height. Such peak height measurements should only be used when all the peaks are very narrow or have similar widths. The tedium and lack of precision associated with non-automated methods of peak area measurements may be overcome using electronic integrators, which are features of most modern instruments. Variation in sample volume is the factor that most affects the precision of quantitative measurements and the use of an injection valve may overcome this and permit the use of external standards. However, it is still often desirable to use an internal standardization procedure as this will reduce the effects of any variation in the detector responsiveness over a period of time. For *external standardization,* replicate standards of known concentration of the pure substance are injected and the height of the resulting peaks measured.

The sample is then injected and the peak height compared with those of the standards to calculate the concentration of the test. In using this technique, replicate injections of both the standards and the test must be made. If the precision offered by the external standard technique is not adequate it is necessary to use an *internal standard.* In this procedure, a known amount of a reference substance, not originally present, is introduced into the sample.

This will result in the appearance of an additional peak in the chromatogram of the modified sample and any variations in the injections volume will equally affect the standard and the test compounds. Because a detector may respond differently to the test substance and the internal standard, it is necessary to determine the response factor (R) of the detector for each test substance relative to the internal standard. This can be done by injecting solutions that contain known amounts of the test substance and the internal standard:

$$\frac{\text{test peak height}}{\text{standard peak heigh}} t \times R = \frac{\text{test concentration}}{\text{standard concentration}}$$

An internal standard can be used in two ways, usually depending upon the nature of the sample. In its simplest form the method uses the equation for calculating the response factor. A known constant amount of the internal standard is introduced into the sample and an aliquot of the mixture is injected. Knowing the concentration of the internal standard (C_s) and the response factor of the test substance (R) the concentration of the test compound is:

$$C_T = \frac{\text{test peak height}}{\text{standard peak heigh}} t \times C_s \times R$$

This method assumes that the response factor is constant over a range of concentrations and it is often more acceptable to determine the response factor for a range of test concentrations.

In this method, a calibration curve is produced by incorporating a fixed amount of the internal standard in samples that contain known amounts of the test compound. For each concentration the ratio of peak heights is determined and plotted against concentration. For quantitation of a test sample, the same amount of the internal standard is introduced in its usual way and the ratio of peak heights for the standard and unknown is used to determine the concentration of the unknown from the calibration curve. This latter method is mainly used when a single substance in a sample is being determined but where the analysis involves the quantitation of many or all of the components of the sample, e.g. in an amino acid analyser, the former method is the more suitable. An internal standard must always be introduced into the sample before any extraction or purification procedures are undertaken as this will compensate for losses in the analysis as a result of these processes.

Choice of Column Materials

HPLC techniques were "initially developed as liquid-liquid chromatographic methods and difficulties in maintaining the stationary phase were resolved by chemically bonding it to the particulate support. Subsequently a whole range of column materials have been developed that enable the basic HPLC instrumentation to be used for the major chromatographic techniques. In selecting a column material for the separation of a specific substance, it is necessary to decide which physical characteristic of the molecule may be useful.

The initial major consideration for small molecules is usually the polarity of the molecule. A choice can then be made from ion-exchange chromatography for ionic species, adsorption chromatography of the molecule. A choice can then be made from ion-exchange chromatography for ionic species, adsorption chromatography for molecules showing moderate degrees of polarity and partition chromatography, which can be applicable to most molecules. For large molecules, gel permeation chromatography should be considered, using non-compressible gels which are effective at the pressures used. The bonding of cyclodextrins to silica has provided a range of media known as chiral stationary phases (CSPs), which are capable of separating some diastereoisomers. Cyclodextrins have rigid, highly defined structures and are able to form inclusion complexes with a range of compounds. The structure of the cyclodextrin is often such that one stereoisomer can fit easily into the molecule but the other isomer cannot.

The latter isomer is therefore eluted from the column first. Such CSPs provide an alternative method to the production of FLEC derivatives for separating stereoisomers. Usually the method is only appropriate for compounds that contain an aromatic group near the chiral centre but development of new chiral media is continually extending the range of applications.

Chemically bonded media provide both polar and non-polar stationary phases for partition methods. Techniques using a non-polar stationary phase with a polar mobile phase are known as reverse-phase systems in contrast to normal-phase chromatography, which uses a non-polar mobile phase. Because the stationary phase is chemically bonded to a supporting medium, the elution of the liquid by the mobile phase is prevented. It is still necessary to avoid extremes of pH to prevent losses due to hydrolytic or other cleaving reactions.

The most common supporting medium for partition methods is silica, to which the stationary phase is linked by either an Si–C bond, or more frequently, by a reaction with an organochlorosilane (R_3SiCl) and a surface silanol group (SiOH) to give $-Si-O-SiR_3$. Variations in the nature of this R group result in a range of stationary phases with considerable differences in polarity.

Table 4.5. Chromatographic media

Name of medium	*Surface group applications*	*Chromatographic*
Adsorption		
Partisil	Silica	Steroids
LiChrosorb Si60	Silica	Alcohols
Nucleosil	Silica	Organic acids
LiChrosorb Alox	Alumina	Vitamins
Spherisorb A	Alumina	Pesticides
Partition (normal phase)		
Partisil PAC	Cyano-amino	
LiChrosorb NH_2	Amino	Sugars
Nucleosil NO_2	Nitrite	Food preservatives
LiChrosorb Diol	Hydroxyl	
Partition (reverse phase)		
Partisil ODS	Octadecylsilane	Barbiturates
LiChrosorb RPB	Octyl	Esters, ethers
LiChrosorb RP 18	Octadecyl	Aromatics
LiChrosorb RP2	Silane	Steroids
Ion-exchange		
Partisil SCX	Sulphonate	Amino acids, nitrogenous bases
LiChrosorb KAT	Sulphonate	Nucleosides
Partisil SAX	Quaternary ammonium	Nucleotides
Nucleosil N $(CH_3)_2$	Dimethyl amine	Organic acids
Gel permeation		
LiChrospher Si	Rigid porous silica	
Macrogel	Semi-rigid polystyrene	Synthetic polymers
Sephacryl S	Dextran-acrylamide	

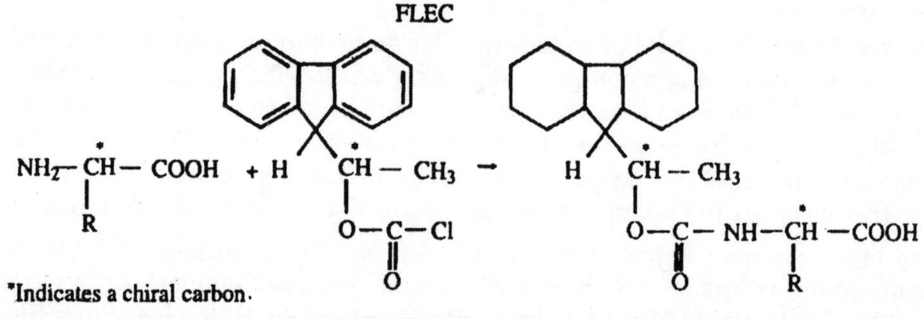

FLEC

*Indicates a chiral carbon.

Fig. 4.12. *FLEC* reacts with alcohols and amines without any racemization of the original sample to produce highly fluorescent diastereoisomers which can be separated by reverse phase chromatography.

The most popular and versatile bonded phase is octadecylsilane (ODS), n-$C_{18}H_{37}$, a grouping that is non-polar and used for reverse phase separations.

Octylsillane, with its shorter chain length, permits faster diffusion of solutes and this results in improved peak symmetry. Other groups are attached to provide polar phases and hence perform normal phase separations. These include cyano, ether, amine and diol groups, which offer a wide range of polarities. When bonded stationary phases are used, the clear distinction between adsorption and partition chromatography is lost and the principles of separation are far more complex.

The Mobile Phase

The great versatility of HPLC lies in the fact that the stability of the chemically bonded stationary phases used in partition chromatography allows the use of a wide range of liquids as a mobile phase without the stationary phase being lost or destroyed. This means that there is less need for a large number of different stationary phases as is the case in gas chromatography. The mobile phase must be available in a pure form and usually requires degassing before use. The choice of mobile phase is influenced by several factors.

Table 4.6. Solve for high performance liquild chromatography

Solvent	*Solvent strength*	*Viscosity*	*Ultravliolet cut-off (nm)*
Pentane	0	0.24	200
Petroleum ether	0.01	0.30	226
Hexane	0.01	.031	200
Carbon tetrachloride	0.18	0.97	263
Toluene	0.29	0.59	284
Benzene	0.32	0.65	278
Diethyl either	0.38	0.24	218
Chloroform	0.40	0.58	245
Dichloromethane	0.42	0.43	245
Ethylene dichloride	0.49	0.79	228
Methylethyl ketone	0.51	0.42	329
Dioxane	0.56	1.44	215
Acetone	0.56	0.30	330
Acetonitrile	0.65	0.34	190
Ethanol	0.88	1.2	210
Methanol	0.95	0.55	210
Water	Large	1.00	200

Compatibility With the Stationary Phase

The mobile phase must not react chemically with the stationary phase or break the bond linking it to the supporting material. For this reason extremes of pH and strong oxidizing agents should normally be avoided. The working pH range of a medium will be quoted by the supplier.

Compatibility With the Detection System

In any chromatographic analysis the method of detection is determined by the nature of the analyte and the mobile phase used must not interfere with this system. The use of ultraviolet absorption detection systems is very common but the solvents used must not absorb significantly at the wavelength used. For instance, absorption at 280 nm is frequently used to detect protein but some solvents, *e.g.* acetone, absorb at this wavelength. Similarly the use of concentration gradients in the mobile phase may present problems with refractive index and electrochemical detection systems.

Pressure Considerations

Small particle media with their desirable large surface areas require relatively high pressures in order to achieve a realistic solvent flow rate. As the viscosity of the solvents will appreciably affect this pressure-to-flow rate relationship a solvent must be chosen that achieves the desired separation without requiring pressures too high for the system.

Polarity

The polarity of a solvent is expressed as *solvent strength* which is a measure of the ability of the solvent to break adsorptive bonds and elute a solute from an absorbent. High values indicate high polarity. The major factor in selecting a mobile phase for separations based on partition or adsorption is the polarity of the analyte. The polarity of the mobile phase should be such that there is an effective partition of the analyte between the two phases, *i.e.* the stationary and the mobile phase, and not a complete affinity for only one.

If a test sample contains more than one component, as is usually the case, it may not be possible to achieve complete separation using a single mobile phase (isocratic separation), such cases gradient elution is required in which the solvent strength of the mobile phase is gradually changed during the separation process by altering the proportion of solvents in the mixture. Reverse-phase chromatography is used mainly for the separation of non-ionic substances because ionic, and hence strongly polar, compounds show very little affinity for the non-polar stationary phase.

However, ionization of weak acids (or weak bases) may be suppressed in solvents with low (or high) pH values. The effect of such a reduction in the ionization is to make the compound more soluble in the non-polar stationary phase but the pH of the solvent must not exceed the permitted range for bonded phases, *i.e.* pH 2–8. While the technique of *ionic suppresson* (or *ionization control*) is only effective with weakly ionic species, *ion-pair chromatography* has been developed for strongly ionic species and again utilizes reverse-phase chromatography.

If the pH of the solvent is such that the solute molecules are in the ionized state and if an ion (the counter-ion) with an opposite charge to the test ion is incorporated in the solvent, the two ions will associate on the basis of their opposite charges. If the counter-ion has a non-polar chain or tail, the ion-pair so produced will show significant affinity for the non-polar stationary phase. Counter-ions which are frequently used include tetrabutylammonium phosphate for the separation of anions and hexane sulphonic acid for cations. The appropriate counter-ions are incorporated in the solvent, usually at a concentration of about 5 mmol l^{-1}, and the separation performed on the usual reverse phase media. This ability to separate ionic species as well as non-polar molecules considerably enhances the value of reverse-phase chromatography.

Derivatives

The process of chemically modifying the test molecules before the separation procedure is known as preparing derivatives or pre-column derivatization. In liquid chromatography this is done to permit the test molecules to be more easily detected after separation and to increase the sensitivity of the detection system and less frequently to alter the separation process. Most derivatives are formed by the introduction into the test molecule of a group that confers either some absorption characteristic, usually in the ultra-violet, or a fluorescent property, thus aiding detection.

Table 4.7. Derivatives uses in liquid chromatography (HPLC)

Derivative	*Suitable for analytes*	*Detection*	
		Method	*Wavelength (nm)*
p-Nitrobenzyl	Carboxylic acids	Absorbance	254
p-Nitrobenzoyl	Alcohols	Absorbance	254
p-Bromophenacyl	Carboxylic acids	Absorbance	260
3.5-Dinitrobenzoyl	Alcohols Amines	Absorbance	254
Dansyl	Amines Peptides Phenols	Fluorescence	360/510

One application in liquid chromatography which does alter the separation process is the use of a specific series of derivatives to enable the separation of chiral (optical isomers) forms of alcohols, amines and amino acids using reverse-phase separation. FLEC is available in the two chiral forms (+)-l-(9-fluorenyl) ethyl chloroformate and (–)-l-(9-fluorenyl) ethyl chlorofor-mate.

Reaction of two stereoisomers of a test compound (*e.g.* T+ and T–) with a single isomer of the derivatizing reagent (*e.g.* R+) will result in the formation of two types of product, T+R+ and T–R+. It is possible to separate these two compounds by reverse-phase chromatography. Derivatives of FLEC are formed without any racemization of the sample and the derivatives of the D and L forms of the test molecules are eluted from reverse-phase columns in sequence. FLEC derivatives are fluorescent, having an excitation maximum at 260 nm and an emission maximum at 315 nm and are particularly useful in the analysis of amino acids.

Gas-liquid Chromatography

GLC depends upon the partition of a solute between two physical states or phases, *i.e.* gaseous and liquid (solution). Hence in GLC there is only one solvent, the stationary phase, which is held immobile in a narrow coiled tube or column. A solution of the test compounds is introduced into the heated column and is blown through the column by the carrier gas.

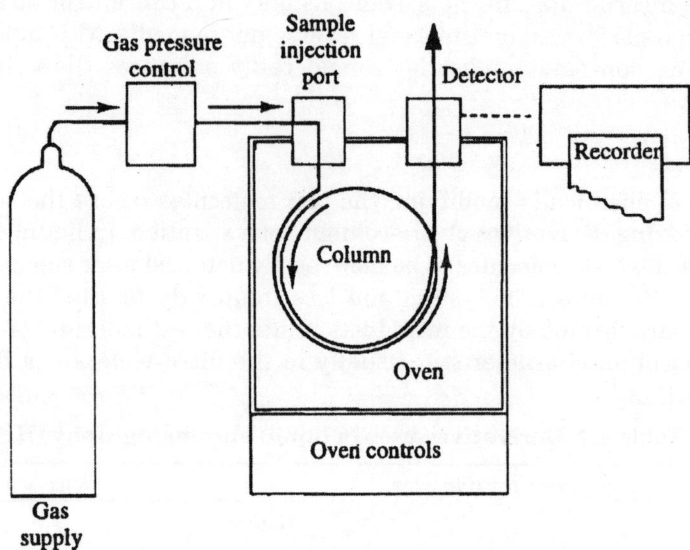

Fig. 4.13. A representation of the components of a gas chromatograph.

Upon initial contact of the solute with the liquid stationary phase, an equilibrium is rapidly established between the amount of solute which dissolves in the liquid phase and the amount of solute remaining as a vapour. The precise equilibrium position is a characteristic of the solute and solvent involved but the equilibrium will always be displaced towards the vapour phase if the temperature of the column is raised. The vapour fraction of the solute is moved down the column by the carrier gas and the equilibrium between the two phases is destroyed. However, in an attempt to re-establish an equilibrium, solute molecules leave the liquid phase restoring the partial pressure above the solution.

The solute vapour which has been moved down the column encounters fresh solvent and a new equilibrium is established. The sample vapours are detected as they leave the column and the compounds which emerge first have either the lowest solubility in the stationary phase or the highest volatility. The detector response is displayed on a recorder and a trace or chromatogram is produced. The impelling force in GLC is the flow of the carrier gas through the column. The gases most frequently used are nitrogen, argon and helium.

Gases vary in their viscosity and the more viscous gases, e.g. carbon dioxide, require higher pressures to maintain the flow rate but compared with the less viscous gases, e.g. nitrogen, restrict the extent of diffusion occurring and so tend to give sharper peaks. It is essential that all gases used are pure and dry and that they are chemically inert towards the solutes and liquid phases being used. The columns used for GLC are several metres in length and are usually coiled in order to save on oven space.

Originally the columns were packed with particles of an inert supporting medium (diatomaceous earth derivatives, porous polymers and occasionally glass beads) which had been coated with a film of the stationary phase. This gives a very large surface area and allows tight packing so that there is very little dead space where solute molecules will not be in direct contact with the liquid. These have now been superseded by capillary columns, which offer greatly improved separation efficiency. Fused silica capillary tubes are used which have internal diameters ranging from 0.1 mm (small bore) to 0.53 mm (large bore) with typical lengths in excess of 20 m.

The wall-coated open tubular (WCOT) columns have the internal surface of the tube coated with the liquid (stationary) phase and no panticulate supporting medium is required. An alter-native form of column is the porous-layer open tubular (PLOT) column, which has an internal coating of an adsorbent such as alumina (aluminium oxide) and various coatings. Microlitre sample volumes are used with these capillary columns and the injection port usually incorporates a stream splitter.

Analytical Procedure

The chromatogram produced by gas chromatography (Procedure 3.2) is comparable to that produced by HPLC and all the considerations regarding peak height and area measurements, and internal and external standards are relevant. The need to use internal standardization and response factors for quantitative work is more important because of the instability of some detectors and the imprecision associated with injection of very small volumes of relatively volatile samples.

Individual components are identified by their retention time, usually measured along the time axis of the chart paper, but the reproducibility of retention times is significantly affected by alterations in the gas flow and column temperature. It is not adequate to rely on quoted values for retention times but it is necessary to determine the values at frequent intervals using identical experimental conditions for tests and reference compounds.

Detectors

The *thermal conductivity detector* or *katharometer* is the simplest detector but is not commonly used nowadays. It measures the variations in resistance of a solid electrical conductor (a length of platinum wire) which are induced by changes in temperature. The wire is suspended in the column and heated electri-cally, and under stable conditions will attain a steady temperature depending upon the current flowing and the loss of heat by radiation and thermal conduction. Any variation in the composition of the gas surrounding the wire will result in a change in its temperature and therefore a change in its resistance.

The resistance of the wire is balanced against the resistance of a similar wire in a reference column (containing only the carrier gas) by means of a Wheatstone bridge arrangement so that variations in resistance can be monitored. A thermal conductivity detector will respond to all compounds and is capable of detecting about 1×10^{-7} mol of substance in the carrier gas but does not show a good linear response to increasing amounts of test substances. The *flame ionization detector* is commonly used and depends upon the thermal energy of a flame causing some ionization of the molecules in the flame.

The ions are collected by a pair of polarized electrodes and the current produced is amplified and recorded. For a particular gas composition, there will be a constant degree of ionization but as the composition of the gas mixture changes due to the presence of the test components so the degree of ionization will change. A supply of hydrogen and oxygen is fed to the detector to produce the flame. Flame ionization detectors are capable

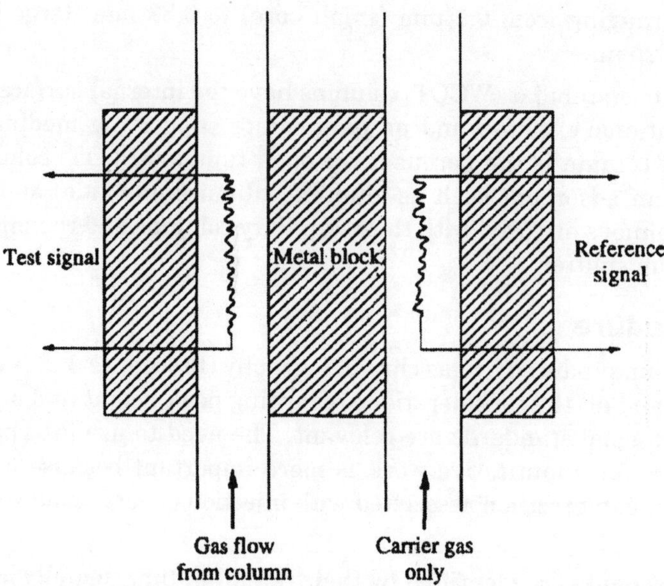

Fig. 4.14. A katharometer. A constant voltage is applied to the detector filaments and the resulting current produces a heating effect. In the reference column (carrier gas only), because the heat losses from the filament are constant, its resistance will also be constant. The heat loss from the test filament will, however, vary with the gas composition and the resulting changes in resistance can be monitored.

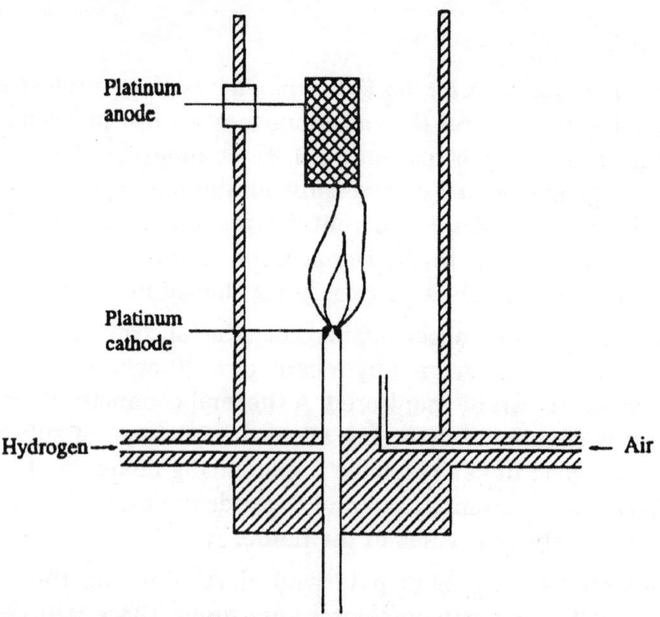

Fig. 4.15. Flame ionization detector. Hydrogen and oxygen are introduced into the gas mixture as it emerges from the column to allow it to be burnt in the detector. Some molecules are ionized in the flame and cause a current to flow between the two polarized electrodes. The degree of ionization varies with the composition of the gas mixture and the resulting changes in current can be monitored.

of detecting virtually all organic compounds and show a lower limit of detection of approximately 1×10^{-9} mol.

They also show good linearity of response and the fact that they do not respond to oxides of carbon or nitrogen or to water makes them particularly convenient for aqueous samples. They have the disadvantage, however, that samples are destroyed unless a stream-splitting device is incorporated. A modification of flame ionization detection involves the introduction into the flame of atoms of one of the alkali metals, e.g. potassium, rubidium or caesium. Such a device, known as an *alkali flame ionization detector* (AFID), gives an enhanced response to nitrogen- or phosphorus-containing compounds, detecting levels as low as 1×10^{-10} mol) The *electron capture detector* also depends upon the ionization of the carrier gases but uses a beta-emitting isotope as a means of ionization.

The isotope, usually ^{63}Ni or ^{3}H, is held on a foil in the ionization chamber through which the emerging gases pass. The ionization of the carrier gas results in the release of electrons and hence a current will flow between two polarized electrodes. If, however, electrophilic compounds are present (*i.e.* those containing oxygen, nitrogen, sulphur or a halogen) they will capture free electrons, reducing the current to some extent.

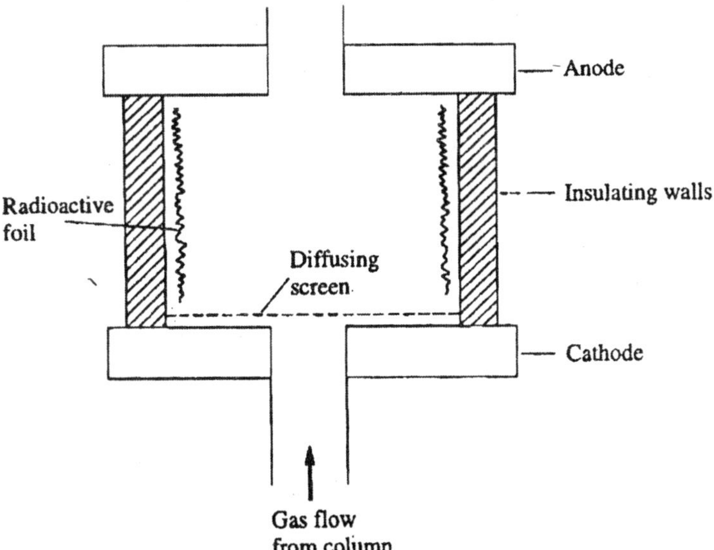

Fig. 4.16. An electron capture detector. The degree of ionization of the emerging guses caused by the emission from a radioisotope is monitored as changes in the resulting current. Electron capture detectors are extremely sensitive (1×10^{-12} mol) but are specific for electrophilic compounds. However, they can be used in parallel with flame ionization detectors to identify specific peaks in a chromatogram.

The Stationary Phase

The major consideration in selecting a stationary (liquid) phase is its polarity relative to the test compounds. A non-polar stationary phase will tend to retain non-polar solutes while a polar liquid will show greater affinity for polar solutes. In non-polar systems, because hydrogen bonding does not occur, elution of the solutes from the column is usually in order of their boiling points.

Hydrogen bonding will occur to varying degrees with polar systems and will appreciably modify the elution order because the molecules involved in hydrogen bonds will be retarded. It is difficult to classify solvents in terms of their polarity and the concept of the selectivity of a solvent was developed initially by Kovats and subsequently by McReynolds, whose classification system is currently the most popular. It is based on Kovats's concept of attributing a retention index (I) to the n-paraffin series of hydrocarbons, selected as reference compounds because of their stability and ease of purification.

For a particular stationary phase, a plot of the logarithm of the retention time against the carbon number results in a straight line graph. This can be used to attribute a retention index value to any compound on the basis of an apparent carbon number determined from the retention time of that compound. For convenience the carbon number determined from the graph is multiplied by 100 to eliminate decimals. This method involves isothermal chromatography but it is not uncommon for specific applications for retention index data to be determined from programmed temperature gradient separations. McReynolds used the retention index of certain solutes to compare different stationary phases and to assess their selectivity compared with a reference liquid phase, squalane.

Squalane is considered to be non-polar and any increase in the retention index of the selected solute on the test column com-pared to squalane may be considered to be due to the greater polarity of that solvent. McReynolds constants have been determined for all stationary phases using a range of solutes of varying polarity and may be used to assist in selecting an appropriate stationary phase.

$$\text{McReynolds constant} = I_{\text{test liquid}} - I_{\text{squalane}}$$

An examination of a table of these constants for different stationary phases and a solute that is most similar to the test substance will indicate the most suitable stationary phase. If it is desired to separate two solutes of differing polarity, a solvent should be selected that shows, a significant difference in the constants for the two most appropriate reference solutes.

Table 4.8. Stationary phases for gas-liquid chromatography

Liquid phase	*Example*	*McReynolds constants for selected test compounds*				
		Benzene	*Butanol*	*Pentanone*	*Nitropropane*	*Pyridine*
Squalane	Reference phase	0	0	0	0	0
Paraffin oils	Nujol	9	5	2	6	11
Apiezon grease	APL	32	22	15	32	42
Fluorinated hydrocarbons	Fluorolube	51	68	114	144	118
Diethyleneglycol succinate	DEGS	492	733	581	833	791
Polyethyleneglycol	PEG 600	350	631	428	632	605
	Carbowax 1000	347	607	418	626	589
Silicones, methyl	OV1	16	55	44	65	42
Silicones, phenyl	OV 17	119	158	162	243	202
Silicones, fluoro	OV 210	146	238	358	468	310
Silicones, cyano	OV275	629	872	763	1106	849

Benzene represents aromatics and olefins. Pentanone represents keto compounds and esters.
Butanol represents alcohols and weak acids. Nitropropane represents nitro- and nitrile compounds.
Pyridine represents N-heterocyles.

Column Conditions

The efficiency of a column can be assessed in a similar manner to that described for HPLC and values for the resolution index of two solutes, the number of theoretical plates and the height equivalent to a theoretical plate may also be calculated. Although it is easier to measure gas pressure, it is the actual gas flow, which is affected by the particle size and compression of the packing, that should be used in column assessment investigations. In many instances, samples contain components with a wide range of volatilities and it may be difficult to separate them quickly and effectively at a fixed temperature; a temperature gradient may be used.

The separation is initiated at the lower temperature for a specified period of time depending upon the retention times of the more volatile components. Subsequently the temperature of the column is raised at a specified rate to speed up the elution of the less volatile components; Such temperature programming can cause some problems but these are usually solved in the design of the instrument. As the temperature rises so the gas flow will fall owing to an increase in the back-pressure within the column, and it is usual to have a system that regulates the gas pressure in order to maintain the flow rate. In addition, the increase in temperature may also cause loss of the solvent (bleeding) from the column, resulting in a baseline increase which must be taken into account in peak height measurements. Temperatures above the specified maximum for the stationary phase must not be used as the column will rapidly deteriorate.

Derivatives

Many substances are not initially appropriate for gas chromatography because of their relatively high boiling points or insolubility. In such cases it is often possible to modify the compound chemically and render it more amenable to separation. In some instances, the chemical modification is used to enable easier detection of the compound, *e.g.* the introduction of a halogen for use with electron capture detectors. There are a wide variety of derivatization reactions but the most frequently used are indicated in Table 4.9.

Table 4.9. Derivatives for gas chromatography

Reaction	*Derivatives formed*	*Reagents used*	*Compounds treated*
Silylation	Trimethylsilyl	BSTFA	Widely used acids,
	R—OH R—OSi$(CH_3)_3$	TMS	phenols, alcohols
Acylation	Trifluoroacetyl	TFAA	Amines, phenols,
	Heptafluorobutyryl		alcohols
Alkylation	Methyl, ethyl or butyl	TMAH	Fatty acids, various drugs

BSTFA, bis(trimethylsilyl)-trifluoroacetamide; TMS, trimethylchlorosilane; TFAA trifluoroacetic anhydride; TMAH, trimethylanililium hydroxide.

Mass Spectrometry

Mass spectrometry is a technique that is widely employed as a detector for GLC and HPLC. It is a powerful tool for the rapid identification and quantification of even femtogram quantities of analyte. Since the early 1980s there has been considerable development of the technique and associated instrumentation and improvements are constantly being introduced. Particularly significant developments have been associated with the computing aspects of mass spectrometry, the means of producing ions from neutral species and the over-all design specification, including the production of *'bench top'* instruments.

The use of mass spectrometry has now been extended to the analysis of large molecules, *e.g.* proteins, and this has considerably enlarged its application in the biochemical field. This is largely due to the widespread use of electrospray and 'time of flight' (TOF) mass spectrometry. In addition, the introduction of mass spectrometry/mass spectrometry (MS/MS) or tandem MS, which are independent of a chromatographic separation process, has opened up a whole new area of analytical applications. There are many types of mass spectrometer, each having special design features, some offering very sophisticated modes of analysis and these details will not be described here. Rather, the fundamental principles and instrumental aspects associated with the technique in general are covered. More detailed information is available in specialist books or instrument manufacturers' publications.

The Mass Spectrum

The production of ions from neutral compounds and the examination of how these ions subsequently fragment is fundamental to mass spectrometry. Neutral sample molecules can be ionized by a variety of processes. The most important of these for the production of positively charged species is the removal of an electron or the addition of one or more protons to give either 'molecular ions' ($M+^{\bullet}$) or *'protonated molecular species'* $(M+nH)^{n+}$. This initial stage of ionization is often followed by fragmentation to produce ionized fragments, *'fragment ions'*. The extent to which further fragmentation pathways now proceed, and hence the fragmentation pattern that is produced, is characteristic of the compound.

The mass spectrometer is designed to separate and measure the mass of the ions using their mass-to-charge ratio (m/z) ratios. Ions are usually formed with a single, positive charge ($z = 1$) and in this situation m/z gives the mass of the ion. In the mass spectrum that is produced the relative amounts of the ions is displayed as their relative abundance on the y-axis and their m/z values on the x-axis. This information is plotted by the computer in alternative ways and it is important for the analyst to be aware of the ways in which the data system is operating.

In the normalized or percentage relative abundance (%RA) method, which is commonly used, the height of each peak is shown as a percentage of the biggest in the spectrum. The total ion current (TIC) is the sum of all the detector responses for each scan plotted against time and this is equivalent to a GLC trace. This information is particularly useful in quantitative analysis.

Instrumentation

There is a variety of differently designed instruments available but they all have the same essential features.

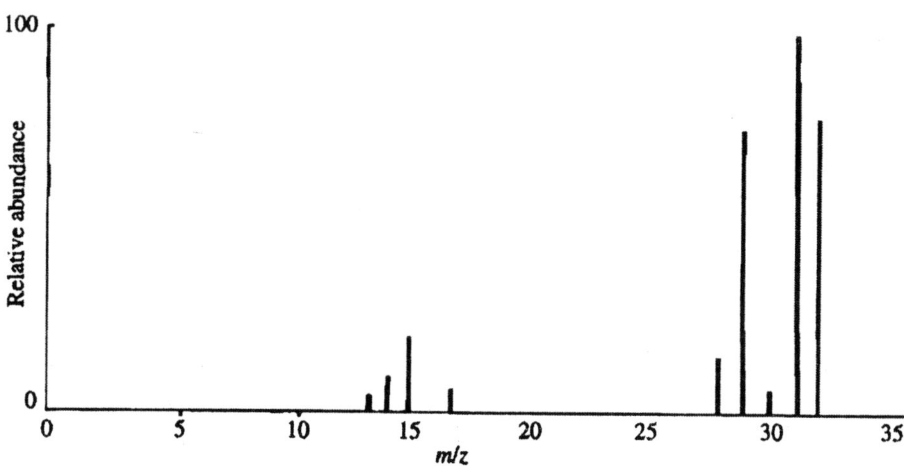

Fragment	*Mass*	*Relative abundance*	
CH	13	0.7	
CH_2	14	2.4	
CH_3	15	13	
OH	17	1	
CO	28	6.3	
CHO	29	64	
CH_2O	30	3.8	
CH_3O	31	100	(base peak)
CH_3OH	32	66	(molecular ion)

The strongest peak is called the base peak and is set to read 100. The signal from the non-fragmented ionized molecule is not usually very strong and is sometimes missing completely.

Fig. 4.17. Mass speetrum of methanol.

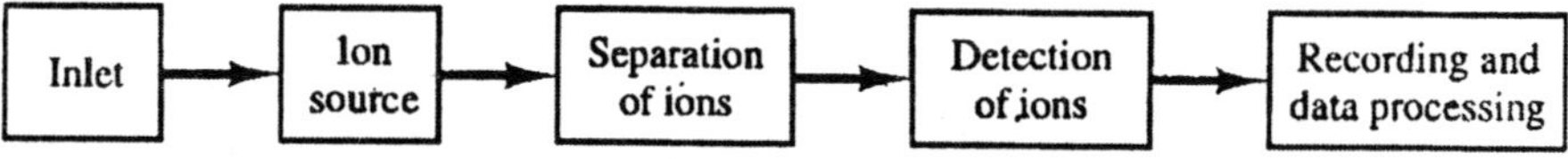

Fig. 4.18. Diagrammatic representation of the components of a mass spectrometer. Often the sample is introduced through the inlet as a vapour and subjected to bombardment with electrons. The products of this process are separated by a variety of means and detected as they exit the system.

Inlets and Ion Sources

Mass spectrometry is traditionally a gas phase technique for the analysis of relatively volatile samples. Effluents from gas chromatographs are already in a suitable form and other readily vaporized samples could be fairly easily accommodated. However the coupling of mass spectrometry to liquid streams, *e.g.* HPLC and capillary electrophoresis, posed a new problem and several different methods are now in use. These include the spray methods mentioned below and bombarding with atoms (fast atom bombardment, FAB) or ions (secondary-ion mass spectrometry, SIMS). The part of the instrument in which ionization of the neutral molecules occurs is called the ion source.

The commonest method of ionization is by electron ionization or electron impact (El) of a volatile sample in an ion chamber maintained under a vacuum. This produces positively charged ions. The energy of the electrons causes an electron to be ejected from the molecule, forming a molecular ion and fragment ions. This technique is commonly used for GC-MS. Several methods of ionization are based on the use of electric fields.

In electrospray (ES) a liquid is pumped through a stainless steel capillary held at a potential of several thousand volts. This results in a spray of highly charged droplets. The droplets are dried using heat and/or a gas flow and after evaporation the charges are retained by the sample molecules. This makes electrospray a suitable ionization method for liquid chromatography/ mass spectrometry (LC/MS). Its applicability to large RMM proteins arises from the fact that in electrospray ions formed are often multiply charged and therefore their m/z values fall within the capabilities of conventional instruments.

Separation of Ions

Sector Instruments

The separation of ions according to their m/z ratios is achieved using electric and/or magnetic fields in a number of ways. The trajectories of ions moving in such fields are determined by their m/z values and these can be monitored to ascertain their mass. Double-focusing instruments use the combined effects of electric and magnetic fields to effect separation in a typical instrument, after the ions have been accelerated away from the ion source through a potential of several kilovolts, they pass through slits to reduce the spread.

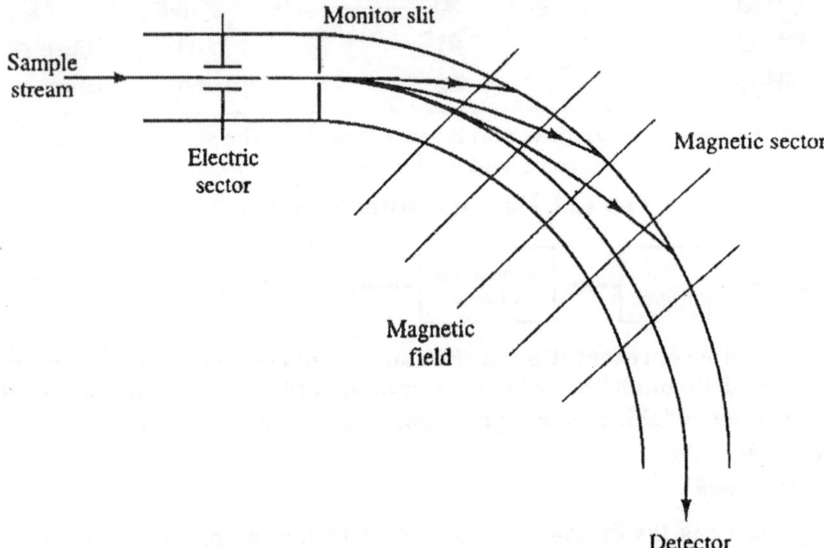

Fig. 4.19. Schematic diagram of a sector mass spectrometer.

The ion beam then passes through an electric sector, formed by applying a potential across two plates. This has the effect of deflecting the beam into an arc, focusing ions of like energies, irrespective of their mass-to-charge ratios, at the monitor slit. Mass separation is effected in the magnetic sector as the mass-to-charge ratios will determine the curvature of the path taken by the ions. Some instruments use an array of multiple detection devices to record the ions of several different *m/z* ratios simultaneously while, more commonly, others scan the magnetic field to record one *m/z* at a time on a point detector.

Quadrupole Instruments

The ions are propelled from the ion source into the quadrupole analyzer by a low accelerating voltage of only a few volts. They enter the space between four or more parallel

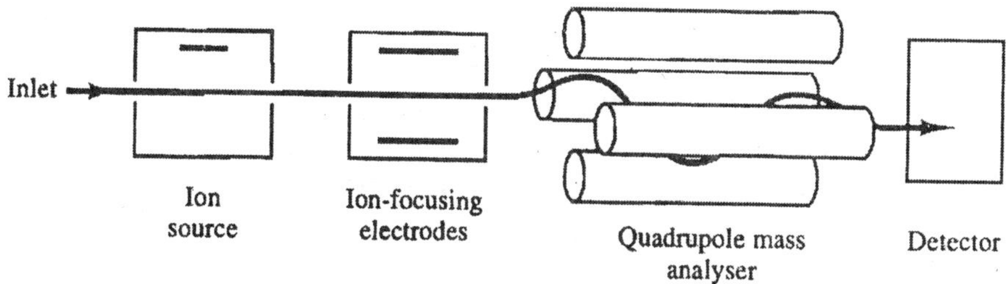

Fig. 4.20. Diagrammatic representation of a quadrupole mass spectrometer system.

rods which are precisely positioned and have a DC voltage and a radio-frequency potential associated with each opposite pair. Under the influence of the combined electric fields the ions oscillate and follow complex trajectories through the rods. The separation of ions according to mass can be effected by changing the DC and radio-frequency voltages while keeping the ratio of the two constant. Although this type of instrument has lower resolution than magnetic sector instruments, it is robust and available in bench top designs, and the rod voltages can be readily changed in order to focus on selected ions (SIM). This makes it very useful for quantitative work. An instrument that stores ions and then ejects those of selected masses is called an ion trap. Like quadrupole analysers such instruments employ electric fields to effect separation and are based on a similar technology. Such systems can be considered as alternatives to tandem MS. Another approach to the separation of the ions is based on differences in their velocity after acceleration through a potential. The mass is related to the time taken to reach the detector. This is known as TOF-MS.

Tandem Mass Spectrometry (MS/MS)

This basically means that two instruments have been linked together. The first analyser can replace the traditional chromatographic separation step and is used to produce ions of chosen m/z values. Each of the selected ions is then fragmented by collision with a gas, and mass analysis of these product ions effected in the second analyser. The resulting mass spectrum is used for their identification. The potential combinations of the various magnetic sector and quadrupole instruments to form such coupled systems is considerable. Ion traps may also be operated in a tandem MS mode.

Computer

Modern mass spectrometers are fully computer controlled for all aspects of operation including automatic scanning, signal processing, data collection, quantitation and, for identification, library searching. This latter function involves comparing the spectrum obtained with that of known compounds held in the computer database.

Quantitation

Internal standards are used in quantitative work in a similar manner to GLC. These are usually homologues or isotopically labelled analogues of the analyte, *e.g.* deuterium (^2H).

METHODS BASED ON IONIC NATURE

Ionizable groups in a molecule give it an ionic character and, as a result, under certain conditions it will carry a charge. However, the intensity of this charge and, in the case of many molecules, the sign (positive or negative) depend upon the pH and composition of the solution. These properties form the basis of several methods of separation, namely ion-exchange chromatography and electrophoresis.

Ion-exchange Chromatography

Although the phenomenon of ion-exchange has been appreciated for many years, it was the development, by D'Alelio in 1942, of synthetic ion-exchange media based on the polystyrene resins that extended the use of ion-exchange as an analytical tool.

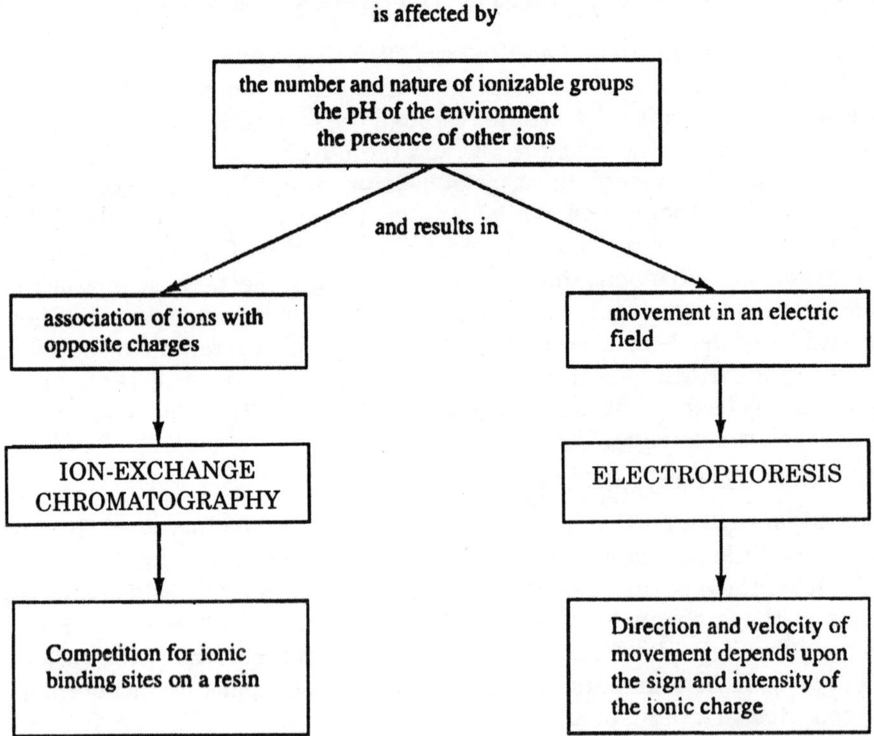

Fig. 4.21. Separation methods in which the ionic nature of the test molecule plays an important part.

An ion-exchange resin consists of an insoluble, porous matrix containing large numbers of a particular ionic group which are capable of binding ions of an opposite charge from the surrounding solution. Hence a cation-exchange resin contains fixed anions and the mobile cations can be any cations from the solution. Ion-exchange chromatography is essentially a displacement technique in which the ions in either the sample or the buffer displace the existing mobile ions associated with the fixed resin ions. Ion-exchange resins are cross-linked polymers which are typically polystyrene, cellulose or agarose based.

Polystyrene is hydrophobic in nature and useful for inorganic ions and small molecules while cellulose and agarose are hydrophilic and more useful for the larger, biologically important molecules, *e.g.* proteins and nucleic acids, which either would be adversely affected by a hydrophobic environment or could not gain access to the small pore structure. The degree of cross-linking of the resin is significant in that it enhances the rigidity and insolubility of, the resin but it also reduces the pore size, and this is undesirable when working with macromolecules. Polystyrene cross-linking is expressed as the percentage of divinyl benzene (the cross-linking group) present in the original preparation and is commonly in the region of 8-10%. Although, theoretically, any ionic group can be used in an ion-exchange resin, in practice the number is limited. For ion-exchange to take place the resin-bound group must be ionized, and its capacity to do this is related to its strength (or ability to dissociate). The dissociation of weak acids is suppressed by the presence of only slightly increased concentrations of hydrogen ions, while the ionic form of a weak base (its conjugate acid) dissociates as the pH increases:

$$NH_2H^+ \rightleftharpoons NH_2 + H^+$$

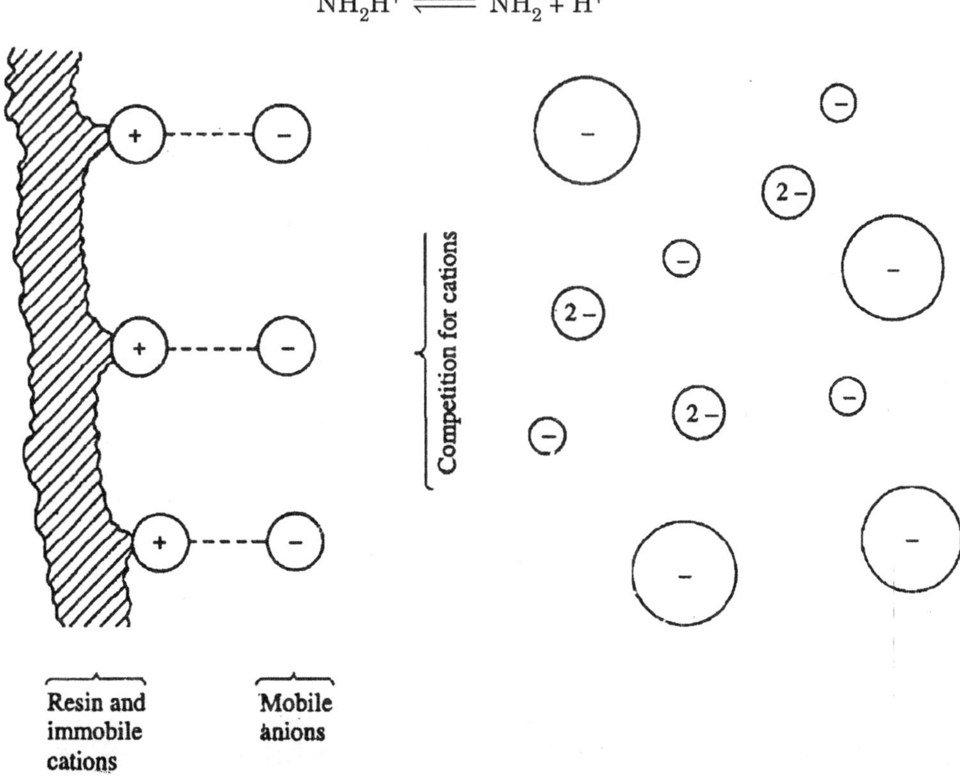

Fig. 4.22. Ion-exchange chromatography. Mobile ions compete for the fixed immobile ions in the resin. The major factors that influence the process are the molar concentration of the ions and the charge that they carry.

Hence an ion-exchange resin cannot be used at a pH that suppresses the ionization of the group, and weak anion exchange resins, for instance, are only effective over the pH range 2–8. The stronger acids and bases, however, are capable of being used over almost the whole pH range. The major consideration in selecting a resin is the charge carried by the test ions. In the case of uncomplexed inorganic ions, it is relatively constant but for many molecules, it can be altered considerably by variations in pH. There are three major factors that determine the binding of ions in such a competitive system.

The size of the charge carried will result in divalent ions showing greater affinity for the resin than monovalent ions. The intensity of the charge is also significant and small monovalent ions, *e.g.* hydrogen, will show greater affinity than large monovalent ions, e.g. potassium. Superimposed on both of these considerations is the effect of the concentration of the ions and this is demonstrated by the fact that a high concentration of a low affinity ion is capable of displacing a low concentration of an ion with a higher affinity for the resin. It is the careful control of these three factors that provides the selectivity in ion-exchange separations.

Methods in ıon-exchange Chromatography

Ion-exchange chromatography may be used for either purification of an individual component or fractionation of a mixture. In order to isolate a particular component from a

Table 3.10. Ion-exchange media

Medium	*Nature*	*Effective pH range*	*Applications*
Anion exchangers			
Quaternary ammonium	Strong	2–11	Strong acids, *e.g.* nucleotides
TTertiary ammonium	Intermediate	2–7	Weakly acids, *e.g.* organic acids
Diethylaminoethyl	Weak	3–6	Weakly polyanionic, *e.g.* proteins
Cation exchangers			
Sulphonated	Strong	2–11	Strong bases, *e.g.* amino acids
Carboxylate	Intermediate	6–10	Weak cations, *e.g.* peptides
Carboxymethyl	Weak	7–10	Weak polycationic, *e.g.* proteins

mixture, it may be selectively retained on a resin while the unwanted constituents, which under the conditions of the analysis should be uncharged or carry the same charge as the resin, will be eluted. Subsequently the required component can itself be eluted in a small volume of an appropriate buffer. Conversely, the resin may be selected so that the required compound does not bind but unwanted ions do and the sample can be eluted quickly from the column without the contaminating ions.

Neutral compounds lend themselves to this technique in which a bed of mixed anion and cation resins can remove unwanted ions and replace them with water (H^+ and OH^-). The technique of fracu'onation is used for samples that contain mixtures of similar ions and permits the separation and quantitation of each component, the various components being eluted sequentially as the solvent composition is changed. It is in this form that ion-exchange chromatography is used as an HPLC technique. The resin should be prepared for use by washing it with a solution containing a high concentration of an ion which has a high affinity for the resin, *e.g.* hydrochloric acid or sodium hydroxide solution (1.0 mol l^{-1}). The hydrogen and hydroxyl ions will displace all other ions present from cation- and anion-exchange media respectively. After washing in water or buffer to remove the excess acid or alkali, the resin is said to be in the hydrogen or hydroxyl form respectively. For some applications it may be necessary to convert it to a different ionic form, *e.g.* the sodium form, that has an affinity for the resin comparable to that of the test ion. The resin is finally equilibrated with the chosen buffer, the pH of which must be carefully selected to ensure the correct ionization of both the test ions and the resin. A small volume of sample is applied to the column and the components of the mixture eluted using the buffer.

A single buffer (isocratic separation) or a gradient technique may be used depending on the complexity of the sample. Gradient elution normally involves either changes in pH (causing alterations in ion affinity for the resin) or changes in the concentration of the buffer (causing displacement of the test ions).

Electrophoresis

Electrophoresis is the term given to the migration of charged particles under the influence of a direct electric current. It is a single-phase system and depends upon the relative mobilities of ions under identical electrical conditions. Moving-boundary electrophoretic techniques, originally demonstrated by Tiselius in 1937, employ a U-tube with the sample occupying the lower part of the U and the two limbs being carefully filled with a buffered electrolyte so as to maintain sharp boundaries with the sample. Electrodes are immersed in the electrolyte and direct current passed between them. The rate of migration of the sample in the electric field is measured by observing the movement of the boundary as a function of time.

For colourless samples differences in refractive index may be used to detect the boundary. Such moving-boundary techniques are used mainly in either studies of the physical characteristics of molecules or bulk preparative processes. Zonal techniques are the most frequently used form of electrophoresis and involve the application of a sample as a small zone to a relatively large area of inert supporting medium which enables the subsequent detection of the separated sample zones.

A wide range of supporting media have been developed either to eliminate difficulties caused by some media (*e.g.* the adsorptive effects of paper) or to offer additional features (*e.g.* the molecular sieving effects of polyacrylamide gel). Although molecular frictional effects impede the movement of molecules through a liquid, an effect that increases with the size of the molecule, the major factors in electrophoresis are the charge carried by the molecule and the voltage applied. The nature of the charge (positive or negative) will determine the direction of migration while the magnitude of the charge will determine the relative velocity. This charge, although initially due to the ionization effects mediated by the pH of the buffer, will be appreciably modified by other features of the buffer composition.

For a colloidal particle in suspension, *e.g.* a protein, its charged surface is surrounded by a layer of oppositely charged ions derived from the solution. There is an immobile, fixed layer of ions adsorbed on to the surface and a diffuse, mobile layer which becomes more mobile with increasing distance from the surface of the colloid. The charge developed by the molecule is due to its chemical nature and the pH of the buffer and is known as the *electrochemical potential.* For colloids, this is reduced as the salt concentration in the buffer increases, due to the partial neutralizing effect of the adsorbed layer of ions.

The net resulting charge is known as the *zeta potential* and is the determining factor in electrophoretic mobility. The combined effect of the fixed and mobile layers of ions also increases the size of the article and reduces its mobility even further. The electrophoretic mobility of an ion is inversely related to the ionic strength of the buffer rather than to its molar concentration. The ionic strength (μ) of a buffer is half the sum of the product of the molar concentration and the valency squared for all the ions present in the solution. The factor of a half is necessary because only half of the total ions present in the buffer carry an opposite charge to the colloid and are capable of modifying its charge:

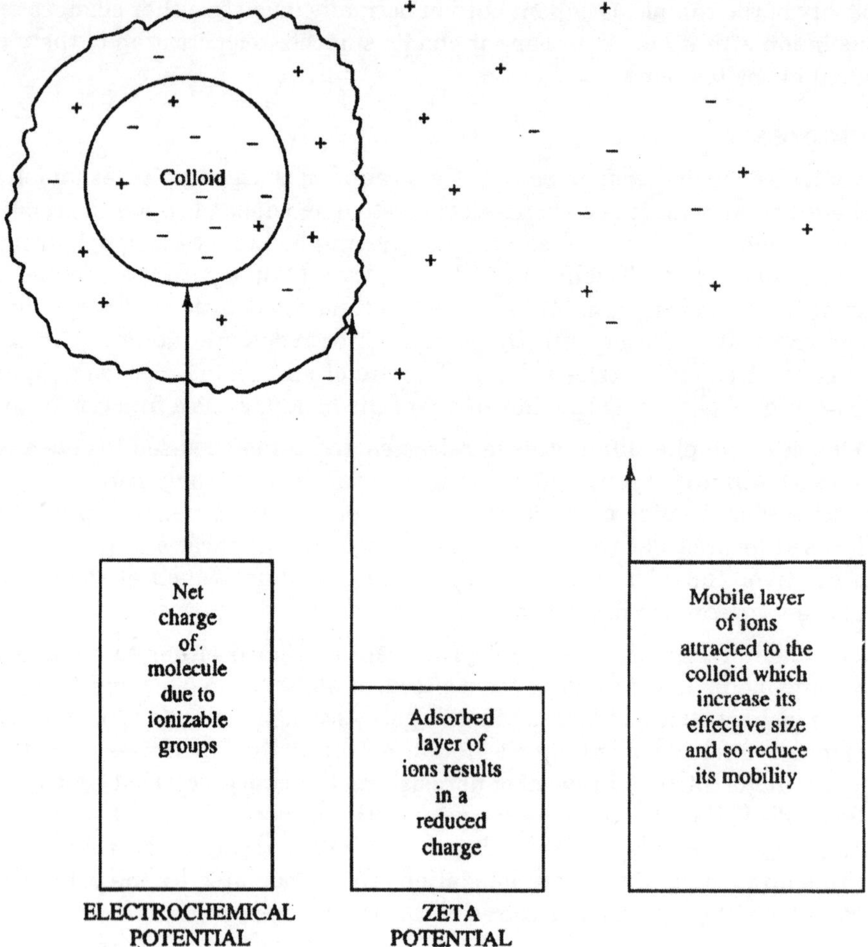

Fig. 4.23. The charge on a colloid. The charge carried by a colloid because of its chemical composition and the pH of the solution (the electrochemical potential) is reduced by the adsorption of ions from the solution and the resulting charge is known as the zeta potential.

$$\mu = \frac{1}{2}\Sigma ct^2$$

The rate of movement of a charged particle is also related to the voltage gradient applied across the supporting medium, which is quoted in volts per centimetre (*i.e.* the distance between electrodes). The current generates heat, which causes an increase in the conductivity of the electrolyte solution and this further increases the current. This rise in temperature will cause some evaporation of the buffer, which in the small amount held in the supporting medium will result in an increase in its concentration. This, will affect both the zeta potential of the colloid and the conductivity of the buffer. Because of these effects it is necessary to control the temperature of the system in some way and to choose between the use of a fixed voltage and a fixed current.

Some power sources have the facility to select a fixed wattage output, which is effectively a compromise position. A additional problem with some supporting media is the phenomenon of *electroendosmosis,* in which the buffer itself moves due to an electrophoretic effect and hence masks the movement of the solute to some extent. However, this feature is exploited in some situations to aid separation. Electroendosmosis is caused by the presence of negatively charged groups on the surfaces of some supporting media, *e.g.* paper, agar, and these induce an opposite charge in the buffer solution. As a result, under alkaline conditions, the buffer moves towards the cathode, the rate being greater for lower concentrations of buffer. A wide range of electrophoretic equipment is available and while it varies considerably in appearance, there is a common basic design.

A well-designed tank should include a cover to prevent any significant evaporation from the supporting medium and an effective cooling system is necessary when gels or high voltage techniques are used. This usually is in the form of an electrically insulated plastic or metal base for the supporting medium, through which cold water is circulated. A stabilized voltage supply is necessary and safety devices on the whole equipment are essential to prevent accidents, particularly if high voltages are used.

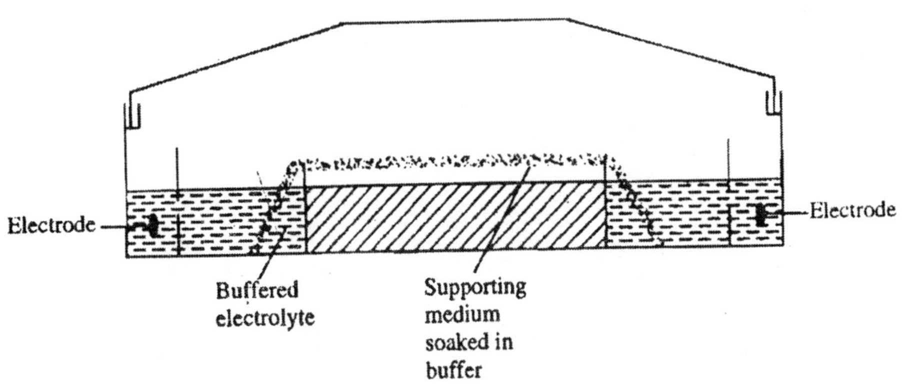

Fig. 4.24. An electrophoresis tank.

Supporting Media

If the supporting medium is a sheet or membrane, it is soaked in the buffered electrolyte, the excess being removed before the sample is applied. Alternatively a gel, such as agar or polyacrylamide, can be prepared in the buffer and a slab or column used for the separation process. *Filter paper* has been an extremely useful and convenient medium for many years and compared with some of the more recent membrane media is capable of handling a relatively large volume of sample, *e.g.* 10 μl. The random structure of the paper, however, tends to result in irregularities in the separation patterns and the relatively polar nature of the cellulose shows some adsorptive effects giving a significant degree of 'tailing' of the zones. Additionally the large buffer volume in the paper, although an advantage from conductivity and evaporation points of view, because of the relatively long separation times (12–18 hours), results in a significant amount of diffusion and hence broadening of the zones. *Cellulose acetate membrane* (CAM) was developed with the aim of reducing the polar nature of paper bylacetylating the hydroxyl groups and producing a more regularly defined pore structure.

For these reasons, it has considerable advantages over paper and shows minimal absorptive ertects with increased separation rates (1–2 hours) but it still shows electroendosmotic effects. Particular care must be taken over handling the membrane, which is very fragile when dry but is much more resilient when wet. Less sample can be applied to cellulose acetate membrane than to paper, volumes of 1–2 μl being common. After locating the separated bands, the membrane can be made transparent for densitometry by soaking in either an oil or an organic solvent mixture specified by the manufacturer. A variety of gels has been used for preparative techniques and to eliminate some of the disadvantages of the solid support media. Purified agar is used at a concentration of 10g l^{-1} in a buffered electrolyte and the samples are introduced in a small well or trough cut out of the gel. It shows an appreciable degree of electroendosmosis and, owing to its polar nature, causes the precipitation of some substances, notably the lipoproteins. In such situations the use of *agarose* (5–8 g l^{-1}), a less polar fraction of agar, eliminates virtually any electroendosmotic effect and gives a molecular sieving effect

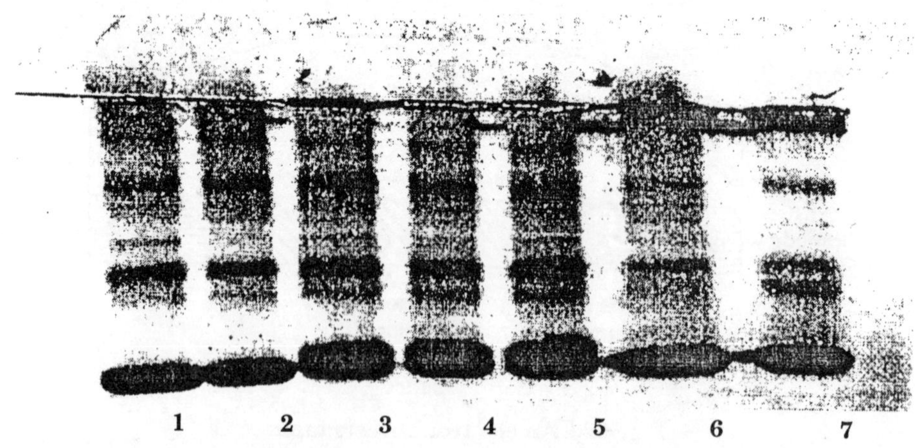

Fig. 4.25. Starch gel electrophoresis of human serum proteins. Samples 1 and 2 are normal while samples 3, 4 and 5 show an extra band 'adjacent to the normal albumin which is due to the presence of bis-albumin. Sample 7 contains a myeloma protein that has remained near the origin.

with large molecules such as proteins and nucleic, acids. The molecular sieving ettect can be increased considerably by using gels with a smaller pore structure. *Starch gel* is prepared by the partial hydrolysis of potato starch; the greater the degree of hydrolysis, the smaller the fragments and, hence, the smaller the pore size of the resulting gel.

In such a medium the larger molecules are impeded by the pore structure and therefore do not move as rapidly as the smaller molecules with the same charge The gel is prepared by heating a slurry of the starch in the appropriate buffer (150 g l^{-1}) in a boiling water bath until the mixture becomes translucent. A vacuum is applied to the flask to remove air bubbles and the solution quickly poured into a heated tray and allowed to cool to form a gel about 5 mm thick. During electrophoresis, the gel must be adequately cooled to prevent it melting and separations may take up to 6 hours at about 150 V. These gels are extremely fragile and the method of hydrolysis is entirely arbitrary; it is difficult to reproduce the pore size in successive preparations. *Polyacrylamide shows* many advantages over starch gel as a medium for high resolution electrophoresis and because of its synthetic nature its pore size can be more easily controlled.

The gel is formed by the polymerization of the two monomers, acrylamide and across-linking agent, *N,* methylene-bis-acrylamide. The proportion of the two monomers and not their total concentration is the major factor in determining the pore size, the latter having more effect on the elasticity and

Acrylamide

```
CH2 = CH — C — NH2
           ‖
           O
```

Methylene-bis-acrylamide

```
CH2 = CH — C — N — CH2 — N — C — CH = CH2
           ‖   |          |   ‖
           O   H          H   O
```

Polyacrylamide

```
— CH2 — CH —   CH2—CH—   CH2 — CH —
        |          |             |
       C=O        C=O           C=O
        |          |             |
       NH2        NH            NH2
                   |
                  CH2
                   |
                  NH
                   |
                  C=O
                   |
— CH2 — CH —   CH2—CH—   CH2 — CH —
        |                        |
       C=O                      C=O
        |                        |
       NH2                      NH2
```

Fig. 4.26. Polyacrylamide.

Preparation of gel

A mixture is prepared containing:

1 volume of the selected buffer,
2 volumes of the selected mixture of monomers,
1 volume of TEMED (3.0 g l^{-1}).

Dissolved oxygen (which tends to inhibit the reaction) is removed under a vacuum. Four volumes of ammonium persulphate (3.0 g l^{-1}) are added and gently mixed before pouring into the mould. The mixture is covered with a thin-layer of water to exclude air and allowed to polymerize (approximately 30 min). Protective gloves should be worn and the preparation of the gel carried out in a fume cupboard because the monomers are toxic.

transparency of the gel. A minimum total concentration of approximately 20g l^{-1} is necessary for gel formation although concentrations in the region of 70 g l^{-1} are frequently used. The pore size of the gel decreases with the increasing proportion of the bis-acrylamide, with a limiting value of approximately 5% of the total, giving minimum pore size. For protein electrophoresis, gels containing 2.5% of the cross-linking agent are often used. Polymerization of the gel may be achieved either by ultraviolet photoactivation with riboflavin, or, preferably, using ammonium persulphate as a catalyst. It is necessary to include an initiator for the reaction, TEMED (tetramethylethylene diamine) being commonly used. Polyacrylamide gel (PAG) electrophoresis is performed either in cylindrical glass tubes or in flat beds. The method is comparable with other electrophoretic techniques but care must be taken to keep the current low to prevent any significant heating effect. PAG electrophoresis is an important technique in many areas.

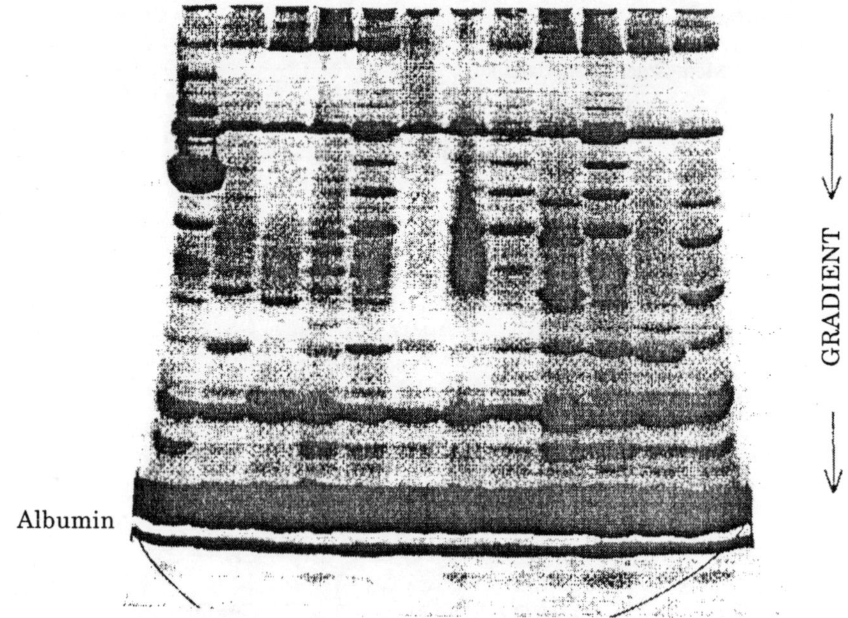

Fig. 4.27. Polyacrylamide gradient gel electrophoresis of human serum proteins. The proteins are separated in a gel which has an increasing concentration gradi-ent with a parallel decrease in pore size, which restricts the movement of the larger molecules. Note the large number of different protein bands that can be demonstrated.

In DNA sequencing it provides a fundamental method of separation, and in protein studies it is important both in the form described here and also in conjunction with the detergent, sodium dodecyl sulphate (SDS), giving a valuable method for assessing the relative molecular mass of a protein.

Iso-electric Focusing

Electrophoresis depends primarily upon the charge carried by a molecule, but at its iso-electric pH (*pI*) a molecule will not migrate in an electric field. If electrophoresis is performed in an electrolyte that has a pH gradient, molecules will migrate to the point where the pH of the solution is the same as their iso-electric pH and remain there as long as the gradient and potential differences are maintained.

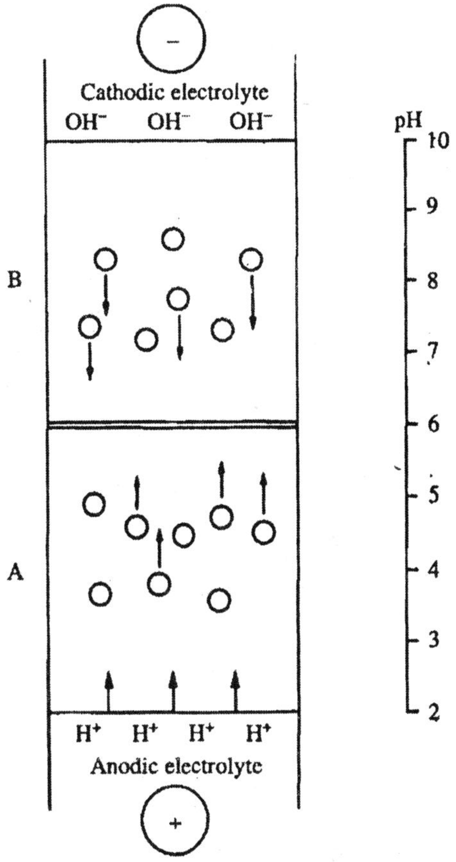

This is the basis of iso-electric focusing techniques. Artificial pH gradients cannot be produced in the same way as concentration gradients but can only be formed electrophoretically. If an acid is used as the anodic electrolyte and an alkali as the cathodic electrolyte, when a voltage is applied the hydrogen and hydroxyl ions will move towards each other. A pH gradient formed in this way would be extremely fragile and has to be stabilized with suitable buffering agents. These will also migrate electrophoretically to their iso-electric pH points and because they will have no charge at that point may also/ show no electrical conductivity, so disrupting the electrophoretic process.

Fig. 4.28 Iso-electric focusing. A protein which has an iso-electric pH of 6, in position A in the pH gradient is on the acid side of its iso-electric pH and will carry a positive (H^+) charge. It will migrate electrophoretically towards the cathode. In position B, the charge on the protein will be negative and it will migrate towards the anode. In both cases, movement is towards its iso-electric pH position, where it will remain.

The use of polyamino-polycarboxylic acids, which show a significant electrical conductivity at their *pI* values, enables the formation of a stable pH gradient. These compounds are commercially available often under the trade name of Ampholine carrier ampholytes.

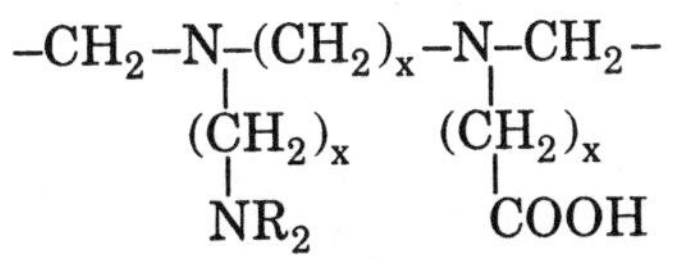

$$
\begin{array}{ccccc}
-CH_2- & N & -(CH_2)_x- & N & -CH_2- \\
 & | & & | & \\
 & (CH_2)_x & & (CH_2)_x & \\
 & | & & | & \\
 & NR_2 & & COOH &
\end{array}
$$

where R = H or $-(CH_2)_x-COOH$

and $x = 2$ or 3

Ampholines are available with a wide range of pI values as a result of their precise chemical nature and this allows the formation of pH gradients varying from 2.5 to 11 or over narrower pH ranges.

They have relative molecular masses below 1000 and hence may be easily separated from the sample macromolecules after electrdphoresis by either dialysis or gel permeation techniques. They show all the qualities necessary for iso-electric focusing techniques, *i.e.* high buffering capacity, high solubility and good, although minimal, conductivity at their pI values. They have generally no toxic effects on cells or enzymic systems nor do they show any antigenic effects and these features are often important when dealing with biological samples. A pH gradient is produced by incorporating a mixture or Ampholines with appropriate pI values in either a polyacrylamide slab or column.

An acid is used at the anode and an alkali at the cathode, the pH of these being approximately the same as the pI values of the two extremes of the Ampholine range. Phosphoric acid and sodium hydroxide are suitable for wide range separations while various amino or carboxylic electrolytes are used for intermediate pH ranges. The mixture of Ampholines will initially show a pH that is approximately the mean of all their pI values but the application cf a voltage will cause each one to migrate towards one of the electrodes but, owing to the pH gradient which also develops, each Ampholine will stop moving electrophoretically at its iso-electric pH.

The gradient will be maintained as long as the voltage is applied but it is necessary to stabilize it using either a gel or a sucrose concentration gradient in liquid column techniquese. Most flat-bed electrophoresis tanks are suitable for iso-electric focusing, provided that adequate support and cooling of the gel is possible. More recent techniques tend to use relatively high voltages (1–2 kV) but satisfactory separations can be achieved using lower voltages of 400–500 V although the separation time will be longer (3–4 hours). The prepared Ampholine gel is set up in the tank and thick filter paper strips are soaked with either the anodic or cathodic electrolyte and placed along the appropriate edge of the gel.

The samples may be applied either to small filter paper squares laid on the surface of the gel or, for bulk preparative work, incorporated in the gel. The appropriate voltage is applied through terminals attached to the electrode wicks and after about 30 min can be switched off to permit the removal of the sample filter papers before continuing the separation.

It is difficult to know when separation is complete but if two samples of coloured protein, *e.g.* haemoglobin, are placed adjacent to each other but in different positions on the gel, they will eventually form sharp bands at the same position, indicating completion of the process. When separation is complete, it is essential to fix the protein in the gel as soon as possible usually by immersion in 5% trichloroacetic acid solution or another suitable precipitating agent followed by an appropriate staining method.

Isotachophoresis

Isotachophoresis or displacement electrophoresis, as it is sometimes called, is a recent development of a principle described in 1920. In this technique the electrolyte between the electrodes is not a uniform single solution but a discontinuous sequence of different electrolyte solutions the ions of which all migrate at the same velocity. Under identical conditions of concentration and voltage, different ions have different electrophoretic mobilities and to cause all the ions to move at the same speed a different voltage gradient is required for each group of ions.

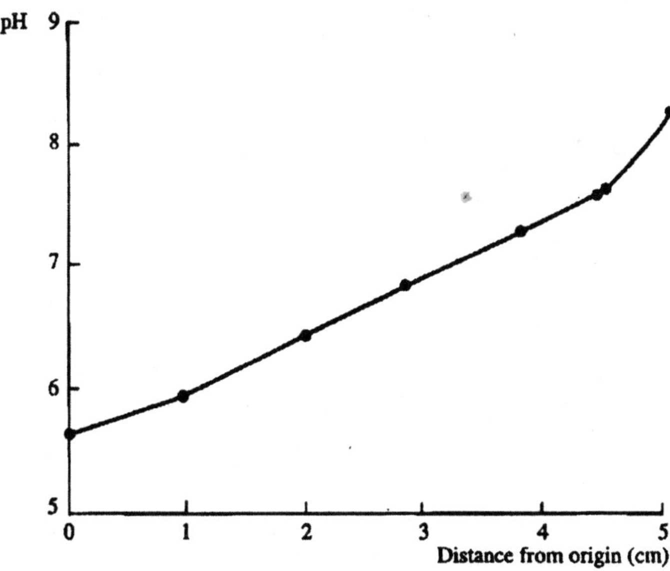

Fig. 4.29. Iso-electric focusing of haemoglobin variants and derivatives.

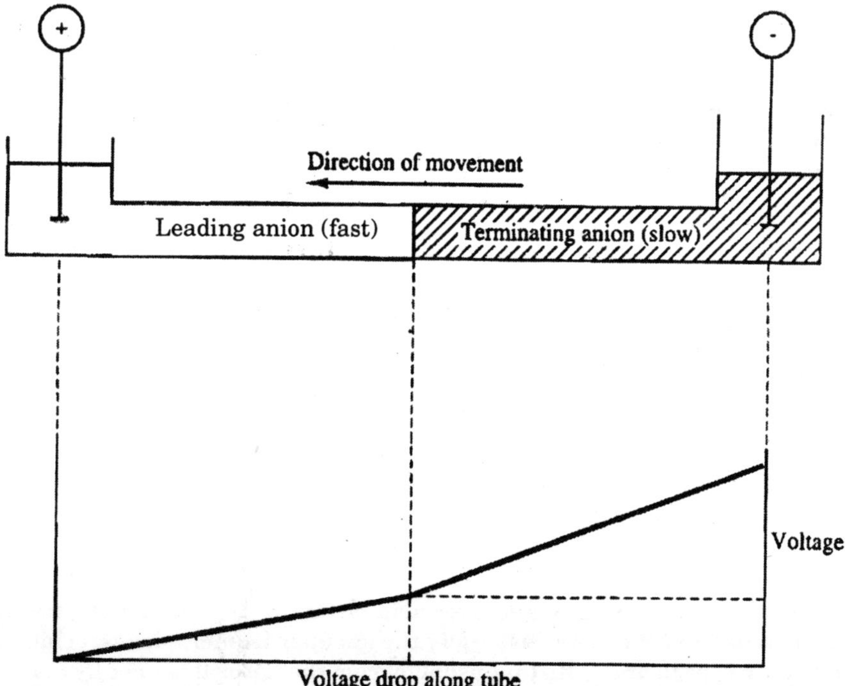

Fig. 4.30. Isotachophoresis. The spontaneous development of different voltage gradients across the two electrolyte solutions has the effect that both types of anion migrate at the same velocity.

These different gradients develop spontaneously during the electrophoretic process. In practice isotachophoresis is usually performed in narrow tubes with electrodes at either end and is one form of capillary electrophoresis. For the separation of a particular type of ion, *e.g.* an anion, two buffered electrolyte solutions are selected that have different anions but a common cation with a buffering capacity. One of the anions (termed the leading electrolyte) should show a greater mobility than the other anion and occupies the anodic end of the tube. The slower anion (the terminating electrolyte) occupies the cathode end of the tube, with a sharp boundary separating the two solutions.

The application of a potential difference across the tube will result in the common buffering cation migrating towards the cathode and the two anions towards the anode. The impossible situation of an ionic gap resulting from the different mobilities of the two anions is prevented by the spontaneous development of a discontinuous voltage drop along the tube with a larger gradient developing across the terminating electrolyte than across the leading elec-trolyte, so maintaining the same velocity for all the ions. If a sample containing various anions is introduced in between the leading and the terminating electrolyte, the different ions will move at a speed that depends upon their ionic mobility and the voltage gradient.

A test ion which is in the terminating electrolyte will move quickly because of the higher voltage but when it moves into the leading electrolyte it will slow down because of the lower voltage gradient. As a consequence the ions will move into zones in decreasing order of their ionic mobilities and a voltage pattern will develop along the tube to maintain a constant velocity of all ions. This forms the basis of the qualitative or preparative form of isotachophoresis. In addition to this qualitative aspect, isotachophoresis may also be used quantitatively.

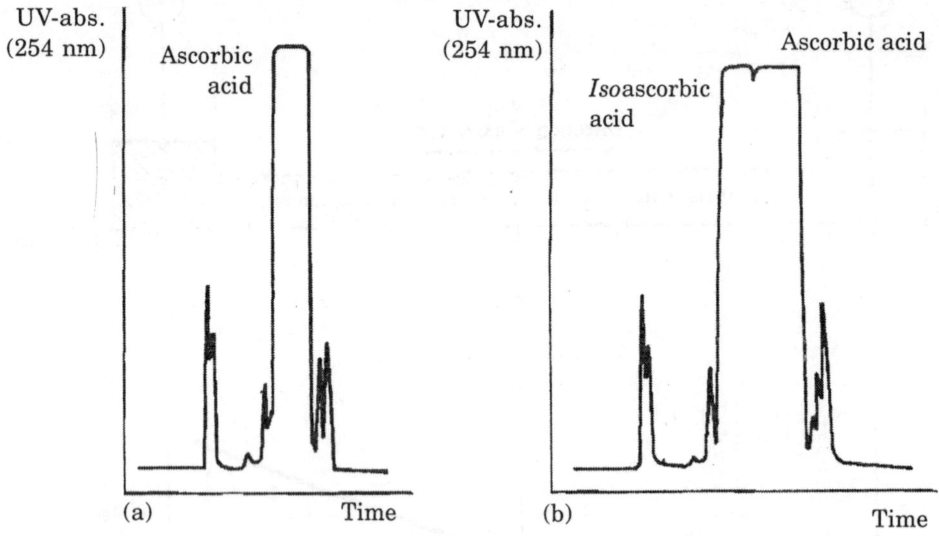

Fig. 4.31. Analytical isotachophoresis. Ascorbic acid (vitamin C) is naturally present in many foods and is often added to others. Occasionally the cheaper isomer, oisoascorbic acid, which has no vitamin action, is used and is distinguishable from the natural isomer by many analytical methods, (a) shows the analysis of a sample of com-mercial fruit juice while (b) shows the same fruit juice to which a known amount (4 nmol) of isoascorbic acid has been added.

Once the sample zones have developed, the conductivity of each zone and hence the voltage drop across it will be related to the concentration of the ions and in order to maintain uniform conductivity the concentration (*i.e.* the volume occupied by each component) will alter. Hence at equilibrium, not only will the components of a mixture be separated from each other but the volume occupied by each component will be proportional to its initial concentration in the sample. Detectors similar to those described for HPLC may be suitable but isotachophoresis is a specialized technique and special instrumentation is necessary. This now resembles that used for other types of capillary electrophoresis.

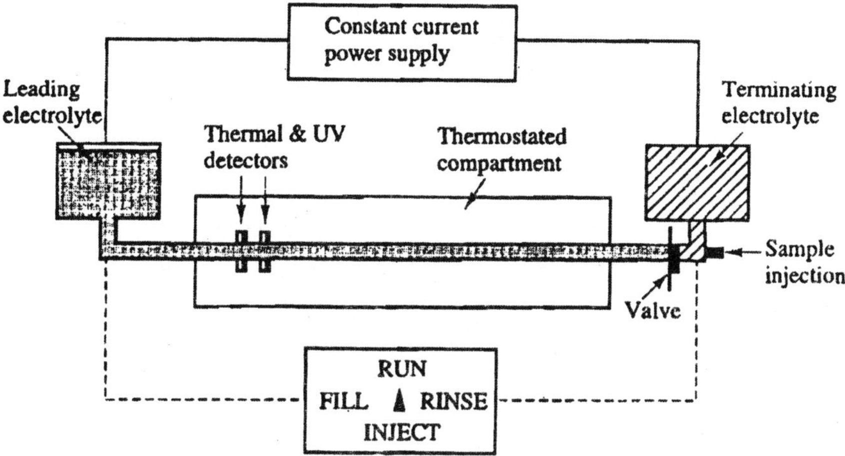

Fig. 4.32. Schematic diagram of the isotachophoresis apparatus.

Capillary Electrophoresis

Capillary electrophoresis (CE) is a family of related techniques that began to gain popularity in the late 1980s, when commercial instrumentation became available. It employs narrow bore (20–100 μm) fused silica capillaries, externally coated with plastic (polyimide) for increased durability in lengths of 10–100 cm. Although high voltages (up to 30,000 V) are used, the heat is dissipated quickly because of the relatively large surface area and cooling is not normally required.

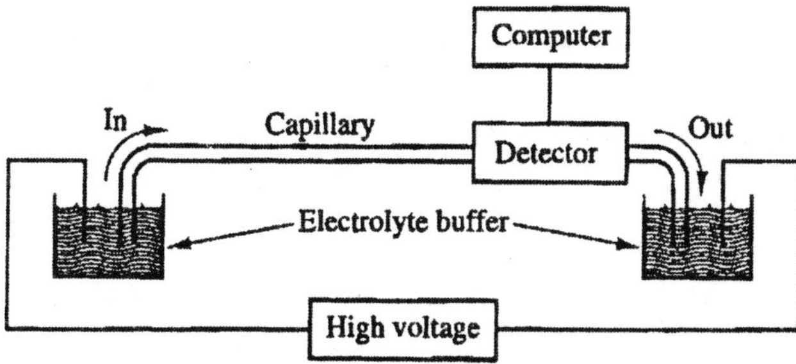

Fig. 4.33. Capillary electrophoresis. The sample is introduced through the inlet, washed in with a wash buffer and then separated under the influence of a potential difference between the two electrodes. The zones are monitored as they pass through the detector and the data captured and computed. The sample is usually introduced into the capillary by dipping it into the sample vial and applying a slight positive pressure. This results in a small volume of the sample entering the capillary.

Depending upon the pressure and the time, the amount can be carefully controlled. The sample is then moved into the capillary as a zone and the entry to the capillary cleaned at the same time by drawing a set volume of a wash buffer. Alternative techniques include suction, siphoning and an electrokinetic technique, but all achieve the same result.

The bands that separate are detected while they are still in the capillary, usually by absorbance, either in the ultraviolet or visible region of the spectrum, sometimes involving a diode array. Alternatively fluorescence or refractive index detectors may be employed and some instruments link the capillary to a mass spectrometer. Capillary electrophoresis has been shown to have several advantages over other separation techniques and since its introduction the number of applications described for it has risen rapidly. It lends itself to the analysis of a wide range of ionic and neutral molecules of varying size and shape. Analysis times are similar to those for HPLC, typically being less than 30 minutes, but separation efficiency is often superior and theoretical plate numbers of greater than 1×10^6 are typical. Nanolitre sample volumes are used, a fac-tor of great importance when limited amounts are available. Reagent costs are low because only microlitre volumes are required for each analysis. The capillaries must be appropriately conditioned or regenerated before use and should be flushed with water and air for storage. Modern instruments can be programmed to perform all the steps in the analysis from the introduction of the sample through to the interpretation of the results, which are presented as a series of peaks on a chart recorder, and preparation of the capillary for the next sample. There are several modes of operation of CE which rely on variations of the principles of electrophoretic separation.

Free solution electrophoresis (capillary zone electrophoresis).

This was the first type of capillary electrophoresis to be developed. Migration of the analytes through a buffer solution, which sometimes contains chemical additives to enhance the separation, is affected by their net charge and the electroendosmotic flow. The magnitude of the electroendosmotic effect is due to negatively charged silanol groups on the capillary wall, which attract positively charged ions from the buffer, and the overall effect is the movement of the buffered solution towards the cathode. Endosmosis is particularly evident with buffers of high pH and low concentration. A variety of internal polymeric coat-ings can be applied to the capillaries to modify the electroendosmotic effect and thus extend the applications of the method.

Micellar Electrokinetic Capillary Chromatography (MEKC or MECC)

This is used to measure small neutral or charged molecules and is applicable to a wide range of analytes including drugs, nucleosides, peptides and vitamins. Surfactants such as sodium dodecyl sulphate (SDS) are added to the buffer solution to produce anionic aggregates called micelles. The surfactants have long-chain hydrophobic tails (between 10 and 50 carbon units) and hydrophilic heads which, when bound in the micelles, point outwards into the solution. Separation depends on the interaction of the analytes with these charged micelles, which modifies their electrophoretic mobility.

When negatively charged surfactants such as SDS are used, cationic analytes are strongly attracted to the micelles in a manner that can be likened to the situation in ion-pair chromatography, and the overall negative charge of the micelle is reduced. Anions and uncharged analytes tend to spend more time in the buffered solution and as a result their movement relates to this. While these are useful generalizations, various factors contribute to the migration order of the analytes. These include the anionic or cationic nature of the surfactant, the influence of electroendosmosis, the properties of the buffer, the contributions of electrostatic versus hydrophobic interactions and the electrophoretic mobility of the native analyte.

In addition, organic modifiers, *e.g.* methanol, acetonitrile and tetrahydrofuran are used to enhance separations and these increase the affinity of the more hydrophobic analytes for the liquid rather than the micellar phase. The effect of chirality of the analyte on its interaction with the micelles is utilized to separate enantiomers that either are already present in a sample or have been chemically produced. Such precapillary derivatization has been used to produce chiral amino acids for capillary electrophoresis. An alternative approach to chiral separations is the incorporation of additives such as cyclodextrins in the buffer solution.

Capillary Gel Electrophoresis

This technique is used for the analysis of large molecules such as proteins and nucleic acids. The molecular sieving property of the gels is important in the separation process. The terms 'chemical-gel' and 'physical gel' are often used to describe the gel matrices. Chemical gels are cross-linked with relatively high viscosity and have a defined pore structure. Polyacrylamide comes into this category. Physical gels are solutions of entangled polymers, *e.g.* alkylcelluloses. They have relatively low viscosity and the contents of the capillary can be easily flushed out after use.

The concentration and chain length of the linear polymers determines the separation characteristics of the gel solution. Capillary procedures offer several advantages, including speed, resolution, sensitivity and technical simplicity, compared with the traditional meth-ods on which they were based. An added advantage for iso-electric focusing in capillaries is the fact that it can be performed without a gel, but a coating on the internal surface of the capillary is usually required to reduce electroendos-mosis. Similarly, isotachophoresis can be conveniently performed in capillary electrophoresis apparatus.

Capillary Electrochromatography (CEC)

This is a more recently developed technique which is a hybrid between HPLC and capillary electrophoresis. The capillary is packed with HPLC media and the mobile phases are aqueous buffers. A voltage is applied to generate an electroendosmotic flow and the analytes separate by interaction with the stationary phase and electrophoretic forces; no pump being required as for HPLC. Improved separation efficiencies have been reported.

METHODS BASED ON SIZE

The separation of macromolecules presents major problems and techniques such as dialysis permit only fairly gross separations to be achieved. However, the development of gel permeation chromatography introduced an extremely valuable method by which molecules may be separated from each other on the basis of their size. A wide range of molecular sizes can be separated by this method but for any particular gel the range is relatively narrow and there are a large number of different gels available commercially. Ultracentrifugal techniques strictly speaking separate on the basis of the mass and not the size of particle.

A variety of ultracentrifugal techniques are available and appropriate to different analytical problems. Preparative techniques generally involve iso-density or maximum velocity ultracentrifugation, whereas it is possible to investigate the relative molecular mass, diffusion characteristics and shape of a molecule using an analytical ultracentrifuge and equilibrium or maximum velocity techniques.

Dialysis and Ultrafiltration

Dialysis a technique for separating macromolecules from smaller solute molecules. The word refers to the diffusion of the solute molecules through a membrane which, owing to its pore size, restricts the movement of large mol-ecules. The passage of the small molecules is due solely to a concentration gradient across the membrane. In *ultrafiltration* the solvent and solutes are forced through the membrane under pressure and the movement of the large molecules is again restricted by the pore size. Cellophane is frequently used for dialysis and it has a pore size of approximately 4–8 μm, which makes it impermeable to molecules with a relative molecular mass in excess of about 10,000.

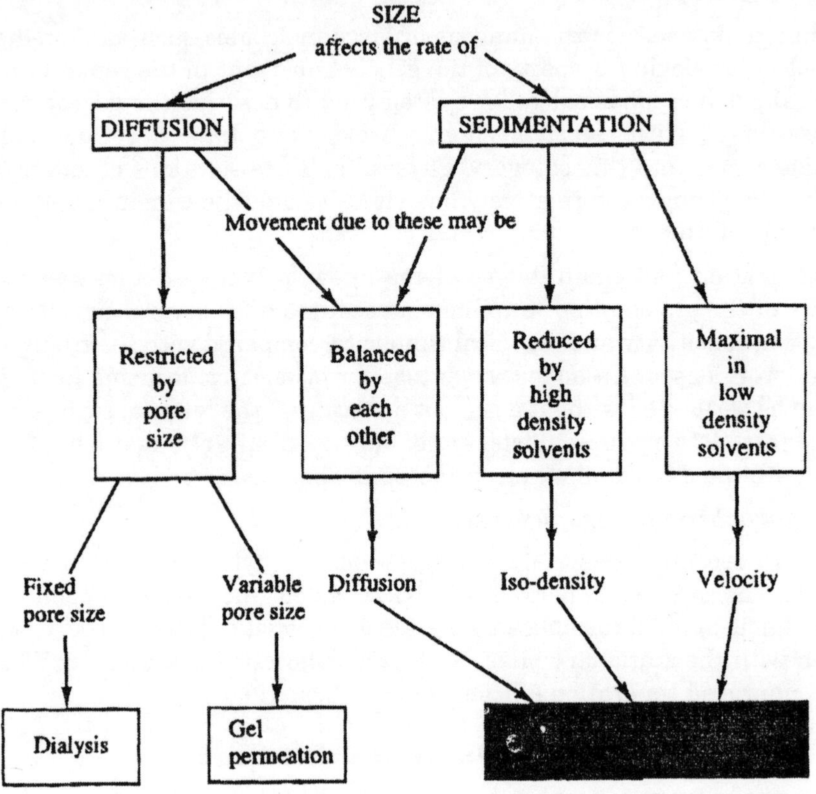

Fig. 4.34. Separation methods in which the size of the molecule plays an important part.

The development of a variety of membrane materials in which the pore size is much more rigorously controlled, has led to wider applications of ultrafiltration. Various cellulose and polycarbonate membranes are available with pore sizes down to 5 run which are capable of excluding molecules with a relative molecular mass of about 50. The internal structure of such membranes, as well as the pore size, determines their exclusion range and as a result precise specifications of membranes vary from one manufacturer to another.

Table 4.11. Applications of ultrafiltration

Particle size (nm)	*Particles*	*Application*
0.1–1.0	Salts, monomers	Water purification
1–100	Proteins, viruses	Fractionation, desalting
100–1000	Bacteria	Sterilization

Gel Permeation Chromatography

Gel permeation chromatography is also known as gel filtration and molecular size exclusion chromatography. The gel structure contains pores of varying diameter up to a maximum size. The test molecules are washed through a column of the gel and molecules larger than the largest pores in the gel are excluded from the gel structure. Smaller molecules, however, penetrate the gel to a varying extent depending upon their size and this retards their progress through the columns.

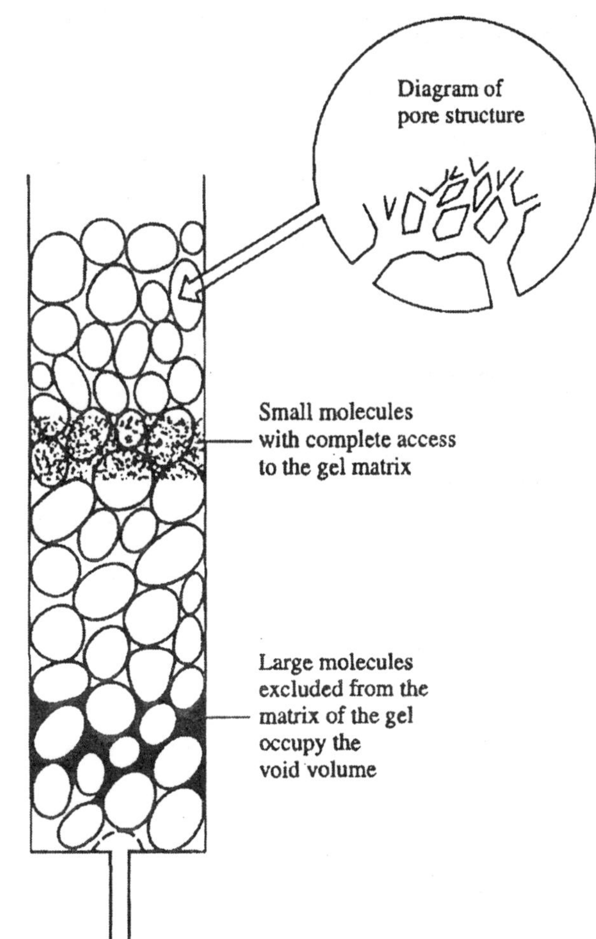

Fig. 4.35. Separation by gel permeation chromatography.

Elution, therefore, is in order of decreasing size. There are a variety of products available for gel permeation chromato-graphy and they are usually classified according to their exclusion limit, i.e. the relative molecular mass above which all molecules are excluded from the gel structure. Sephadex, the original medium, is based on dextran (a linear glucose polymer) that is modified to give varying degrees of cross-linking, which determines the pore size of the material. It is a strongly hydrophilic polymer which swells considerably in water and before a column is prepared, the gel must be fully hydrated with a large volume of liquid. Polyacrylamide beads are also available with a wide range of pore sizes and are prepared commercially in a manner similar to that described for polyacrylamide electrophoresis. They are also hydrophilic and swell significantly in aqueous solutions but are chemically more stable than the dextran gels.

Agarose gels are particularly useful when gels with a very large pore size are required. They are also hydrophilic but are usually sold in the swollen form. A major problem is the fact that they soften at temperatures above 30 °C. Mixed gels of polyacrylamide and agarose are available in which the polyacrylamide provides a three-dimensional structure which supports the interstitial agarose gel. Polystyrene gels are hydrophobic, and as a result are used primarily with non-aqueous solvents and for organic chemical applications rather than with biological samples.

Table 4.12. Range of gel permeation media

Trade name and manufacturer	*Chemical nature*	*Fractionation range (RMM)*
Pharmacia		
Sephadex G 10	Cross-linked dextrans	Up to 700
to		
G 200		5000–800 000
Sepharose 6B	Agarose	Up to 4×10^6
to		
2B		Up to 40×10^6
Sepharose Cl–6B		Up to 4×10^6
to		
Cl–2B		Up to 40×10^6
Biorad		
Biogel P2	Cross-linked polyacrylamide	200–2500
to		
P300		100 000–400 000
Biogel A 0.5 Agarose	10 000–500 000	
to		
A 150		1×10^6–150×10^6
Biobeads S × 1	Polystyrene	600-14 000
to		
S × 12		Up to 400
LKB		
Ultragel AcA22	Agarose and polyacrylamide	100 000–1.2×10^6
to		
AcA54		5000–70 000

Column Conditions

Columns for gel permeation chromatography must have both a minimal dead space volume at the outlet, to prevent broadening of the zones developed during the separation, and a suitable bed support, *e.g.* nylon mesh, which will retain the gel without becoming blocked. The length of the column affects the resolution, while the diameter influences the amount of sample that can fee applied without reducing the resolution. Gels with a small bead size result in columns with lower void volumes and hence improved resolution.

However, columns of small beads give a high resistance to liquid flow resulting in slower separations. For preparative work, therefore, larger beads are preferable. The major technical difficulty in gel permeation chromatography involves the careful packing of the column to ensure a uniform structure and it is absolutely essential that the level of the solvent is never allowed to fall below the level of the gel otherwise disruption of the gel bed occurs. Additionally, because most gels are compressible, only an absolute minimum of pressure should be applied to prevent alteration of the pore characteristics. In practice hydrostatic pressures greater that 30–40 cm should not be used and pumps should be used only with non-compressible gels.

Elution Parameters

The effluent volume is the most convenient parameter to measure in gel permeation chromatography because flow rates are often variable, making the use of retention times unsuitable. The sequence of different solutes emerging from a column will therefore be reported as the total volume of solvent that has emerged from the column when the substance appears in the effluent (V_e). Molecules larger than the exclusion limit of the gel will be eluted when a volume of solvent equal to the void volume (V_o) leaves the column and similarly those molecules that can gain access to all parts of the gel matrix will have an effluent volume

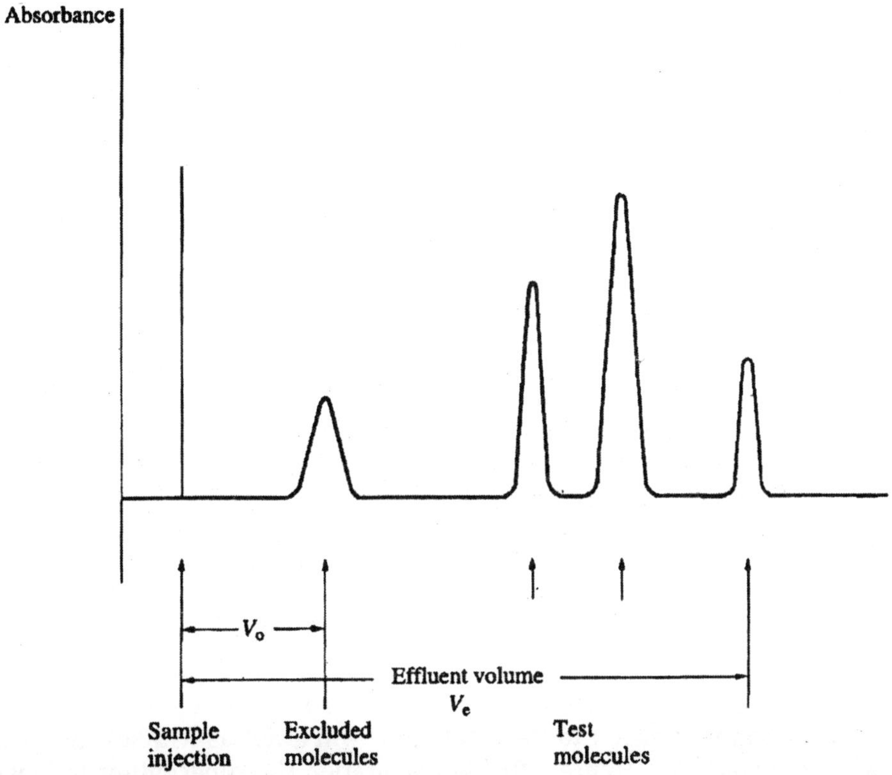

Fig. 4.36. Gel permeation chromatogram. All molecules larger than the exclusion limit of the gel appear at V_o (the void volume). Molecules which can gain access to the gel structure to varying degrees are eluted in order of decreasing size.

equal to the total bed volume (V_t). Molecules of an intermediate molecular size will be eluted in volumes between V_t and V_o. The measurement of effluent volume is not very reliable because of the effect of the geometry and packing characteristics of any column.

It is often more useful to use a reduced parameter, such as V_e / V_o, which is not so dependent upon column characteristics and is comparable with the calculation for R_F values in thin-layer chromatography.

It is possible to calculate a partition constant (K_d) for a solute in terms of the access gained to the gel matrix:

$$V_e = V_o + K_d \times V_i$$

where V_i is the gel inner volume. Hence

$$K_d = \frac{V_e - V_o}{V_i}$$

However, because it is difficult to measure V_i precisely, the equation may be modified to determine the 'available' part of the resin, K_{av}

$$K_{av} = \frac{V_e - V_o}{V_t - V_o}$$

In order to separate two solutes satisfactorily, their values for K_{av} must be significantly different from each other.

Applications of Gel Permeation Chromatography

Gel permeation chromatography is commonly used in the fractionation of mixtures of substances that vary in their relative molecular mass and it is extremely useful for labile molecules, such as enzymes. The elution volume of a solute is determined mainly by its relative molecular mass and it has been shown that the elution volume is approximately a linear function of the logarithm of the relative molecular mass.

It is possible to determine the relative molecular mass of a test molecule using a calibration curve prepared from the elution volumes of several reference substances of known relative molecular mass. This should be done using the same column and conditions and hi practice it may be possible to calibrate the column and separate the test substance at the same time by incorporating the reference compounds in the sample. Such a method is rapid and inexpensive and does not demand a highly purified sample, provided that there is a specific method for detecting the molecule in the eluate. Solutes of low relative molecular mass may be removed from preparations of macromolecules, often called desalting, using a small column of a gel which excludes the macromolecules but retains the small solutes.

The macro-molecules will be eluted in the void volume, which for a small column will be only 2–3 ml. A *dry gel* may also be used to concentrate a dilute aqueous solution of a macromolecule. When it is added to the solution, it will swell as it takes up water but will exclude the larger molecules. Hence, after sedimentation the supernatant fluid will contain the macromolecule, leaving the majority of the water in the pores of the gel. Such a technique is quicker and more effective than dialysis.

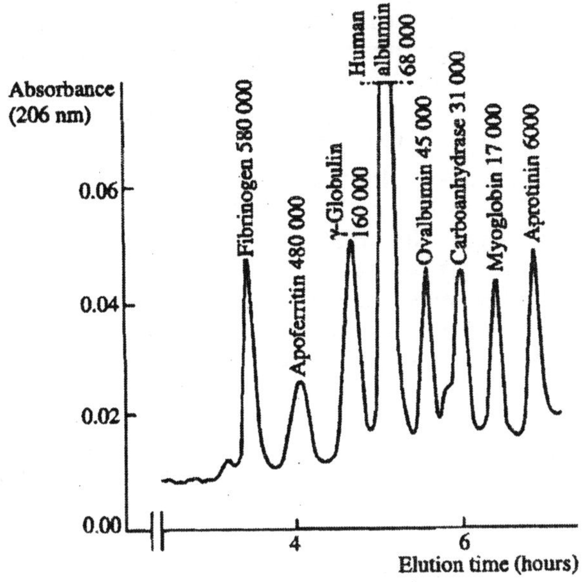

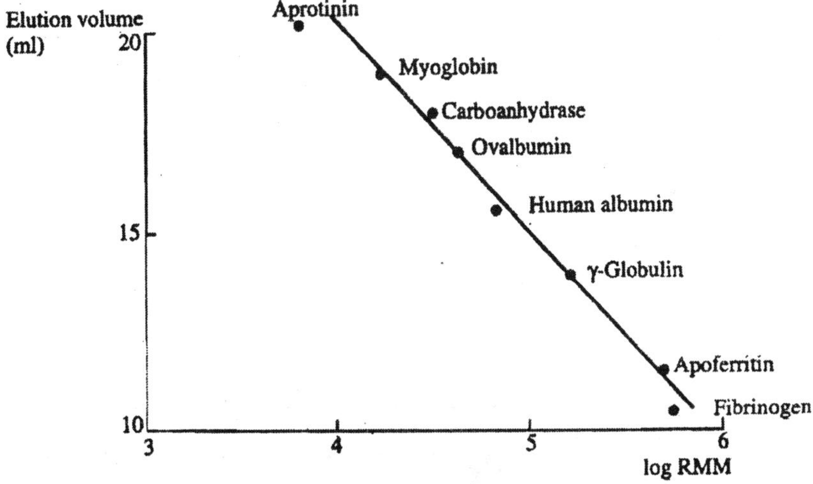

Fig. 4.37. Determination of relative molecular mass by gel permeation chromatography. A mixture of proteins (approximately 10 μg of each) was separated on an UltroPak TSK SW column and the elution volume for each protein plotted against the logarithm of its relative molecular mass (RMM).

Ultracentrifugation

Although the use of high rotor speeds is not always necessary, most ultra-centrifuges are designed to operate safely at rotor speeds of up to 60 000 or 100 000 rpm. Despite the use of high speed, the sedimenting forces involved are only small and the technique is extremely gentle, providing a valuable method for the separation of large, labile molecules or particles.

Preparative ultracentrifugation uses rotors with relatively large capacities to separate various components of a mixture, while analytical ultracentrifugation involves the measurement of sedimentation velocities in small volumes under carefully controlled conditions. Because particles sediment in a tube during centrifugation, it may appear that a force is acting radially outwards, whereas in fact there is no such force.

Any particle that is rotating in a circle will always tend to move away at a tangent to the circle and to prevent this a force has to be applied towards the centre of rotation. This is known as the centripetal force and is provided in a centrifuge by the rotor arms and spindle. Particles suspended in solution, when rotated in a centrifuge tube, are not connected to the axis of rotation and therefore tend to move at a tangent to the circle. Their movement, however, relative to the centrifuge tube is radially down the tube and appears to be determined by a force equal and opposite to the centripetal force. It is convenient, therefore, in ultracentrifugation to assume that this movement in an apparently stationary tube is due to centrifugal force and to compare this force with that due to gravity by quoting the intensity of the centrifugal force with respect to that of gravity. This is known as the *relative centrifugal force* (RCF) and enables quick and simple comparisons of the forces applied by different centrifuges and different rotors.

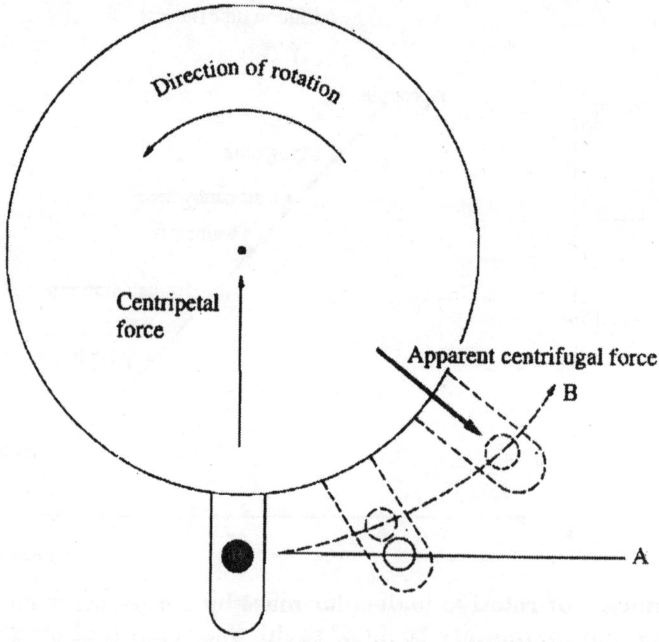

Fig. 4.38 Centrifugal effects. The path of a particle unimpeded by factional effects is at a tangent (A) to the circle described by the rotation of the centrifuge tube. When impeded by friction, its path (B) more closely resembles circular movement with a reduced rate of sedimentation in the tube.

The effective centrifugal force may be calculated as:

$$F = M\omega^2 r$$

where M is the mass of the particle, ω is the angular velocity in radians per second and r is the radius of rotation in centimetres. The effect of centrifugation is therefore dependent upon the mass of the particle and the convenience of using the relative centrifugal force is that the value for the mass of the particle is eliminated from the equation, permitting more general comparisons of centrifuge performance:

$$\mathrm{RCF} = \frac{f_c}{f_g} = \frac{M\omega^2 r}{Mg} = w^2 r \times g^{-1}$$

It is more usual to monitor the speed of a centrifuge in terms of revolutions per minute (rpm) rather than in radians per second. Hence:

$$\mathrm{RCF} = \left(\frac{2\pi \text{ rpm}}{60}\right)^2 r \times g^{-1}$$

In comparing the relative centrifugal force developed by different centrifuges it is important to appreciate that the value varies considerably from the top to the bottom of a centrifuge tube.

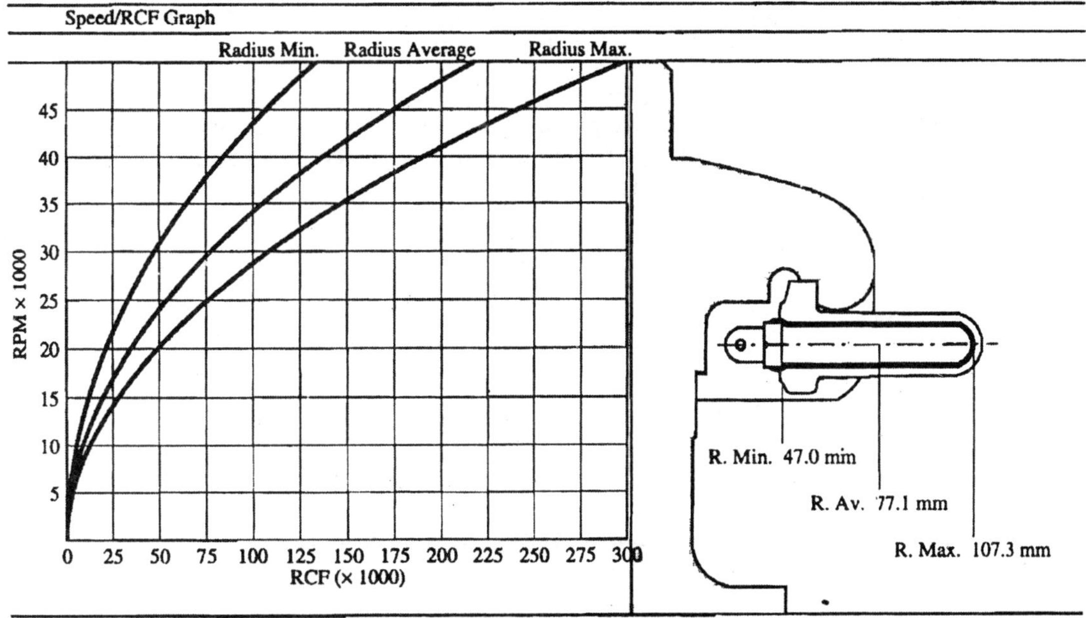

Fig. 4.39. Relative centrifugal force. The relative centrifugal force developed by any rotor can be calculated but will vary depending upon the distance from the centre of rotation.

For a particle to sediment it must displace an equal volume of the solvent from beneath it. This can only be achieved by centrifugation if the mass of the particle is greater than the mass of the solvent displaced. Hence, density as well as mass is an important consideration and explains the phenomenon of low density particles floating rather than sedimenting during centrifugation. When a particle is moving it will be necessary to pass solvent molecules regard less of the direction of movement and the resulting frictional effects will always oppose the movement.

This effect is proportional to the velocity of the molecule and is affected by its size and shape as well as the viscosity of the medium:

$$\text{Friction} = fv$$

where f is the frictional coefficient of the particle in the solvent, and v is the velocity of the particle.

Because of the centrifugal force a particle will accelerate, its velocity increasing until the frictional force equals the centrifugal force. Under these conditions the net force acting on the particle is zero and the particle will not accelerate any further but will have achieved a maximum velocity:

$$\text{Net force} = (M_p - M_s)\omega^2 r - fv$$

where M_p is the mass of the particle and M_s is the mass of equal volume of solvent.

It is convenient to classify particles by their rate of sedimentation per unit centrifugal field, a parameter known as the sedimentation coefficient (s), which has units of seconds. A Svedberg unit (S) is defined as a sedimentation coefficient of 1×10^{-13} seconds.

Maximum Velocity Methods

Rotor speeds that are high enough for all the test particles to achieve their maximum velocity are used in preparative and analytical separations.

Moving-boundary Ultracentrifugation

A centrifuge tube is filled with the sample and centrifuged at a speed high enough for all the components of the sample to achieve their maximum velocity. The components begin to separate out, leaving clear solvent (supernatant)

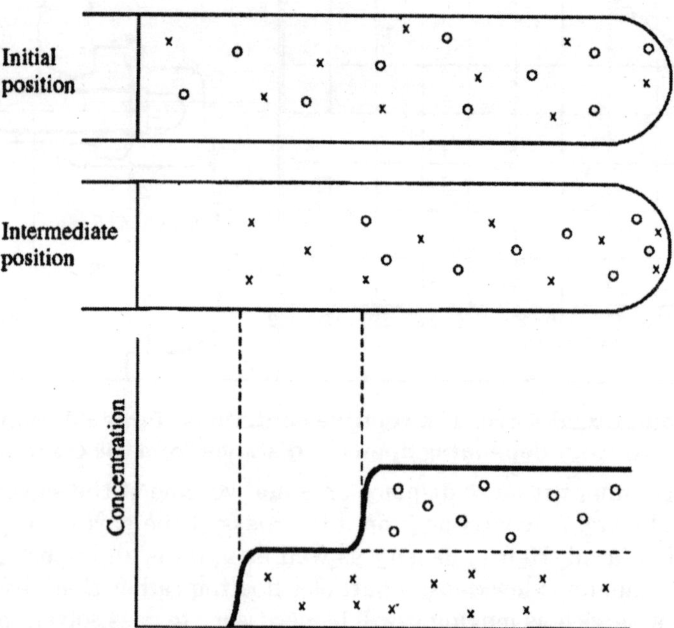

Fig. 4.40. Moving-boundary ultracentrifugation. The heavier particles (o) sediment more rapidly than the lighter particles (x) and as a result concentration boundaries develop as the supernatant fluid is cleared of particles. Monitoring the movement of a boundary will permit the determination of sedimentation coefficients.

at the top of the tube, and a series of concentration boundaries developing upon the number of components in the sample. The movement of a boundary is characteristic of the particle in the bulk of the solution ahead of the boundary and although it may not be possible to purify a particular component by this technique it is possible to measure its rate of sedimentation by following the movement of the boundary. Separation into components can only be achieved by stopping the process when sedimentation of the desired component has occurred. The sediment is then resuspended in fresh solvent and centrifuged at a lower speed, when the heavier particles will sediment leaving the component in suspension. Such a method is known as differential sedimentation and is particularly useful for the fractionation of cellular components.

Moving-zone Ultracentrfugation

A layer or zone of the sample is superimposed on the solvent in the centrifuge tube and again, under the influence of a centrifugal force, the particles sediment and discrete zones or bands develop depending upon the relative mobility of each particle. As centrifugation continues, so each zone will pellet at the bottom of the tube in sequence, resulting finally in a mixed pellet.

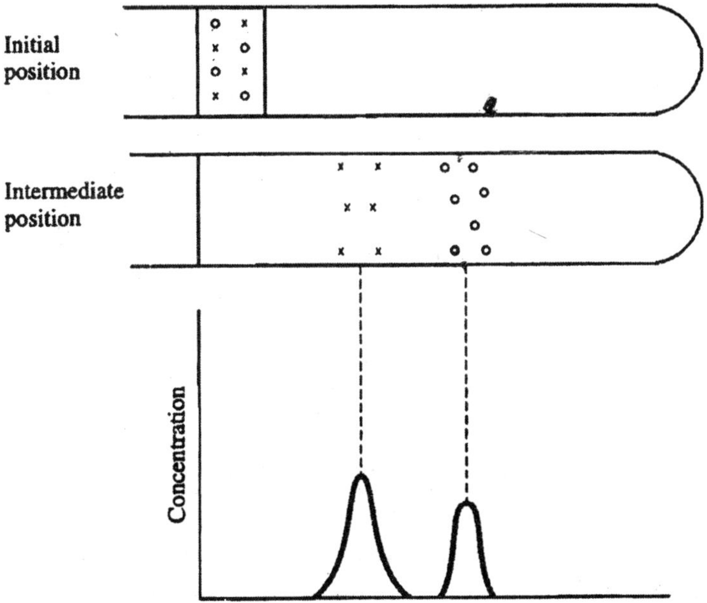

Fig. 4.41. Moving-zone ultracentrifugation. Zones develop during centrifugation, the lower ones being heavier particles.

Centrifugation is stopped before this occurs and the different zones can be removed. Zonal techniques are almost solely preparative and the resolution of the zones will be maintained more effectively if a slight density gradient is used. Zonal techniques may be used for the separation of a wide range of particles and macromolecules, *e.g.* mitochondria, nucei, ribosomes and proteins. The technique may be used for bulk preparative work using a zonal rotor which is filled with a solvent gradient while running at a slow speed.

The sample is similarly introduced and the rotor speed is then increased to the desired value. After centrifugation is complete, the contents are drawn off while the rotor is running slowly by displacing them with a more dense solution.

Iso-density Methods

Regardless of the rotor speed and maximum velocity, sedimentation (or flotation) will not occur in a solution of equal density to the sample. Iso-density methods use this lack of movement in a manner comparable to a pH gradient in iso-electric focusing techniques. The methods are a combination of sedimentation and flotation, achieved by using a density gradient that straddles the density to the panicles concerned.

On centrifugation, the particles sediment until they reach a solvent zone with the same density. This results in the development of a zone for each type of particle present in the sample.

Pre-formed Density Gradients

The centrifuge tube is filled with a solvent in a graded range of concentrations prepared from mixtures of two different stock solutions which determine the limits of the gradient. While it is possible to produce a stepwise gradient by layering decreasing concentrations of the solution carefully on top of each other, it is preferable to produce a smooth gradient using a gradient former.

Solutions of either sucrose or caesium chloride are usually used and the sample is layered on the surface and centrifuged. Under the influence of the centrifugal force, each component will move to an area of iso-density, the fastest particle not necessarily being the most dense. Density gradient techniques are particularly useful in the separation of cellular components and macromolecules but are not generally as efficient as differential sedimentation. Table 4.13 indicates the order of separation of cellu-lar organelles after ultracentrifugation to equilibrium on a density gradient of sucrose ranging from a high density of 1.25 ρ at the bottom of the tube to a minimum value of 1.10 ρ at the top.

The cell extract is prepared in a sucrose solution with a density of 1.05 ρ and layered on top of the gradient. Some organelles, such as mitochondria, show fairly consistent density values but others do vary considerably depending upon the cellular source and the manner of preparation of the extract. Density values for rough endoplasmic reticulum vary consider-ably depending upon the proportion of dense ribosomes present.

Table 4.13. Density gradient ultracentrifugation of sub-cellular organelles

Organelle	*Density* (ρ)
Plasma membrane	1.14
Rough endoplasmic reticulum	1.16
Mitochondria	1.17
Lysosomes	1.20

A cell extract was separated on a sucrose gradient ranging from 1.10 ρ at the top to 1.25 ρ at the bottom and centrifuged to equilibrium.

Equilibrium Density Gradients

This is a method for the separation of molecules with very similar densities and depends upon the formation of a density gradient by the effect of centrifugal force on the supporting solute molecules but because of their low mass the technique demands a long period of centrifugation, *e.g.* 2–3 days. The test compounds are suspended initially in a solution of selected density determined usually by preliminary preformed gradient technique.

The gradient forms and the molecules separate into zones simultaneously during the long period of centrifugation. The materials usually used to produce such gradients are salts of heavy alkali metals, caesium chloride being the most frequently used. This salt has a high solubility and its low relative molecular mass permits rapid diffusion enabling the gradient to be formed reasonably quickly. Concentrations of salt up to about 2 g ml^{-1} can be used and are chosen on the basis of information about the density of the test particles or macromolecules. The technique is used frequently in the separation of viral particles and nucleic acids.

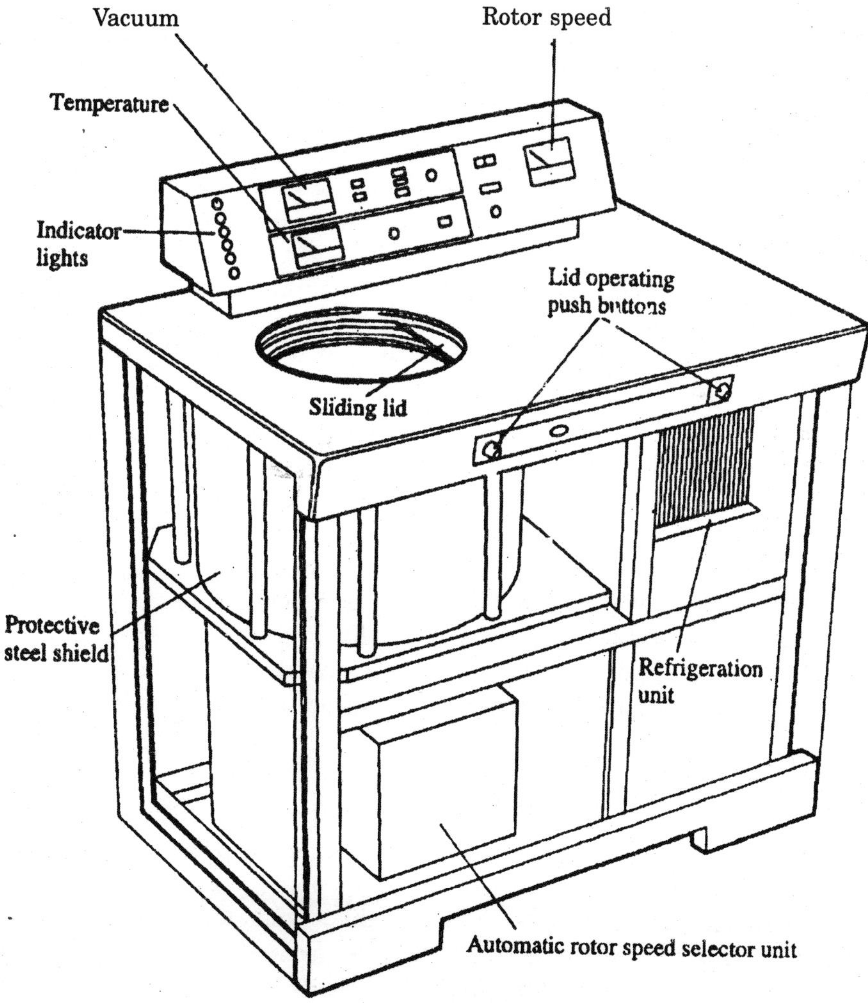

Fig. 4.42. An ultracentrifuge.

Equilibrium Methods

Centrifugation of a suspension of macromolecules at relatively low speeds will not result in complete sedimentation, because of the effect of diffusion, but a concentration gradient will be produced owing to the equilibrium between sedimentation and diffusion. Such a gradient may be studied in the analytical centrifuge and the resulting measurements can be used to determine the relative molecular mass of the molecule. It is not a preparative technique and requires very long periods of centrifugation for the equilibrium to be established.

Instrumentation

The major difference between a preparative and an analytical ultracentrifuge lies in the design of the rotors and the need for an optical monitoring system in the latter. A vacuum is necessary to minimize the factional effects of high speed rotation and a cooling system is essential to protect the high speed bearings, etc., and any labile materials being studied. Rotors are generally made of a aluminium or titanium, the latter being more durable and safer at high speed.

They are protected by a hard anodized film to prevent corrosion but have a limited life owing to metal fatigue and careful records of their use must be kept. Preparative rotors may be either the swing out or fixed angle type of varying sample capacity. The large capacity rotors have a lower maximum safe speed, with swing out heads having lower limits than the corresponding fixed angle rotors. Centrifuge tubes are cylindrical and usually made of polypropylene or polycarbonate with a capping or sealing device. It is essential that tubes should be filled to capacity to prevent collapse and that they are carefully balanced before centrifugation. Analytical rotors are usually solid with holes where the cells containing the samples are inserted, the simplest type of rotor incorporating the cell together with a reference.

These cells are usually made of quartz or sapphire and are sector shaped to prevent distortion of the boundaries that develop during sedimentation. The optical system required to monitor the movement of the boundary during centrifugation presents design problems in maintaining a vacuum in the centrifuge while also monitoring the rapidly rotating sample cell. Absorption of ultraviolet radiation may be used to detect zones of protein, nucleic acid, etc., the zones being detected either photographically or absorptiometrically. A reference cell is necessary to compensate for absorbance by the solvent. Variations in concentration that are produced as a boundary develops result in changes in the refractive index along the length of the cell.

In Rayleigh interference optics, a fringe pattern produced by the superimposition of two parallel rays of light is distorted by a change in refractive index. The degree of distortion is proportional to the change in the refractive index and hence to the change in concentration. The Schlieren optical system utilizes the same phenomenon but results in a trace that represents the change in the refractive index gradient and shows the boundary as a peak. While fractions or samples can be removed from the centrifuge tubes using a pipette it is possible with appropriate equipment to pierce the base of a plastic tube to allow the contents to drip slowly out.

Alternatively the contents of the tube may be displaced by slowly injecting a dense solution to the bottom of the tube or by puncturing the tube in the appropriate position and withdrawing the desired band with a syringe.

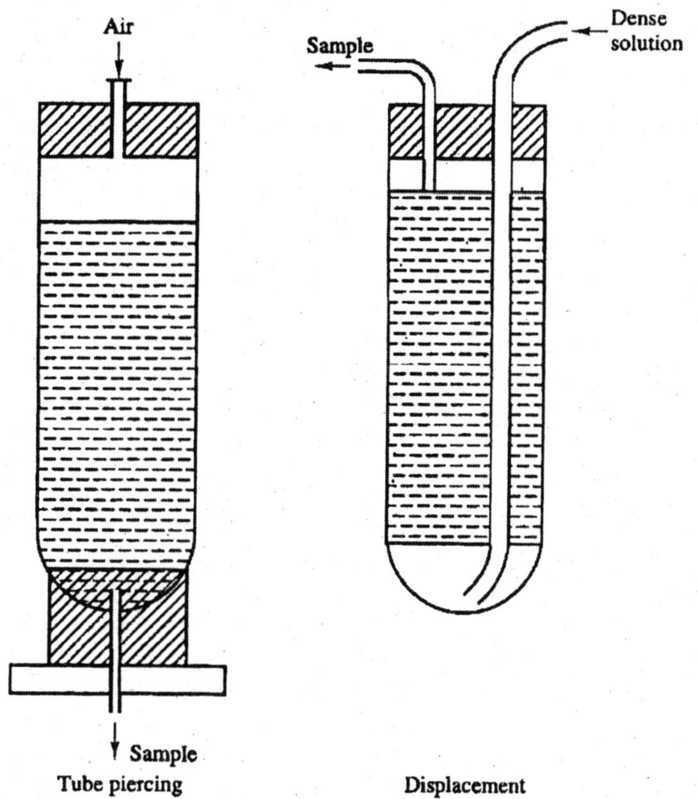

Fig. 4.43. Methods of sampling from plastic centrifuge tubes.

Determination of Relative Molecular Mass

During ultracentrifugation, when the frictional force equals the centrifugal force and the particle achieves maximum velocity, the relative molecular mass of the particle may be represented by the Svedberg equation:

$$M = \frac{RT}{(1-\nabla\rho)} \times \frac{s}{D}$$

where s is the sedimentation coefficient, R is the gas constant, T is the absolute temperature, D is the diffusion coefficient, ρ is the density of the solvent and ∇ is the partial specific volume of the solute.

The partial specific volume is the increase in the volume of a solution caused by the presence of the solute and is defined as the volume occupied by 1 kg of the solute in 100 kg of the solution. The value may be calculated using the densities of both the solvent (ρ) and the solution (ρ_s):

$$\overline{V} = \frac{\dfrac{100}{\rho_s} - \dfrac{100-n}{\rho}}{n}$$

where n is the percentage concentration of the solute in the solvent (w/w).

The value of the partial specific volume for proteins is approximately 0.7 while for nucleic acids the value varies around 0.5.

The sedimentation coefficient may be determined by measuring the velocity of the particle at a fixed centrifuge speed (ω, in radians per second) and, from a series of observations; plotting the logarithm of the distance moved *(x)* against the time taken in seconds *(t)*. The relationship is expressed by the equation:

$$s = \frac{\ln x}{\omega^2 t} = \frac{2.303 \log_{10} x}{\omega^2 t}$$

and the value for s can be calculated from the slope of the resulting graph. Although relative molecular mass may be calculated using the Svedberg equation, the calculations are complicated by difficulties in determining the diffusion coefficient and correcting for differences in viscosity and temperature.

For these reasons the sedimentation coefficient of a particle is often used instead of its mass, the two being proportional to one another. Meniscus depletion is an alternative method for the determination of the relative molecular mass. If a homogenous sample is centrifuged at a speed great enough to cause the particles to move away from the meniscus, resulting in a solute concentration of zero at the meniscus and producing a concentration gradient due to the solute, the relative molecular mass is proportional to the slope of a plot of the logarithm of concentration against the distance squared.

This relationship is not absolute and depends upon measurements being taken early in the process and assumes that the density of the solution is constant throughout the solution despite the very slight concentration gradient that does develop.

The concentration of solute may be determined from the steepness of the interference fringes formed in a Rayleigh pattern.

The relative molecular mass (RMM) may then be calculated using the equation:

$$\text{RMM} = \frac{4.606RT}{(1-\overline{V}\rho)\omega^2} \times \frac{\text{log concentration}}{\text{distance}^2}$$

The basis of many biochemical processes within a cell lies in the shape rela-tionships that exist between the reacting molecules, *e.g.* an enzyme active site and its substrate. The affinity and specificity that such molecules show for each other form the basis of methods such as immunoassays, and they can also be exploited in affinity chromatography.

Affinity Chromatography

A binding molecule (or ligand) is linked to an insoluble polymer in such a way that the affinity of the ligand for its complementary molecule is not significantly altered. The polymer is subsequently used in a manner comparable to adsorption or ion-exchange media but has the additional advantage of specificity. The process may also be used in a reverse manner by insolubilizing the complementary molecule and using it to separate a specific binding molecule, *e.g.* the isolation of specific antibodies from an antiserum. Antibodies probably provide the largest source of ligand molecules but enzymes, transport proteins and membrane receptors are all used.

Methods of Insolubilization

A variety of techniques has been used varying from simple adsorption to entrapment within a polymer matrix but the commonest method is the covalent linking of the ligand to a suitable polymer, such as dextran, agarose or polyacrylamide. The polymer should be hydrophilic and of a pore size large enough to allow the test molecules to gain access. If the ligand group is linked closely to the polymer backbone it is possible that steric interference will impair its binding ability and it is desirable that a relatively long spacer arm (6–8 atoms) is used in

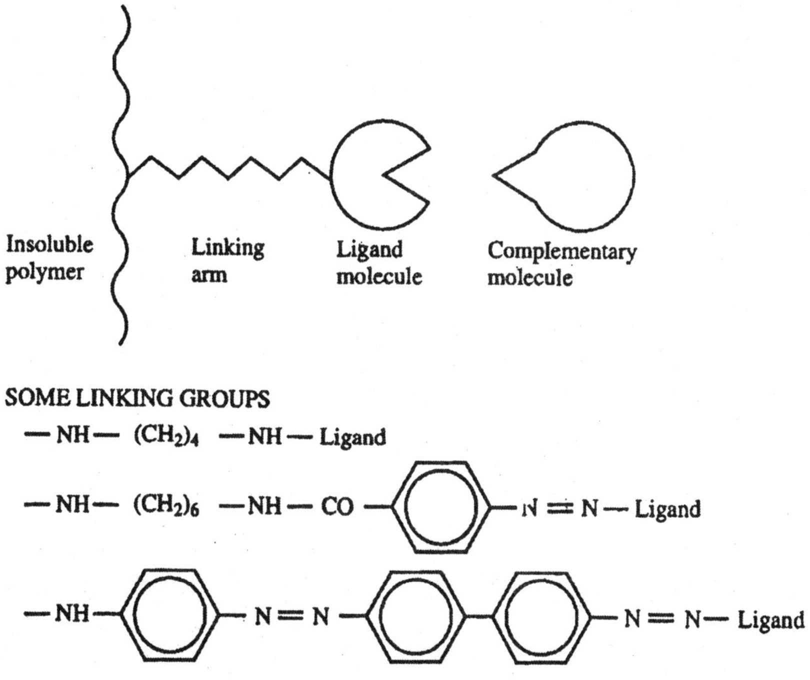

Insoluble polymer

Linking arm

Ligand molecule

Complementary molecule

SOME LINKING GROUPS

—NH— $(CH_2)_4$ —NH— Ligand

—NH— $(CH_2)_6$ —NH— CO —⟨benzene ring⟩— N = N — Ligand

—NH—⟨benzene ring⟩— N = N —⟨benzene ring⟩—⟨benzene ring⟩— N = N — Ligand

Fig. 4.44. Affinity chromatography.

the linking reaction. Several linking reactions have been described but the two common examples both involve aromatic amino groups in the polymer.

These can be converted to reactive diazo groups by the action of nitrous acid and will subsequently react with free amino groups in the ligand proteins. Alternatively, they can be converted to isothiocyanate groups by thiophosgene and will then react with tyrosine and histidine residues in the proteins. Other functional groups in the polymer can be used in a similar manner. A range of prepared gels is available commercially and the best gel for a particular situation must be determined by trial and error.

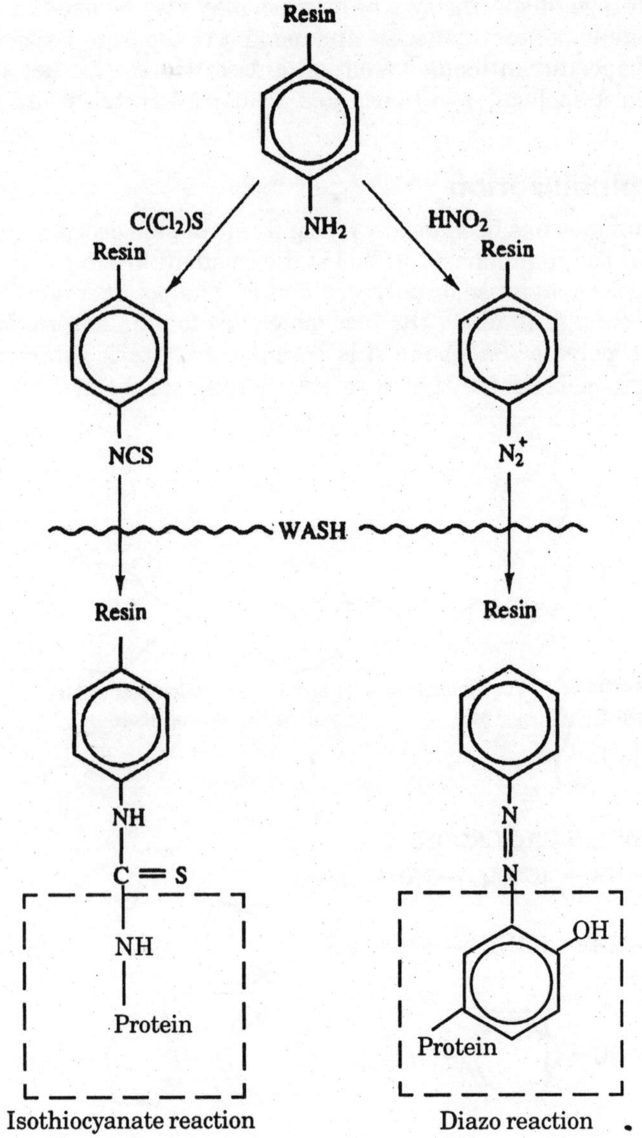

Fig. 4.45. Examples of reactions used for insolubilizing proteins using aromatic amino resins.

Methods of Affinity Chromatography

Affinity chromatography techniques are not generally appropriate for the fractionation of mixtures because of the very positive and specific affinity between the ligand and solute molecules. Separation may be achieved in a column where the solute molecules bind to the matrix, and, after washing to remove unwanted substances, the test substance can be eluted. Alternatively, a batch procedure may be used in which the affinity medium is mixed with the sample solution and allowed to react.

The solid-phase complex is then removed from the bulk solution by filtration or centrifugation and after washing, the purified solute is separated from the complex. In order to elute molecules which are strongly bound to the ligand group it is necessary either to reduce their affinity for each other or to introduce molecules that are more strongly bound or in greater concentration and will therefore be able to displace the test molecules.

Generally, non-specific methods are preferred and involve altering either the ionic strength or the pH of the buffer. These will result in conformational changes in the proteins and hence their binding characteristics. Changes in the dielectric constant of the solvent caused by the introduction of organic solvents will result in altered hydrophobic bonding and again aid the dissociation of the complex. It is possible to use specific agents which compete for the binding sites, such as alternative substrates and inhibitors for enzymes.

Applications

The nature of the binding between the ligand and its complementary molecule restricts affinity chromatography to a particular type of biological compound and some examples are given below.

1. *Enzymes.* The specificity of an enzyme for its substrate, coenzyme or competitive inhibitor provides the basis for many affinity chromatographic separations. Enzymes may be extracted and purified using insolubilized substrates, coenzyme or inhibitors. Less frequently, enzymes are used as the ligands.
2. *Antibodies.* The reaction between an antibody and its antigen does not result in the chemical modification of the antigen compared with the action of an enzyme and provides the basis for producing chromatographic media capable of selecting the complementary molecules. Either the antigen is insolubilized and used to isolate and purify the appropriate antibodies or with the increased availability of monoclonal antibodies, the reverse procedure is used.
3. *Lectins.* Some proteins extracted from certain seeds are capable of binding compounds containing carbohydrate groups. These proteins are known as phytohaemagglutinins or lectins. Affinity chromatographic media using such lectins have been used to investigate cell membrane structures and aid in the study of cell interactions. They are also used in conjunction with quantitative column chromatographic methods and in some electrophoretic separations of carbohydrate-rich proteins.
4. *Receptor proteins.* Hormones act on cells via specific membrane receptors and can be used to purify these receptors from cell homogenates. Receptors for compounds such as insulin, oestrogens and acetylcholine among others have been purified in this way.
5. *Nucleic acids.* Immobilized polynucleotides can be used to extract nucleic acid binding proteins as well as complementary strands of nucleic acids.

CHAPTER 5

Rapid Flow Diagnosis

One of the main goals of investigators in the fields of (clinical) diagnostics and contamination-control programs consists of providing new analytical strategies to obtain accurate data as quickly as possible. When a new screening test is requested, the method has to be sensitive, specific, fast, cheap, and easy to perform. In general, immunological techniques (immunoassays) can fulfill these high requirements to a great extent. Over the past decade, single-use lateral flow immunoassays have been extremely successful in the laboratory and in outpatient clinic and primary care environments. In this type of assay, all reaction components are impregnated or immobilised on a porous solid phase, usually a nitrocellulose membrane, and are brought into contact with the sample in sequence after addition of a diluent. This immunoassay format is also known as *strip test, one-step strip test, immunochromatographic test. rapid flow diagnostic. Rapid immunoassay (test), lateral flow immunoassay (LFI) on-site test (assay)* or *near-patient test (NPT)*. Immunoassays are analytical measurement systems that use antibodies as test reagents.

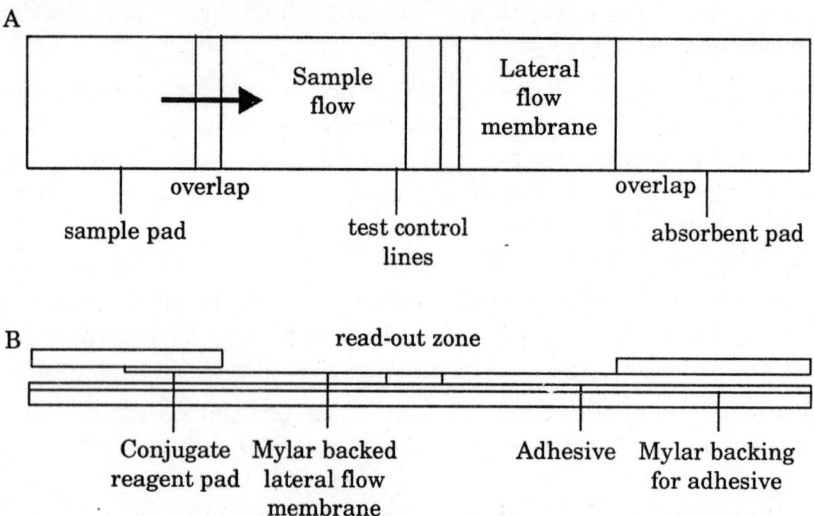

Fig. 5.1. Schematic of a typical test strip, including a Mylar backing plate with adhesive, a sample pad a conjugate pad an absorbent pad and a membrane that incorporates the capture reagents. A. Top view, B. Side view. A complete strip test device consists of a test strip as shown here, packed in a plastic cassette (housing).

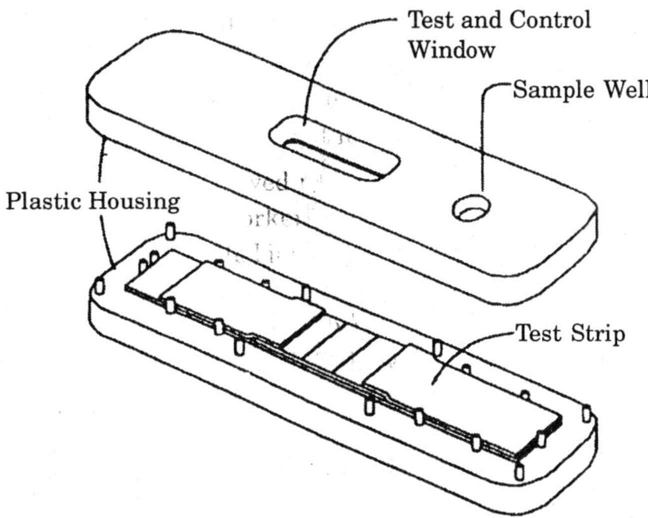

Fig. 5.2. Example of a plastic cassette (housing) used in a strip-test device.

The antibodies are attached to some kind of label and then used as reagents to detect the substance of interest. This label can be either an enzyme for colorimetric detection or a colored collodial particle such as gold (red) carbon (black), silica (several colours), or latex (several colours) for direct visualization of the immunoreaction. In practise, collodial gold particles or gold nanoclusters having a diameter of 25 to 40 nm are probably the most commonly applied labels in strip tests. For this reason, we will focus on the use of such gold nanoclusters. A typical strip test device consists of a test strip that is mounted in a plastic cassette (housing).

The test strip is made up of a number of components, including a sample pad, a conjugate pad, an absorbent pad, and a membrane that contains the capture reagents. The main purpose of the housing is to fixate the several components of the test strip and to keep them in close contact with each other. Moreover, the housing determines the dimensions of the sample well and contains the viewing window with readout indications. When describing the principle of the strip test, one has to distinguise between the direct assay to detect high-molecular-mass components, usually proteins and the indirect or competitive assay to detect low-molecular-mass analytes such as drug residues, antibiotics, hormones, etc. In both type of tests, the user dispenses a liquid sample (buffer extract, milk, urine, serum, plasma, whole blood, etc.) onto the sample pad.

The sample then flows through the sample pad into the conjugate pad, where it releases and mixes with the detector reagent.

In the direct assay format for detecting high-molecular-mass analytes such as proteins the test detector reagent consists of gold nanoclusters coated with specific antibodies to the protein of interest. When that particular protein is present in the sample it will react with the test detector reagent. The formed protein-antibody-gold complexes are mobile and are able to move freely from the reagent pad into the membrane with the flow of the fluid. At the test line, the complexes will be captured by immobilized anti-protein antibodies. Thus, the presence of the protein of interest in the sample will result in a colored test line.

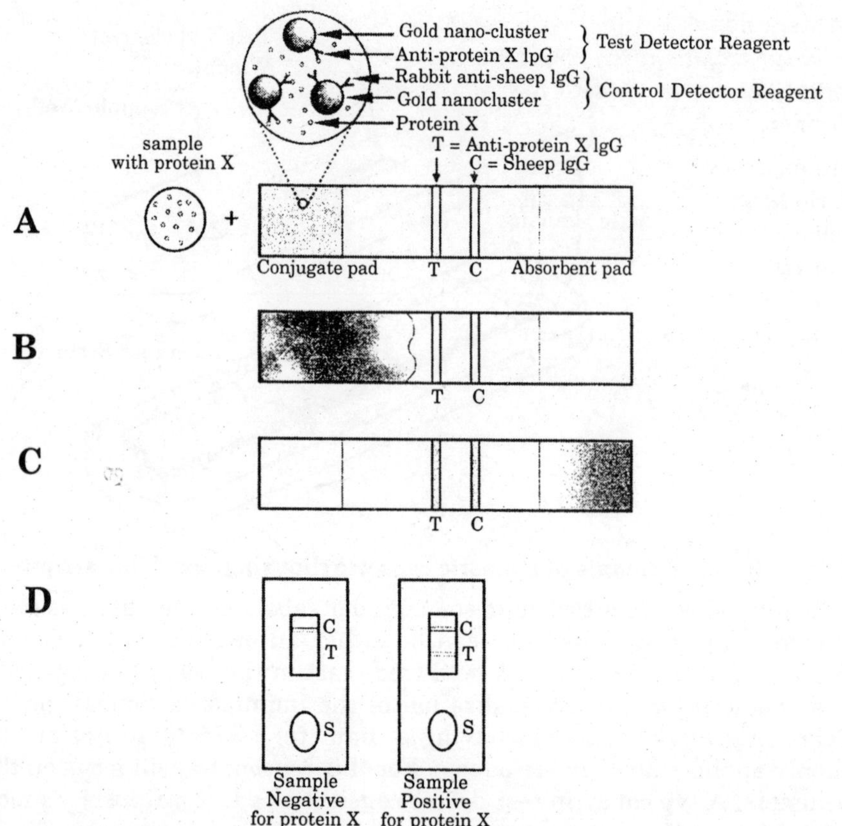

Fig. 5.3. Principle of a strip test for the detection of high-molecular mass-analytes (proteins). A. Fluid sample is brought onto the sample pad. B. The control detector reagent. C. Sample passes over the capture zones. D. The immobilized capture antibodies. E. Reading the results:

The colour intensity of the test line is proportional to the concentration of the analyte in the sample. When the concentration of the protein of interest is lower than the lowest detection concentration or when the protein of interest is completely absent, no test line will be visible. Excess sample that flows beyond the test and control lines is taken up in the absorbent pad In the competitive assay format for detecting low-molecular-mass analytes the test detector reagent consists of gold nanoclusters coated with antibodies to the analyte of interest. The test line here consists of a protein-analyte conjugate. The more analyte present in the sample, the more effectively it will compete with the immobilised analyte on the membrane for binding to the limited amount of antibodies of the detector reagent. Thus, the absence of the analyte in the sample will result in a coloured test line whereas an increase in the amount of analyte will result in a decrease of signal in the readout zone.

At a certain concentration of analyte in the sample, the test line will be no longer visible. The lower detection limit (LDL) is defined as the amount of analyte in the sample that just causes total invisiblity of the test capture line. Strip-test devices are commercially available for an increasing number of antigens (high and low molecular mass). The first major target analyte for this test format was human chorionic gonadotropin (HCG) for the detection of pregnancy. At present, several test strip assays are available *e.g.,* for the detection of hormones (pregnancy, fertility, ovulation, menopause, sexual disorder, thyroid functions), tumour markers (prostate, colorectal, etc.), viruses (HIV, hepatitis B and C), bacteria *(Streptococcus* A and B, *Chlamydia trachomatis, Treponema pallidum, Heliobacter pylori,* etc.), IgE (allergy), and troponin T in cardiac monitoring. All these analytes are measured on the basis of their presence or absence.

An extensive review of near-patient testing in primary care has been published by Hobbs et al. So far, these tests are intended mainly for human diagnostics where, especially for the low-molecular-mass analytes, relatively high concentrations of analytes are measured. In food diagnostics, however, much lower detection levels are required. For the detection of veterinary drug residues, for example, detection levels at ppb level (ng analyte/g of sample) are necessary as for many of these drugs a maximum residue level (MRL) has been established.

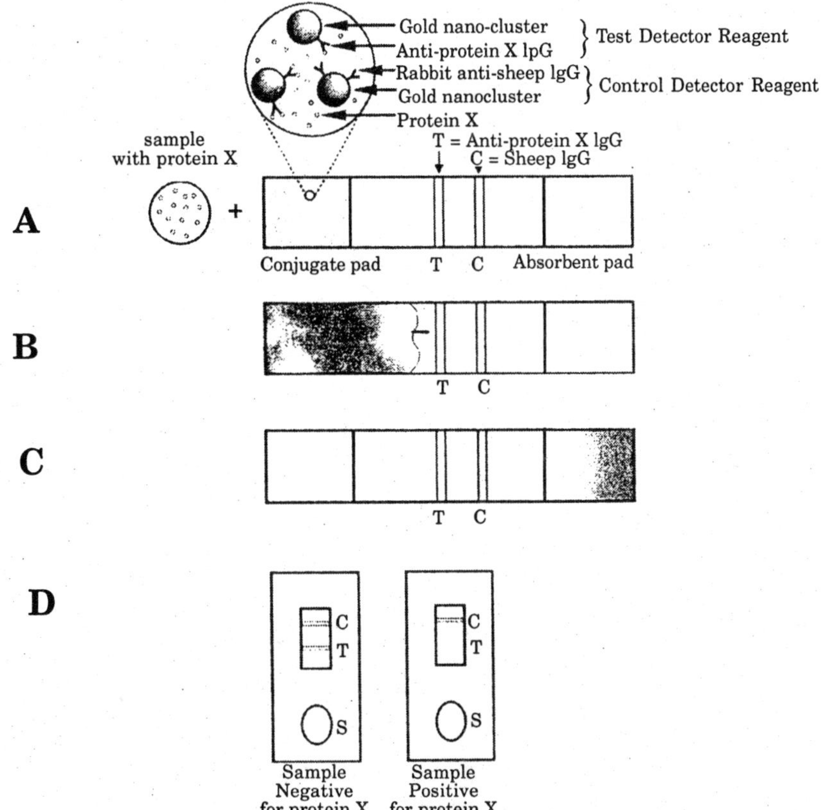

Fig. 5.4. Principle of a strip test for the detection of low-molecular mass-analytes. A. Fluid sample is brought on to the sample pad. B. The control detector reagent. C. Sample passes over the capture zones.

Although strip tests appear simple, complex interactions among their various components lead to a number of challenges in both the development and manufacturing environments. These challenges are even greater for quantitative tests where the colour intensity of the test line must be repeatable between production lots. From a developmental perspective, creating a successful test system means optimizing the interactions among its raw materials, component design, and manufacuring techniques. It is beyond the scope of this manual to exhaustively describe all possible methods and materials that can be used for the development and manufacturing of strip tests. We will merely describe our experience in making good and reliable strip tests for the detection of both low- and high-molecular-mass analytes. Because of this limitation, some additional reading on new and improved methods is recommended.

MATERIALS

Materials

Amersham Pharmacia Bietech AB (Uppsala, Sweden):

— CNBr-activated Sepharose 4B (17–0430-01)

— HiTrap Protein G (1 × 5 ml) (17–0405–01)

Amicon (Beverly, MA, USA):

— Centriprep–30 concentrator (4306)

BioDot Inc. (Irvine, Ca, USA):

— BioDot Surfactant Starter Kit

— BioDot Diagnostic Materials Kit

Bio-Rad Laboratories B.V. (Veenendaal, The Netherlands):

— Two-way stopcock (732–8102)

G&L (San Jose, Ca, USA):

— 0.01" white matte vinyl GL-187 (6.3 cm x 30 cm) (991093)

Gelman Sciences (Ann Arbor, MI, USA):

— 5 μm Acrodisc filter (4489)

Greiner Bio-One B.V. (Alphen a/d Rijn, The Netherlands):

— 50 ml test tubes (210261)

Kenosha C.V. (Amstelveen, The Netherlands):

— Plastic cassettes (housings): 16 × 71 mm for test strips of 5 × 63 mm

— Foil pouches: OPA15/PE15/ALU12/ LDPE75 MU (80 × 144 mm); 403389

Multisorb Technologies Inc (Buffalo, NY, USA):

— MiniPax (sachet containing 1 g of silica gel dessicant)

Millipore Corporation (Bedford, MA, USA):

— MilliQ 185 Plus water purification system

Nalge Nunc International (Rochester, NY, USA):

— Drop-Dispenser bottle (15 ml; 2411-0015)

Pall Gelman Sciences (Champs-sur-Marne, France):

— Cellulose type 133 (8" × 10") (AC1071)

— Cytosep 1660 (8" × 10") (AC1081)

Pierce (Rockford, IL, USA):

— BCA protein assay reagent (23225)

— PharmaLink Immobilization Kit (44930)

Schleicher & Schuell (Dassel, Germany)

— Nitrocellulose membrane strips (25 × 300 mm) AE 100 12 μm on Mylar 5 backing

— Cellulosic paper GB002 (20 × 20 cm) (10 426 681)

— Cellulose acetate filter FP 030/3 (0.2 μm) (462 200)

Serva Feinbiochemica (Heidelberg, Germany):

— Visking dialyzing tubing; 20/32, ∅ 16 mm (44110)

Varian (Harbor City, CA, USA):

— Varian Bond Elut Reservoir (8 ml) (1213-1015)

Whatman International Ltd (Maidstone, UK):

— Coarse PVA Bound Glass Fibre, grade F 075-17 (A4-size; 676475)

— Cellulosic paper 3MM Chr (20 × 25 cm; 3030-866)

— Cellulosic paper 3MM D28 (80048)

Chemicals

Caltag Laboratories (Burlingame, Ca, USA):

— Affinity purified goat anti-rabbit IgG (H + L) antibodies (1.25 mg/ml)

Marvel, Premier Brands UK (Moreton, UK):

— Skimmed milk powder

Merck (Darmstadt, Germany): -

— Acetic acid (100%) (CH_3COOH; 100063)

— Disodium hydrogen phosphate dihydrate ($Na_2HPO_42H_2O$; 106580)

— Glutaraldehyde (25% w/w solution in water; 820603)

— Glycerol (extra pure $C_3H_8O_3$; 104093)

— Hydrochloric acid (32%) (HCl; 100319)

— Methanol (CH_3OH; 106009)

— Potassium carbonate (extra pure K_2CO_3; 104924)

— Potassium dihydrogen phosphate (KH_2PO_4; 4873)

— Polyvinylpyrrolidone (PVP; 107370)

— Sodium acetate (CH_3COONa; 106268)

— Sodium chloride (NaCl; 106404)

— Sodium dihydrogen phosphate monohydrate ($NaH_2PO_4.H_2O$; 106346)

— Sodium ethylene diamine tetra acetate (Na-EDTA, $C_{10}H_1N_2Na_2O_8.2H_2O$; 108421)

— Sodium hydrogencarbonate ($NaHC0_3$; 106329)

— Sodium hydroxide (NaOH; 106495)
— Tri sodium citrate dihydrate ($Na_3C_6H_5O_7.2H_2O$; 112005)
— Tris(hydroxymethyl)aminomethane ($C_4H_nNO_3$; 108382)
— Tween 20 (Polyoxyethylene sorbitan monolaurate; 822184) Sucrose ($C_{12}H_{22}O_n$; 107651)

Pragmatics Inc (Elkhart, In, USA):

— Surfactant 10G (p-Isononylphenoxypolyglycidol; 50% aqueous solution)

Serva Feinbiochemica (Heidelberg, Germany):

— Sodium dodecylsulfate (SDS, $C_{12}H_{25}O_4S.Na$; 20760)

Sigma Chemical Company (St Louis, MO, USA):

— 1,4-Butanediol diglycidyl ether ($C_{10}Hi_8O_4$; B 1029)
— Bovine Gamma Globulins (BGG; G 5009)
— Bovine Serum Albumin, fraction V (BSA; A 7888)
— Sodium azide (NaN_3; S 2002)
— Ovalbumin grade V (A 5503)
— Sulfadimidine (SDD, 4-amino-N-[4, 6-dimethyl-2-pyrimidinyl] benzene- sulfonamide = sulfamethazine; S 6256)
— Streptomycin sulfate (S-6501)
— Tetra-chloroauric[III] acid trihydrate ($HAuCl_4.3H_2O$; G 4022)

Equipment

The BioDot system (BioDot Inc., Irvine, Ca, USA) consisted of two BioJet Quanti3000 dispensers attached to a BioDot XYZ3000 dispensing Platform. An AZCON Sur-Size™automatic guillotine cutter (Model SS-4) was supplied by AZCON (Elmwood Park, NJ, USA). Sealing equipment (Magneta, Model 421) was purchased from Kenosha (Amstelveen, The Netherlands).

Solutions, Reagents and Buffers

Solution 1	Phosphate Buffered Saline (PBS): 5.39 mM Na_2HPO_4; 1.29 mM KH_2PO_4; 153 mM NaCl; pH 7.4
Solution 2	Saturated ammonium sulfate solution: 80 g $(NH_4)_2SO_4$ in 100 ml of water of 20 °C
Solution 3	IAC coupling buffer: 0.1 M $NaHC0_3$ pH 8.3 containing 0.5 M NaCl
Solution 4	IAC blocking buffer: 0.1 M Tris-HCl buffer pH 8.0
Solution 5	IAC washing buffer 1:0.1 M Tris-HCl buffer pH 8.0 containing 0.5 M NaCl
Solution 6	IAC washing buffer II: 0.1 M sodium acetate-acetic acid buffer pH 4.0 containing 0.5 M NaCl
Solution 7	IAC elution solution: 0.1 M acetic acid
Solution 8	Retentate mixing buffer in synthesis BSA-streptomycin conjugate (0.1 M potassium phosphate buffer pH 8.0): 26.5 ml 0.2 M KH_2PO_4 and 473.5 ml 0.2 M K_2HP0_4 is adjusted to 1 1 with water
Solution 9	Strip-test membrane blocking buffer: 0.9 g/1 $NaH_2PO_4.H_2O$ pH 7.5 containing 2% (w/v) skimmed milk powder and 0.02% (w/v) s'odium dodecylsulfate (SDS)
Solution 10	Strip test membrane washing buffer: 0.9 g/1 $NaH_2PO_4.H_2O$ pH 7.5 containing 0.01% (v/v) Surfactant 10G (S24 from the BioDot Sur factant Starter Kit)

Solution 11 Extraction buffer for food products and plant leaves: PBS pH 7.4 containing 0.5% (v/v) Tween 20 and 0.5% (w/v) polyvinylpyrrolidone (PVP)

METHODS

Antibodies

Antibodies form the heart of the immunological strip-test device. In practise, they will be obtained either from an animal serum (polyclonal antibodies) or from a culture supernatant (monoclonal antibodies). Both types of antibodies should be purified at least to some extent before being applied in a strip test. For the control capture antibody, an ammonium sulfate precipitation to isolate the total immuunglobulin fraction (Ig fraction) often will be sufficient. For the test and control detector reagents, as well as for the test capture reagent in the direct assay format, however, merely antibodies that have been purified by immunoaffinity chromatography (IAC), should be used.

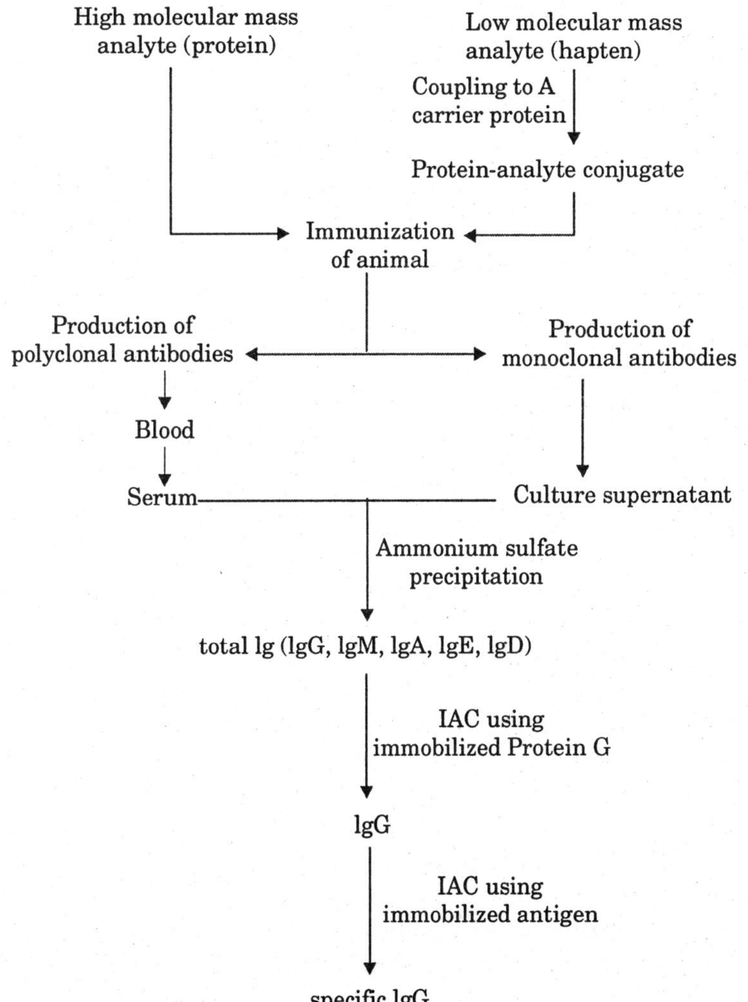

Fig. 5.5. Flowchart showing the various steps in the production and purification of both polyclonal and monoclonal antibodies.

For an even better performance of the test, one may decide to use antigen-specific purified antibodies as test reagents. These can be prepared by additional IAC of the IgG fraction using a column in which the antigen has been immobilised onto a chromatographic support. Fig. 5.5 shows the various steps in the production and purification of both polyclonaland (mouse) monoclonal antibodies. It is important to determine the cross-reactivities of the antibodies with sample matrices to be used in the assay.

Moreover knowledge about the behaviour and properties of the antibodies in a specific sample matrix is essential in order to obtain an optimal performing strip test. When testing urine samples, for example, one has to know the binding properties of the antibodies at the wide range of possible pH values in such samples. It goes without saying that in a direct assay where monoclonal antibodies are used for the detector and capture reagent, each of the two antibodies should be directed against a different epitope on the antigenic protein. When using polyclonal antibodies, however, the same pool of antibodies may be applied for both the detector and the capture reagent.

Isolation of Total Ig by Ammonium Sulfate Precipitation

Protocol 1: Isolation of total Ig by ammonium sulfate precipitation

1. When starting from a crude rabbit or sheep serum, 10 ml of serum is first diluted with 20 ml of PBS (Solution 1), after which 30 ml of a saturated ammonium sulfate solution (Solution 2) is added drop-wise with constant stirring at ambient temperature.
2. When starting from a culture supernatant, an equal volume of a saturated ammonium sulfate solution is added directly with constant stirring at ambient temperature.
3. After standing for 30 min., the antibodies are collected by centrifugation for 10 min. at 10,000 × g at ambient temperature. The use of a swinging-bucket rotor is recommended.
4. After discarding the supernatant as completely as possible, the pellet is dissolved in as small a volume of PBS as possible (a few ml).
5. The solution is then exhaustively dialysed against PBS at 4 °C for at least 48 h.
6. The dialysed fraction can then be concentrated by using a centriprep-30 concentrator according to the manufacturer's user guide.
7. Finally, the protein content is determined, *e.g.*, by performing a BCA protein assay according to the manufacturer's user guide and using bovine gamma globulins (BGG) as standard.

Notes: When isolating sheep IgG for the control capture line, the dialysed fraction is now purified sufficiently and can be stored in small aliquots at –20 °C until used. For other applications the antibodies should be further purified by IAC. Before use, some dialysis tubing has to be boiled for 5–10 min. in the presence of sodium ethylene diamine tetra acetate (Na-EDTA). The obtained sterile tubing can be stored for several months at 4 °C in 50% (v/v) glycerol in water.

Isolation of IgG by Immunoaffinity Chromatography (IAC)

HiTrap Protein G columns (Amersham Pharmacia Biotech) contain Protein G immobilised to Sepharose High Performance as chromatographic support. Like Protein A, Protein G binds specifically to the Fc region of IgG. However, the range of polyclonal IgGs that bind strongly to protein G is much wider, and adsorption takes place over a wider pH range. This makes IAC using Protein G ideal for fast purificaton of specifically the IgG fraction from a serum or culture supernatant. For the use of HiTrap Protein G columns, the manufacturer's user guide should be followed.

Note: For the production of (mouse) monoclonal antibodies, the hybridoma cells are often cultured in the presence of 1% to 10% foetal calf serum. This implies that the antibody fraction obtained after IAC purification on a HiTrap Protein G column of such samples will contain a relatively large amount of bovine IgG as well.

Isolation of Specific Antibodies by IAC

IAC Using Immobilised Antigenic Proteins

CNBr-activated Sepharose 4B is a preactivated gel for immobilization of ligands containing primary amines such as proteins. Such a gel with immobilized proteins can be used to isolate specific antibodies to the proteins from a serum or culture supernatant.

Protocol 2: Preparation of the column

1. 2g freeze-dried powder of CNBr-activated Sepharose 4B is swollen in 15 ml 1 mM hydrochloric acid for about 10 min.
2. The swollen gel is transferred into an 8 ml Varian column with a porous bed support (frit) at the bottom of the colomn and sealed with a two-way stopcock.
3. The gel is subsequently washed with 400 ml 1 mM hydrochloric acid and 100 ml coupling buffer (Solution 3).
4. The gel is transferred into a 50 ml Greiner tube and mixed with 7.5 ml coupling buffer containing 20 mg of the proteins of interest (5-10 mg protein/ml gel).
5. The coupling solution is then gently rotated end over end for 2 h at ambient temperature or overnight at 4 °C. Do not use a magnetic stirrer!
6. After coupling, the gel is either centrifuged for 2 min. at 200 x g or allowed to settle out for 5-10 min.
7. The supernatant is removed and the gel is washed three times with 30 ml coupling buffer, each wash followed by either centrifugation or standing at the bench for 5-10 min.
8. After removing the last washing solution, any remaining active groups are blocked by adding 15 ml blocking solution (Solution 4) to the gel and gently rotating the mixture end over end for 2 h at ambient temperature.
9. The gel is then transferred to the column, after which the upper-frit is placed.
10. The gel is washed with three cycles of alternating pH. Each cycle consists of a wash with 20 ml washing buffer I (Solution 5) followed by a wash of 20 ml washing buffer II (Solution 6).
11. Finally, the column is washed with 30 ml PBS, pH 7.4.

Protocol 3: Binding of antibodies

1. The column is loaded with ammonium-sulfate-precipitated and HiTrap Protein G immunoaffinity-purified antibodies in PBS, pH 7.4. The capacity of the column is high enough to cope with an amount of total IgG originating from 25 ml crude animal serum. The sample is loaded three times onto the column.
2. The column is washed with 20 ml PBS, pH 7.4

Protocol 4: Elution of antibodies

1. Bound antibodies are eluted with 15 ml elution solution (Solution 7).
2. The eluate is immediately brought at neutral pH by addition of 1 M sodium hydroxide.
3. The eluate is concentrated to about 5 ml by using a centriprep-30 concen-trator, according to the manufacturer's user guide.
4. Finally, the protein content is determined, *e.g.*, by using the BCA protein assay according to the manufacturer's user guide.

Storage:

— The column is stored at 4 °C in PBS, pH 7.4, containing 0.05% (w/w) sodium azide.

Notes:

— Be sure that none of the reagents in the coupling reaction contains sodium azide, as this will seriously disturb the reaction.

IAC Using Immobilised Small Ligands

For the purification of antibodies to low-molecular-mass ligands, the ligand of interest has to be coupled to a chromatographic support. Because the chemical nature of small ligands usually varies widely, there is no such thing as a general immobilization protocol. Often, coupling of small ligands to a support appears to be far more difficult than coupling of a protein as described in the section above.

An overview of small ligand immobilization techniques is described in The PharmaLink Immobilization Kit enables the coupling of a large variety of difficult-to-immobilize compounds. It utilizes the Mannich reaction in which certain active hydrogens of the ligand are condensed with formaldehyde and an amine-containing support (the PharmaLink gel). Particular hydrogens in ketones, esters, phenols, acetylenes, α-picolines, quinaldines, and a host of other compounds can be aminoalkylated using the Mannich reaction.

For an overview of the possibilities of the kit and a step-by-step guide to the procedure, see the manufacturer's instruction manual. When the ligand of interest contains an alifatic amine group, coupling can be performed easily by using CNBr-activated Sepharose 4B (protocol 2), now adding 1 to 10 umoles of the ligand/ml gel. When the ligand contains an aromatic amine group, however, coupling is more successful when using a modified PharmaLink protocol as described by Verheijen *et al.* The example describes the immobilization of sulfadimidine (SDD; sulfamethazine), a sulfonamide drug that is coupled by its aromatic amine group.

Note:

— Be sure that none of the reagents in the coupling reaction contains sodium azide, as this will seriously disturb the reaction.

Preparation of 40 nm Gold Nanoclusters (G40)

Preparation of gold nanoclusters is based upon reduction of $HAuCl_4$ and has been described for a wide variety of reagents, including formaldehyde, white phosphorus, citric acid, ascorbic acid, tannic acid, and hydrogen peroxide. The size of the individual nanoclusters can be manipulated easily and depends on several factors such as the chemical nature of the reducing reagents, temperature, pH, and concentration of the reagents. Protocols have been published for the preparation of colloidal gold particles with a diameter ranging from 3 to 150 nm. The most commonly used method is the reduction of gold chloride with sodium citrate as described by Frens.

The procedure for preparing a gold nanoclusters with an average particle diameter of 40 nm (G40) is as follows:

Protocol 5: Preparation of gold nanoclusters (40 nm)

1. 100 ml 0.01% (w/v) tetra-chloroauric [III] acid trihydrate ($HAuCl_4.3H,O$) is heated to boiling under reflux conditions.
2. 1 ml of 1% (w/v) trisodium citrate dihydrate is added to the boiling solution under constant stirring.
3. In about 25 s the slightly yellow solution will turn faintly blue (nucleation). After approximately 70 s the blue colour then suddenly changed into dark red, indicating the formation of monodisperse sperical particles.
4. The solution is boiled for another 5 min. to complete reduction of the gold chloride. The optical density, measured at 540 nm (OD540), of such G40 suspensions will be about 0.9.
5. Supplemented with 0.05% (w/v) sodium azide, the obtained G40 suspensions can be stored at 4 °C for several months.

Notes:

All reagents should be prepared in highly purified water, *e.g.,* by using a Milli-Q 185 Plus water purification system (Millipore) yielding water with a resistivity of > 18.2 MΩ.cm.

Another frequently used particle size of gold nanoclusters is that of 25 nm.

For the preparation of such particles, the amount of added trisodium citrate solution should be increased to 1.5 ml. A G25 suspension will be somewhat lighter red in colour than a G40 solution.

Suppliers of gold nanoclusters are:

British BioCell International, Golden Gate, Ty Glas Avenue,Cardiff CF4 5DX, UK, http://www.british-biocell.co.uk, e-mail: info@britishbiocell.co.uk; and Aur-ion, Costerweg 5, 6702 AA Wageningen, The Netherlands,

http://www.aurion.nl, e-mail: info@aurion.nl

Preparation of the Detector Reagents

The test and control detector reagents consist of G40 nanoclusters coated with the appropriate antibodies. The amount of antibody attached to the individual gold nanocluster depends upon many factors, including the size of the gold nanocluster, the pH at which the coupling is performed, the concentration of the antibody added, and the electrical charge of the antibody. Coupling will take place only if the negative charge of the unstabilized gold nanocluster can be compensated by a net positive charge of the antibody. Because of the lability of unstabilized gold nanoclusters and its sensitivity to salts, the salt concentration in the coupling solution should be kept as low as possible.

The amounts of electrolytes necessary to induce flocculation depend upon the valences of the particular ions. In order to eliminate the same amount of negative surface charge of colloidal gold and to induce flocculation, cations such as K^+, Ba^{2+}, and Al^{3+} have to be added in equivalents amounts of 1000:10:1 to cause the same effectr. The minimal protecting amount (MPA), *i.e.*, the minimal amount of protein needed to protect the sol from salt-induced precipitation, can be determined by the procedure of Horisberger and Rosset. In this procedure, small amounts of unstabilized gold nanoclusters (0.5 to 1.0 ml) are added to a series of increasing concentrations of the protein of interest in a constant volume (100 μl). After 5 min., 100 μl of a 10% (w/v) sodium chloride solution is added to the mixture. If the colour changes from red to violet and finally to blue, the protection was incomplete and aggregation, followed by flocculation of unstabilized gold nanoclusters, was induced by the added salt.

Spectrophotometric analysis of a suspension of gold nanoclusters at 520 nm results in a more precise estimate of the degree of stabilization of the sol. In general, a final concentration of 10% to 100% above the MPA is used for coating. When using the G40 nanoclusters made in our lab, the MPA was always found to vary around 5 μg antibody/ml G40 suspension. Coating was therefore performed with 6.0 to 7.5 μg antibody per ml of G40 suspension.

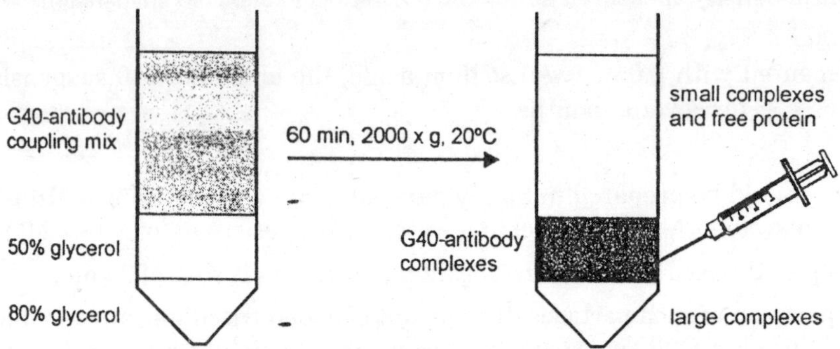

Fig. 5.6. Isolation and purification of the detector reagents.

The general procedure for coating G40 nanoclusters with antibodies is as follows:

Protocol 6: Coating of gold nanoclusters

1. 50 ml G40 suspension (OD540 ~ 0.9) is adjusted to pH 8.5 with a 0.2 M potassium carbonate solution (K_2CO_3).
2. Coating of the G40 clusters is performed by incubating X μg purified antibody (X > MPA) per ml G40 for 20 min. at ambient temperature while being gently swirled on a shaking platform. While adding the antibodies, the sol should be stirred vigorously in order to obtain as uniform a coating as possible.

3. After coating, the sol is further stabilised by adding 5 ml 1% (w/v) skimmed milk powder that has been adjusted to pH 8.5 with 0.2 M potassium carbonate.
4. The mixture is gently swirled for another 60 min. at ambient temperature on a shaking platform.
5. The suspension is split in half, and each half (about 27.5 ml) is centrifuged for 60 min. at 2500 x g and 20 °C in a 50 ml Greiner tube over a discontinuous glycerol gradient consisting of 5 ml 80% (v/v) glycerol, of pH 8.5, and 7.5 ml 50% (v/v) glycerol, pH 8.5, using a swinging-bucket rotor. Under these conditions the antibody-coated G40 nanoclusters will be found almost entirely in the 50% (v/v) glycerol layer.
6. Particles are harvested by piercing the centrifuge tube with a syringe needle and removing the dark-red layer sideways out of the tube. In this way, the antibody-coated G40 nanoclysters will not be contaminated with free protein (antibodies and blocking proteins) and very small complexes present in the upper water layer and/or with large protein cluster complexes present in the lower 80% glycerol layer (Fig. 6).
7. The obtained 4-5 ml of coated G40 nanoclusters are diluted to 15 ml with water that has been adjusted to pH 8.5 with 0.2 M potassium carbonate.
8. The solution is then transferred into a centriprep-30 concentrator.
9. The glycerol is removed in about five centrifugational runs in which the retentate after each run is filled up to 15 ml with water of pH 8.5.
10. In an additional run, the suspension is then further concentrated to 2 ml.
11. After addition of 400 μl of a 6% (w/v) skimmed milk powder solution in water of pH 8.5 (1% (w/v) end concentration) and 12 ul of a 10% (w/v) sodium azide (0.05% (w/v) end concentration), the suspension is stored at 4 °C until used.

Notes:

When using other volumes and/or different sized gold nanoclusters, one has to change the centrifugal speed and/or the running time such that the purified detector reagent is found almost entirely into the 50% glycerol layer.

The washes to remove (most of) the glycerol are necessary to prevent problems with drying of the conjugate pads. Moreover, glycerol reduces the flow of the detector reagent over the membrane.

Preparation of the Capture Reagents

Control Capture Reagent

The control capture reagent consists of a total Ig fraction which can be prepared by ammonium sulfate precipitation of sheep serum.

Test Capture Reagent in the Direct Assay

In the direct assay, the test capture reagent consists of polyclonal or monoclonal antibodies to the antigenic protein of interest that have been purified by immunoaffinity chromatography (HiTrap Protein G).

Test Capture Reagent in the Indirect (Competitive) Assay

In the competitive assay, the test capture reagent consists of a ligand-protein conjugate in which the ligand of interest has been coupled to a protein, usually ovalbumin or bovine serum albumin (BSA). Similar to the immobilization of low-molecular-mass ligands to a chromatographic support a successful coupling of small ligands to a protein largely depends on the available reactive groups on the ligand. Moreover, the low solubility of a ligand in an aqueous solution may be an additional problem. A large variety of coupling protocols have been described. One of these procedures is the use of glutaraldehyde as linker molecule between an amine group of the ligand and the ε-aminogroup of the lysin residues of the protein. As an example, the coupling of the sulfonamide drug SDD by its aromatic amine group to ovalbumin will be described.

Protocol 7: Coupling of a sulfonamide drug to ovalbumin

1. 0.1 mg SDD/ml aqueous solution is prepared by dissolving 10 mg of SDD in 1 ml of methanol which is then added to 100 ml of PBS (Solution I).
2. Solution II is prepared by dissolving 300 mg of ovalbumin in 30 ml of PBS, after which the solution is filtrated over a 5 μm Acrodisc.
3. The reaction mixture consists of 30 ml of Solution I, 30 ml of Solution II and 30 ml of PBS, to which 900 μl 25% (w/w) glutaraldehyde solution is added slowly under vigorous constant stirring.
4. Approximately 10 min. after adding the glutaraldehyde, the clear solution will take on a yellowish appearance. The mixture is stirred for 4-5 h at ambient temperature.
5. The conjugate mixture is exhaustively dialysed against PBS for 4-5 days at 4 °C.
6. The dialysate is then concentrated to 50 mg/ml using a centriprep-30 concentrator and stored in 1 ml aliquots at -20 °C.

Another useful method is the coupling of the analyte to an epoxy-activated protein. As an example, the synthesis of a BSA-streptomycin conjugate is described in which the antibiotic streptomycin is coupled to BSA by using the two epoxy groups of 1,4-butanediol diglycidyl ether as reactive groups. First, one epoxy group is coupled to the ε-amino group of lysine residues as well as to the hydroxyl group of tyrosine residues. Thereafter, the remaining epoxy group of the obtained epoxy-activated BSA is coupled to the streptomycin molecule. In this way, 1,4-butanediol diglycidyl ether provides a 12-atom spacer between BSA and the streptomycin molecule. The procedure is as follows:

Protocol 8: Coupling of epoxy-activated BSA to streptomycin

1. A solution of 630 mg BSA in 10 ml water is brought on pH 11.2 with 0.5 M sodium hydroxide.
2. After adding 152 mg 1,4-butanediol diglycidyl ether, the solution is gently stirred for 20 h at 27 °C.
3. During the first few hours of the incubation, the pH has to be checked every half hour and kept between 11.1 and 11.2 by adding small amounts of 0.5 M sodium hydroxide.
4. After the incubation, the brownish solution is transferred into a centriprep-30 concentrator and centrifuged for 15 min. at 1500 × g and ambient temperature.
5. The retentate is mixed with retentate mixing buffer (Solution 8) to an end volume of 15 ml and centrifuged again in a centriprep-30 concentrator.

6. After five such washing cycles, the remaining 10 ml of epoxy-activated BSA solution is allowed to react with 376 mg streptomycin sulfate at pH 10.8 during an incubation of 48 h at 27 °C under constant stirring. During this incubation, a yellow precipitate is formed.
7. The conjugate mixture is exhaustively dialysed against PBS for at least 48 h at 4 °C.
8. The yellow precipitate is spun down by centrifugation for 10 min. at 3500 × g and 4 °C.
9. The yellow supernatant is concentrated to approximately 3.5 ml, divided into small aliquots and stored at -20 °C.

Choice of Membrane

A proper choice of membrane is essential for the good performance of a strip-test. Several different strip test membranes are available, varying in capillary flow rate (sec/4 cm), tensile strength (the presence or absence of a plastic membrane support), thickness, and surface quality. The parameters that affect capillary flow rate are the membrane's pore size, pore size distribution, and porosity. Millipore gives an overview of how these parameters are related. The thickness of the membrane is of influence on the bed volume, the dispensed line-width and the tensile strength. The advantage of a membrane that has been cast directly on a plastic support is the large increase of the membrane's tensile strength properties, which makes its handling much easier.

Moreover, the presence of a membrane support prevents the migration of adhesives from the backing plate into the membrane layer. On a direct-cast membrane, however, it is not possible to apply reagents to the belt side, *i.e.,* the side of the membrane that is the smoothest and is relatively defect free. Nitrocelluose is probably the most commonly used polymer for strip test membranes. The pore size of such membranes varies between 5 and 20 μm, which is large compared to a pore size of 0.2 to 1.2 μm of a nitrocellulose membrane used for protein blotting. A large pore size implies a small membrane surface area and, consequently, a low-protein binding capacity, *i.e.,* 20 to 30 μg IgG/cm^2 instead of the 110 μg IgG/cm^2 for a blotting membrane with a pore size of 0.45 μm. Nitrocellulose membranes bind proteins electrostatically; the strong dipole of the nitroester interacts with the strong dipole of the peptide bonds of the protein. Because nitrocellulose membranes are completely neutral, their binding properties are independent of the pH of the immobilization solution.

However, the pH might have an effect on both the solubility and immobilization efficiency of a particular protein. when using nitrocellulose membranes that bind proteins electrostatically, detergents such as Tween-20 and Triton-X 100 should not be used or should be used only in very low concentrations (< 0.01% v/v). Ionic detergents, such as sodium dodecylsulfate (SDS), are compatable with this type of membrane at concentrations up to 0.5% (w/v). Depending on the choice for a batch or a continuous manufacturing process, membranes are available in different formats, i.e., as small strips (30 × 2.5 cm), sheets (20 × 20 cm), or rolls (30 cm × 100 m).

Immobilization of the Capture Reagents into the Membrane

A crucial step in manufacturing strip tests is the proper application of reagents to the membrane.

For the preparation of our striptest devices, a BioJet Quanti3000 attached to a BioDot XYZ3000 Dispensing Platform is used for dispensing the test and control capture antibodies onto nitrocellulose membrane AE 100 (12 μm) strips (300 × 25 mm) on a Mylar 5 support. Prior to use, the antibody stock solutions (5 mg IgG/ml) are filtered over a 0.2 μm FP 030/3 disposible cellulose acetate filter.

A typical amount of antibody used for both the control and test capture reagent is 0.5 μg/cm membrane. The distance between control and test line largely depends on the readout indications on the viewing window of the housing, but is usually about 5 mm.

Drying of the Membranes

After dispensing the sample and/or control lines, the membrane is dried for 1 h at 50 °C. Drying a membrane completely is essential in order to obtain a proper fixation of the proteins that have been deposited. Drying efficacy is a function of temperature, humidity, and time. If the membrane is backed, then the drying times must be extended. The conditions that are typically required to achieve complete drying of a membrane are 0.5 h, 1–2 h, and > 16 h for drying temperatures of 15–25 °C, 35–40 °C, and 45–50 °C respectively.

Blocking and Washing of the Membranes

Blocking is often necessary in order to prevent non-specific binding of the analyte and/or detector reagent to the membrane. After drying the membrane strips, blocking is performed by incubating the strips for 30 min. in blocking buffer.

Because an excess of blocking agent can interfere dramatically with the membrane's capillary flow properties, unadsorbed blocking agent has to be removed by washing the membrane in a weak buffer solution.

To promote uniform re-wetting of the blocked membrane, the presence of a low concentration of surfactant in the washing solution is recommended. In case one would like to compare the effects of various surfactants, the BioDot Surfactant Starter Kit may be very useful. Using the nitrocellulose membrane AE100, washing can be performed three times for 5 min. each with washing buffer (Solution 10). After drying for another 2 h at 50 °C, the membranes can be either further processed or stored under dry conditions, *i.e.,* sealed in plastic in the presence of desiccator, until use. Storage may be performed at ambient temperature (for a few weeks) or at –20 °C (for a longer period).

Note:

The term surfactant is an acronym for surface active agent and generally refers to welters, solubilizers, emulsifiers, dispersers, or, detergents. Surfactants almost always act to reduce attractive interactions between "like" particles and bring them to "unlike" surfaces. An alternative for the above-described blocking of the test strip directly after dispensing of the readout lines is the "locking on the fly" technique, by which the blocking agent is included in with the conjugate pad impregnation mixture. When the sample then enters the conjugate pad, the solubilized blocking agent will run with the front along the membrane, binding to non-specific sites.

Choice of the Conjugate Pad

The conjugate pad performs multiple tasks, the most important being the delivery of the assay detector reagent, which is deposited into the pad material typically in conjugation with some combination of blocking agents and/or surfactants. When sample flows from the sample pad into the conjugate pad, the detector reagent should be released consistently, quickly, and quantitatively and then float with the sample front. Such an ideal-acting conjugate pad thus consists of a well-balanced combination of the proper conjugate pad material and the proper surfactants. Materials that are commonly used to make conjugate pads include glass fiber filters, paper (cellulosic) filters, and surface-treated (hydrofilic) propylene filters.

A number of these materials can be found in the BioDot Diagnostic Materials Kit, whereas the BioDot Surfactant Starter Kit contains a variety of surfactants that one may try for his own specific application. In our strip tests, Whatman glass fiber Paper with binder F 075–17 was chosen to be the most appropriate material for a conjugate pad because of the ease with which the material is wetted, the high capacity to absorb fluid, the uniform distribution of the detector reagent on the pad, and the low amount of detector reagent that remains on the pad after running the assay.

Composition of the Detector Reagent on the Conjugate Pad

For a membrane strip of 5 mm in width, a conjugate pad of 5 x 5 mm may be applied. Using Whatman Glass Fiber Paper with binder F 075–17, such a pad can be loaded with 12 to 13 µl of conjugate pad impregnation mixture. Besides the two detector reagents, the choice of other key ingredients appears to be a matter of trial and error. Moreover, the required concentration of the test and control detector reagent in the mixture largely depends on the quality of these reagents.

As an example, the composition of a conjugate pad impregnation mixture is given for a test that has been developed in our laboratory for the detection of Pepino Mosaic virus (PepMV) in tomato plants:

- 215 µl rabbit anti-PepMV IgG coated G40 nanoclusters (test detector reagent)
- 45 µl rabbit anti-sheep IgG coated G40 nanoclusters (control detector re-agent)
- 130 µl 40% (w/v) sucrose in water
- 10 µl 1% (v/v) Surfactant 10G in water

Each conjugate pad (5x5 mm) was loaded with 13 µl of this mixture. After drying for 2 to 3 h at 50 °C, the conjugate pads were stored at –20 °C until used.

Another example is the composition of a conjugate pad impregnation mixture for a test that has been developed in our laboratory for the detection of (dihydro) streptomycin in milk:

- 70 µl of anti-(dihydro)streptomycin IgG coated G40 nanoclusters (test detec-tor reagent)
- 150 µl of rabbit anti-sheep IgG coated G40 nanoclusters (control detector reagent)
- 70 µl of 12% (w/v) sucrose in water
- 10 µl of 1% (w/v) Surfactant 10G in water
- 100 µl of water

Each conjugate pad (5 x 5 mm) was loaded with 10 µl of this mixture. After drying for 2 to 3 h at 50 °C, the conjugate pads were stored at –20 °C until used.

Choice of the Absorbant Pad

Absorbent pads are placed at the end of the test strip. The major advantage of using an absorbent pad is that assay sensitivity can be increased by virtue of the increase in the total volume of sample that is analyzed. Without an absorbant pad, the total volume of sample analyzed is determined by the bed volume of the membrane. Most absorbent pads are made from paper filters. Cellulose type 133 (Gelman), cellulosic paper 3MM (What man), or cellulosic paper GB002 (Schleicher & Schuell) are all excellent materials for use as absorbent pad.

Choice of the Sample Pad

Sample pads are placed at the beginning of the test strip. Their major task is to absorb the sample and provide a uniform flow of the sample fluid from the sample pad via the conjugate pad onto the membrane. Moreover, the sample pad acts as a filtration device by removing particles, *e.g.*, fat particles in milk, from the sample solution. The pad may be pre-treated with a special buffer solution in order to adjust the pH or viscosity of the sample solution.

Most sample pads are made from paper filters. When using aqueous sample solutions like urine, buffer, etc., cellulose paper type 133 (Gelman), 3MM Chr (Whatman), 3MM D28(Whatman), or GB002 (Schleicher & Schuell) can be used as sample pad. However, these materials are pretty thick and therefore the sample pads will be easily clogged when using samples with high viscosity and/or that contain a high concentration of particles or sample debris. For milk samples as well as for food and/or plant extracts, cytosep 1660 (Gelman) is a good choice. When using whole-blood samples, special blood separation filters will be needed as sample pad.

A partial list of blood separation media manufacturers is provided below: Furthermore, there are two interesting patents in this field that should be mentioned. US5915521 Lateral flow filter devices for separation of body fluids from particulate materials. Describes the use of an asymmetric filter wherein red cells and plasma flow laterally along the filter in order to improve the volume of plasma obtained without clogging the filter. PCT WO97/34148A1 Immunoassay Device. Describes various devices where the analytical membrane also acts to separate plasma from whole blood however, a certain amount of analyte in the sample will result in the appearance of a test line. Originally, strip tests were designed to be judged by the naked eye. However, one tends to apply optical reading equipment for interpretation of the test results.

An example of such equipment is the BioDot Test Strip Reader (Model TSR3000), which is a reader designed for the qualitative and/or quantitative measurement of a test strip itself or within a housing. The system has a CDD array camera to take a high-resolution image of the test strip. The user can then define a window and measure the intensity of light reflected from the test strip. The great advantage of such a system is that the results can be stored, analyzed, and compared in a spreadsheet format. See also the article by Tisone *et al* on image analysis for rapid-flow diagnostics.

Patents Related to Strip Tests

Finally, it should be mentioned that there are many patents that cover a number of technologies and materials that are commonly used in strip-test devices. Patents are a vital aspect of research, particularly if commercial interests are involved. When preparing this manuscript it became clear that writing a meaningful overview of patents in this field would be difficult and complex because of technical and legal implications.

However, in order to comply with international trade and licensing agreements and to prevent possible legal problems after product launch, it would be prudent to review the patent literature prior to commercialization. A useful source of patent information can be the Diagnostic Club patent abstract service (contact Tony Towen, Diagnostics Club Administrator; e-mail: towen@waitrose.com).

TROUBLE SHOOTING

In developing a strip test, multiple reagents, materials, and techniques are brought together in order to create a sensitive, specific, fast, cheap, and easy to perform assay for the detection of either low-molecular-mass analytes (hormones, antibiotics, drugs of abuse, etc) or high-molecular-mass components (antibodies, bacteria, viruses, cancer markers, food allergens, etc.). Moreover, the test should be applicable for a wide variety of sample matrices such as whole blood, plasma, serum, urine, saliva, milk, food extracts, plant leaves, and even sweat. All these aspects in developing a strip test rely upon one another for their efficacy; often changing just one of these items can lead to completely un expected and/or undesired results.

The possible causes of a poorly performing strip test are numereous. The most common difficulties encountered in creating a strip test are listed in Table 5.1 which consists of the troubleshooting list by Schleicher & Schuell, extended with our own experience on the subject.

Table 5.1. The most common difficulties encountered in creating a strip test and the possible solutions. Note that various problems mentioned in the table are related to each other, *e.g.*, the item "No sample front visible" is a more profound definition of the item "Migration problems".

Problem	*Possible solutions*
Nonspecific binding	Evaluate absorption
(Heterophile reactions)	Add IgG
	Add surfactants
	Antibody fragmentation
	Change pH
	Add buffers
General nonspecific reactions	Add a blocking agent
	Change the blocking agent
	Add or change surfactant
	Smaller particles production
	Change pH
	Change buffer make up
	Decrease reagent/strength
	Change conjugation procedure
Lack of sensitivity	Increase quantity capture material
	Increase quantity of the conjugate

Problem	*Possible solutions*
	Increase strength of conjugate
	Increase particle size
	Decrease or change blocking agent
	Decrease flow rate
	Decrease rate of conjugate release
	Use higher affinity antibodies/stronger antigens
	Repurify reagents to up activity
	Add activity enhancer
	Check for analyte absorption
	Increase capacity of the absorbant pad
Lack of specificity	Repurify the reagents
	Try new reagents
	Check pH of the sample
Migration problems	Try smaller particles
	Try other type of particles
	Change production protocols
	Change or increase blocking agents
	Add a surfactant
	Block membrane directly
	Look for leeching of reagents from one component to another
	Check the viscosity of the sample
Stability issues	Change the conjugate pad material
	Change preparation of conjugate pad
	— increase or add surfactants
	— increase or add polymers
	— change pH of buffers
	— add or increase sugars to conjugate mixture
	Look for moisture problems
	Try new membrane materials
	Add preservatives to membrane block
	Add preservatives to capture protein prior to dispensing
	Look for leeching effects
	Look for degradation of materials on the sample pad
No control line visible	Flooded test. Repeat with new device and add fewer drops
	Check the control capture reagent
	Check the control detector reagent
	Check pH of the sample

Problem	*Possible solutions*
Faint lines	Check dispending equipment and/or dispensing procedure Check the capture reagents Check the detector reagents Check membrane blocking, drying and washing procedure Look for membrane anomalies Check pH of the sample
No sample front visible	Sample volume is too small, add more sample Make sure the sample pad is not clogged by sample material (fat, debris)
Detector reagents are not completely reased from the conjugate pad	Change conjugate pad material Increase sample volume Add or increase sugar concentration on conjugate pad

APPLICATIONS

In principle, there are no limitations for the type of analyte that can be detected by using a strip test. This is demonstrated by the large number of commercial strip tests for the detection of a wide variety of analytes. Most strip tests, with special emphasis on the pregnancy tests, find their application in human clinical diagnostics. A systematic review of near-patient tests in primary care has been published by Hobbs *et al.* This lists several hundred commercial devices.

Furthermore, strip tests have been developed in the field of veterinary diagnostics food testing, and plant pest and disease diagnosis. In most human diagnostic applications, relatively high concentrations of analyte are measured. It has been shown, however, that strip tests are also applicable in the low ppb (ng analyte/g sample) detection region, which is often required for the detection of veterinary drug residues in food diagnostics. As for pesticides and other environmental contaminants, LDLs are required in the ppt (1 ng/kg) region rather than in the ppb region; to our knowledge, no strip tests have been described for the detection of analytes at such low concentration. Multianalyte devices also have been developed for the simultaneous detection of several analytes, *e.g.,* drugs of abuse.

REMARKS AND CONCLUSIONS

The proper performance of a strip test for the detection of either low-molecular-mass analytes (indirect or competitive assay) or high-molecular-mass analytes (direct assay) can be summarized as being a delicate balance between all of the applied components. Changing just one of these components can lead to completely unexpected and/or undesired results. The sample matrix also will greatly influence the performance test. The strip test has some important advantages, *i.e.,* the test is cheap, fast (test results can be obtained in 5 to 10 min.), and easy to perform. Moreover, all reagents are included in the test device and no expensive equipment is required for using the test.

A disadvantage of the strip test is the fact that large number of samples can not be tested simultaneously, as in an ELISA. Also, the high specificity of a strip test may be considered a disadvantage as well. Especially for general screening purposes in food diagnostics, there is a demand for (immuno)assays with which one is able to detect a whole group of analytes, *e.g.*, several aminoglycosides or sulfonamides, rather than one specific member of such a group. A possible solution for this problem is the application of generic antibodies that are directed against the generic part of a specific group of analytes. In principle, there are no limitations for the type of analyte that can be detected by using a strip test. However, when detection limits in the ppt region are required, the strip test in its present form is simply not sensitive enough.

Another drawback of the strip test is its nonquantitative character. In practise, interpretation of the test results will allow merely two possible conclusions: "the analyte is detectable" or "the analyte is not detectable". For analytes where an MRL has been established, the LDL of the strip test should be at least as high as or power than the MRL value. Despite the above-mentioned limitations, the strip test has its own specific place in the field of immunological screening assays. The development of strip tests and related assays is increasing and constantly improving.

A most interesting development in the field of rapid assay devices is provided by the Wolfson Applied Technology Laboratory (WATL) of the university of Birmingham, UK. Their work involves devices based on flow in porous media such as nitrocellulose membranes. These devices are classified in first- and second-generation devices. The first-genera-tion devices have channels formed by printing so that liquid flows in two dimensions, distinct from conventional strip devices in which the flow is essentially unidirectional (patent US 5354538; Liquid Transfer Devices). Their second-generation devices additionally include a switchable barrier to control the flow in the channels (patent WO 01/25789A1; Fluid-flow Control Device). This is made by printing down a resin, typically polycoumarone-co-indene resin, and switching with a surfactant such as octyl-β-D-glucopyranoside. Undoubtly, applications of this type of rapid assay device will be heard of in the near future.

CHAPTER 6

Electrical Analysis

The subject of electrochemistry deals with the study of the chemical interaction of electricity and matter generally, but it is the interaction with solutions that is of particular value in analytical biochemistry. The electrical properties of a solution depend upon both the nature of the components and their concentration and permit qualitative and quantitative methods of analysis to be developed. These electrical properties of solutions are measured using electrodes in arrangements known as electrochemical cells. There are two main types of electrochemical cell:

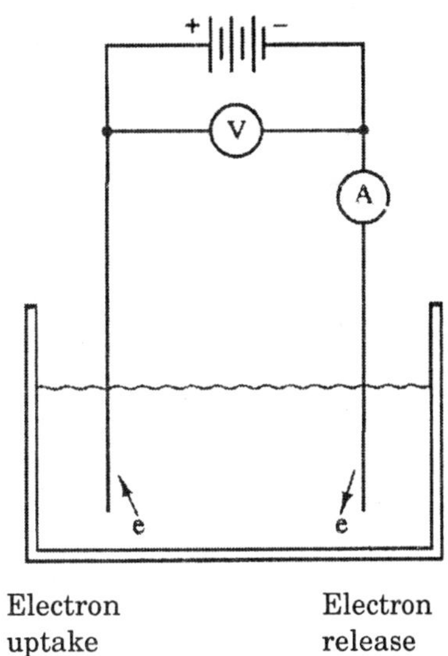

Fig. 6.1. An electrolytic cell. The current which flows between the electrodes depends not only on the voltage that is applied but also on the electrical properties of the solution.

1. *Electrolytic cells :* When a potential difference is applied across two electrodes that dip into a solution, a current will flow between them. The amount of current that flows depends upon the voltage applied and the electrochemical properties of the solutions. This provides the basis for conductimetric and polarographic methods of analysis. In a similar manner, the total amount of chemical change which takes place at an electrode is related to the total amount of current. This forms the basis of coulometric methods of analysis.

2. *Voltaic or galvanic cells :* In some cells, the chemical nature of the electrodes and solutions results in a chemical reaction taking place at the electrodes with the production of electrical energy (batteries) but without the need for an external voltage. This type of cell is used in potentiometric methods in which no voltage is applied to the cell and although no current actually flows between the electrodes (which are said to be unpolarized), they develop a potential relative to the solution due to the nature of the electrodes and the solution. This potential can be measured and related to the concentration of the ions in the solution.

POTENTIOMETRY

The potential developed by a single electrode in a solution is caused by the tendency of the solution either to donate or accept electrons and can be Calculated using the Nernst equation :

$$E = E^0 - \frac{2.3026RT}{nF} \times \log a$$

where E is the electrode potential at the specified concentration,

E_0 is the standard electrode potential,
R is the gas constant,
T is the absolute temperature,
n is the number of electrons involved,
F is the Faraday constant,
A is the activity of the ion.

For measurements made at 25 °C the equation simplifies to:

$$E = E^0 - \frac{0.059}{n} \log a$$

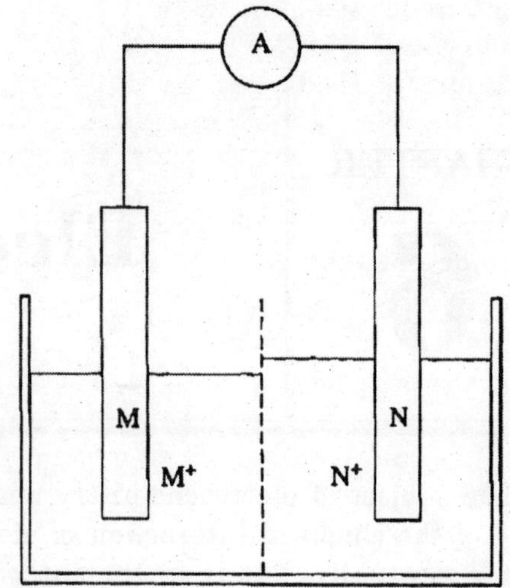

Fig. 6.2. A galvanic cell. A reaction taking place at the electrodes results in a potential difference developing between them and, in some cases, a measurable current will flow between them.

There can be no chemical reaction in such a system without a complementary electron donor or acceptor to complete the process. Each of these electrode systems is known as a half-cell and the potential developed by a half-cell cannot be measured in absolute terms but only compared with that of another *half-cell*. The chemical reaction occurring at each half-cell is known as a half-reaction.

An active electrode consists of an element (M) in its uncombined state which is capable of establishing an equilibrium with a solution that contains its ions:

$$M = M^+ + e^-$$

The ionization of atoms or molecules results in a potential being developed by such an electrode, the intensity of the potential being related to the concentration of the ions. The effective concentration of the ions (known as the activity of the ions) is more significant than the molar concentration.

The values for activity and concentration are only the same in very dilute solutions. Inert electrodes, such as silver, platinum, and carbon, are used solely to make electrical contact with the solution and only reflect the potential of the solution. They are used to measure the potential of solutions containing mixtures of ions which have a tendency to transfer electrons between them, *e.g.*, ferric and ferrous ions:

$$Fe^{3+} + e^- = Fe^{2+}$$

Such reactions are know as redox reactions and, in this case, the potential developed by the electrode system depends on the tendency of ferrous ions to donate an electron compared with the tendency of the ferric ions to accept one. The standard electrode potential of an element is defined/as its electrical potential when it is in contact with a molar solution of its ions.

For redox systems, the standard redox potential is that developed by a solution containing molar concentrations of both ionic forms. Any half-cell will be able to oxidize (*i.e.* accept electrons from) any other half-cell which has a lower electrode potential.

Table 6.1. Standard Electrode Potentials at pH 7.0

Half-reaction	E'_0
Oxygen/water	0.81
Ferric/ferrous	0.77
Ferricyanide/ferrocyanide	0.36
Oxygen/hydrogen peroxide	0.30
Cytochorome *c* ferric/ferrous	0.22
Dehydroascorbic acid/ascorbic acid	0.08
Pyruvate/lactate	–0.19
FAD/$FADH_2$	–0.22
NAD^+/NADH	–0.32
$2H^+/H_2$	–0.42
Ferrous/iron	–0.44

It is impossible to measure the potential of a half-cell directly and a reference half-cell must be used to complete the circuit. The hydrogen electrode is the standard reference electrode

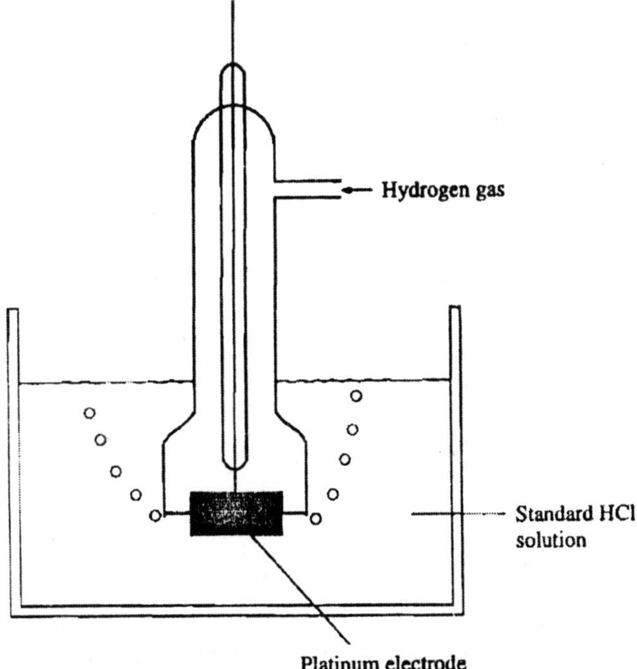

Fig. 6.3. The hydrogen electrode. An electrode of platinum foil, covered with platinumoTack, dips into a solution of hydrochloric acid (1.0 mol l^{-1}). Hydrogen gas at a pressure of 1 atm (101 kPa) is bubbled over the electrode and is absorbed by the platinum black. The half-reaction for the electrode can be represented as:

against which all other half-cells are measured and is arbitrarily attributed a standard electrode potential of zero at pH 0. Because it is difficult to prepare and inconvenient to use, the saturated

$$2H^+ + 2e^- = H_2$$

calomel electrode is frequently used instead. This has a standard electrode potential of +0.242 V relative to the hydrogen electrode. It consists of a mercury electrode in equilibrium with mercuric ions in the slightly soluble salt, mercuric chloride. A high concentration of chloride ions is maintained by using a saturated solution of potassium chloride, which also provides a means of ensuring electrical contact with any other half-cell used.

A platinum wire makes electrical contact with an electrode which is composed of a paste of metallic mercury, mercuric chloride (calomel) and potassium chloride. A saturated solution of potassium chloride completes the half-cell and provides electrical contact through a porous plug.

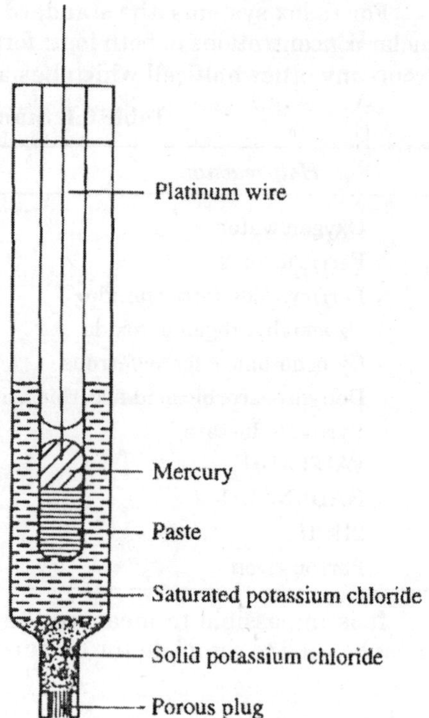

Fig. 6.4. The saturated calomel electrode.

Potentliometric Measurements

The sensitivity of instruments using low resistance circuits is determined primarily by the sensitivity of the *galvanometer*. Electrode systems that have a high resistance, *e.g.*, glass electrodes, require a high impedance voltmeter, which converts the potential generated into current which can be amplified and measured. Such instruments are commonly known as pH meters but may be used for many potentiometric measurements other than pH. Titrations can often be conveniently followed potentiometrically and in many cases it is not the actual value of the electrode potential that is important but the pattern of changing potential as the composition of the solution varies –pH and redox measurements are particularly well suited to such methods.

In many instances the equivalence point will be indicated by a significant change; in potential but sometimes the change at the equivalence point is difficult to detect and it might be more convenient to use a derivative plot. Instead of plotting potential (E) against volume of titrant (V), a graph is plotted of the change in potential for the unit change in volume $\Delta E/\Delta V$ against volume. Such a first-derivative plot indicates the equivalence point as a peak or spike.

The voltage from the test circuit is balanced against a known voltage by means of a variable resistance using a galvanometer to indicate the position at which no current flows in either direction.

Acetic acid (10 ml of a 0.1 mol l^{-1} solution) was titrated with a sodium hydroxide solution (0.2 mol l^{-1}) and the pH of the resulting solution plotted against the amount of alkali added.

If two identical electrodes are placed in separate solutions that are similar in every way except for the concentration of the test ions, the potential developed between the electrodes will be related to the ratio of the two concentrations, *e.g.*, Ag—AgCl electrodes in solutions containing chloride ions.

A calibration curve of potential developed in a series of known concentrations of test ions can be used in the analysis of unknown samples: Often it is advisable to plot the graph as potential versus logarithm of concentration to give the straight line relationship as indicated by the Nernst equation. It is necessary to add a constant amount of a high concentration of a non-reacting electrolyte to give all solutions tested the same ionic strength. This is because the potential depends upon the activity of the ions rather than concentration.

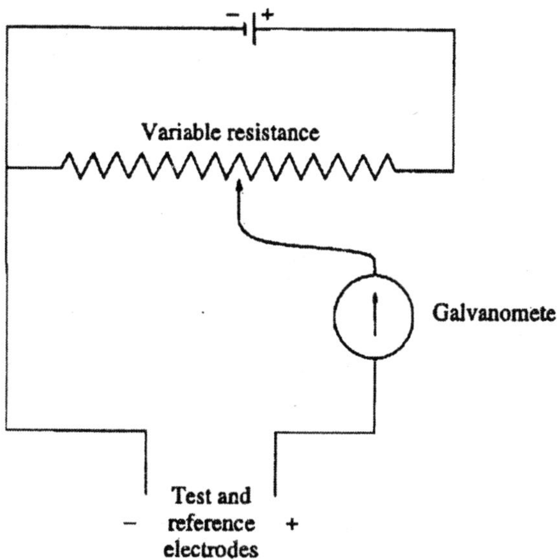

Fig. 6.5. A potentiometer circuit.

Measurement of pH

The glass electrode is most commonly used for routine measurement of pH because the use of the hydrogen electrode is impracticable. It consists of a silver-silver chloride electrode in a reference solution of hydrochloric acid (usually 0.1 mol l^{-1}) contained in a glass membrane, (The membrane is made of a special glass, usually a hydrated aluminosilicate containing sodium or calcium ions. It is selectively permeable to hydrogen ions and the potential that develops across the membrane depends upon the hydrogen ion concentration of the test solution compared with the reference acid solution inside the electrode. This potential can be measured against a reference calomel electrode using a high impedance voltmeter. For ease of operation the glass electrode is often combined with a reference calomel electrode in a single probe. Glass electrodes must be stored in water between use in order to keep the glass membrane fully hydrated. There is an almost linear relationship between potential and pH over the pH range of 2–10 but at extreme pH values special glass electrodes must be used.

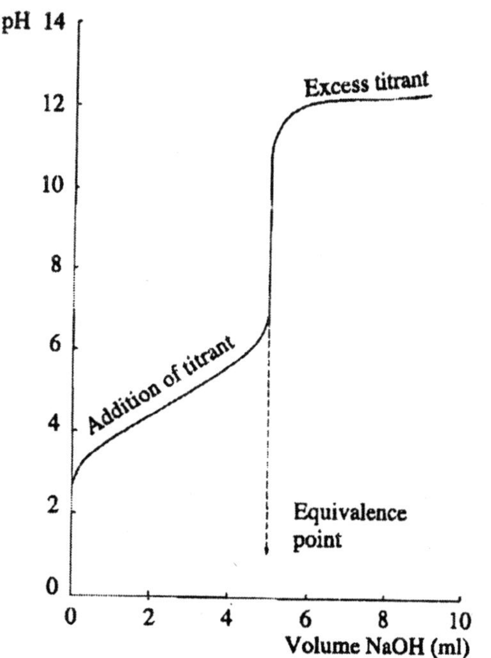

Fig. 6.6. A titration curve.

Calibration of the instruments is essential and may be achieved with either buffers whose pH has been previously measured using a hydrogen electrode or solutions of very pure chemicals prepared in specific concentrations, *e.g.* a solution of potassium hydrogen phthalate (0.05 mol l^{-1}) has a pH of 4.00 at 15 °C. For measurements over a range of pH values it is necessary to standardize the instrument on at least two standard buffer solutions which cover the required range, making sure that the correct temperature setting is used.

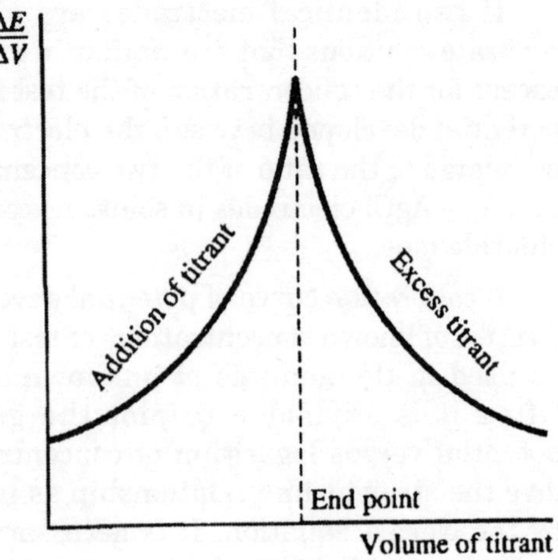

Fig. 6.7. A first derivative plot of a potentiometric titration curve.

Ion-selective Electrodes

The membrane of the glass electrode used for pH measurements is selectively permeable to hydrogen ions and from this basic concept a whole range of ion-selective electrodes have been developed. Varying the composition of the glass membrane can change the permeability of the glass and several cation-sensitive electrodes have been developed in this way. Although such electrodes may be described as specific for ions such as Na^+, K^+, Ca^{2+} and NH_4^+, they are in fact only selective and not specific and often show significant interference from other ions, particularly hydrogen ions. In all cases the basic design of the electrode is the same but the nature of the ion-selective membrane varies.

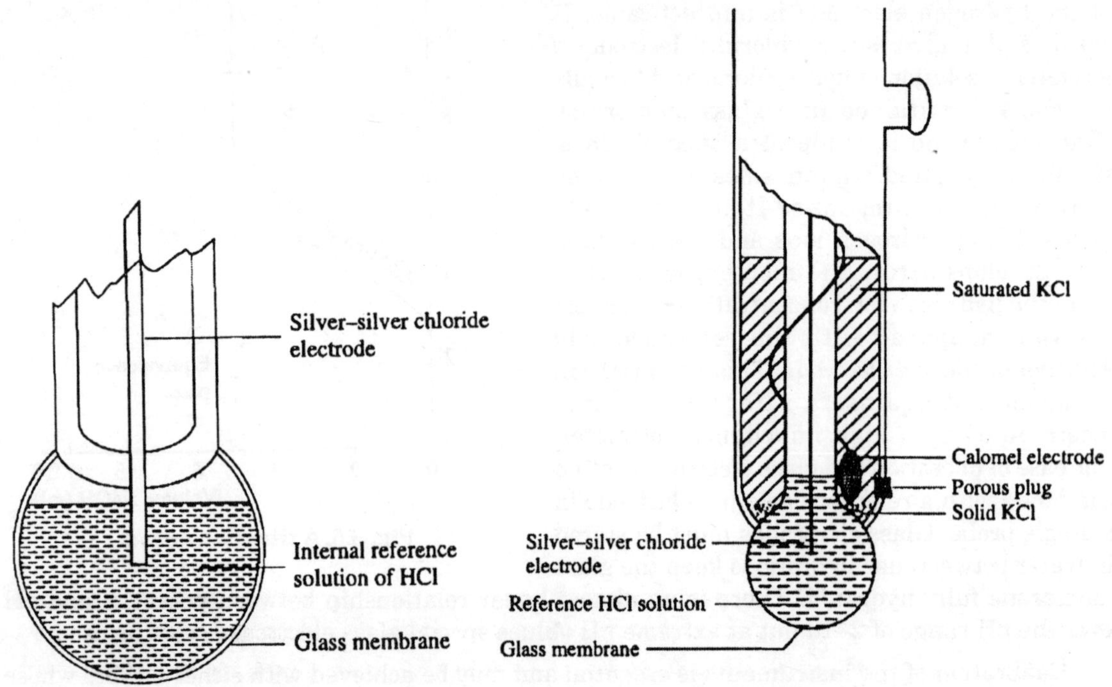

Fig. 6.8. The glass electrode.

Fig. 6.9. A combined pH electrode.

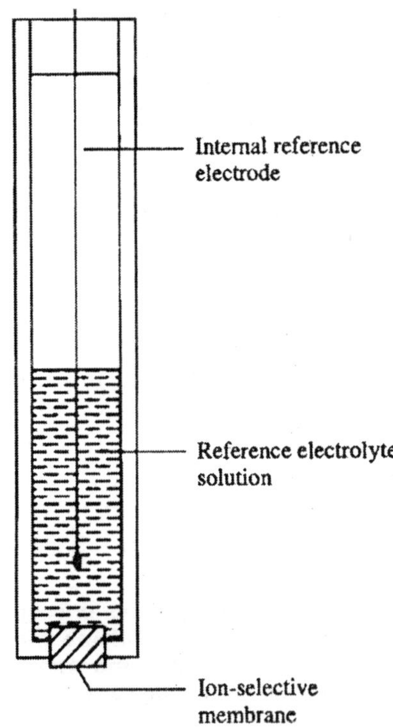

Fig. 6.10. The basic design of an ion-selective electrode.

Table 6.2. Ion-selective Electrodes.

Test ion	*Membrane material*	*Major interfering ions*
Solid-state electrodes		
Fluoride	LaF	I^- Br^- Cl^-
Chloride	$AgCl/Ag^2S$	S^- I^-
Bromide	$AgBr/Ag_2S$	S^- I^-
Iodide	AgI/Ag_2S	S^-
Sulphide	AgI/Ag2S	
Cupric	Ag_2S/CuS	Hg^+ Ag^+
Lead	Ag_2S/PbS	Hg^+ Ag
Cadmium	Ag_2S/CdS	Hg^+ Ag^+ Cu^{2+}
Silver	Ag^2S	
Liquid-membrane electrodes		
Potassium	Valinomycin in diphenyl ether	
Ammonium	Macrotetrolides in tris(2-ethylhexyl)phosphate	
Calcium	Calcium dialkylphosphate in dioctylphenylphosphonate	
Calcium/magnesium	Dialkylphosphoric acid in aliphatic alcohol	

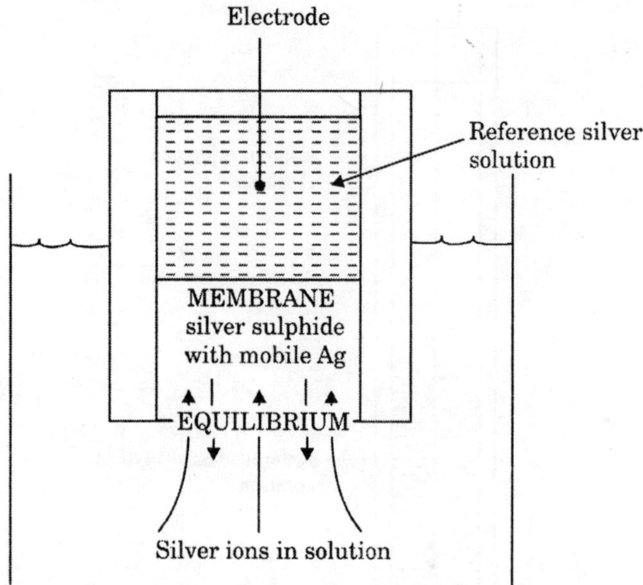

Fig. 6.11. A solid-state electrode showing a first-order response. An electrode designed to measure the activity of silver ions uses a crystalline membrane of silver sulphide. An equilibrium between the mobile silver ions of the membrane and the silver ions in the solutions results in the development of a potential difference across the membrane.

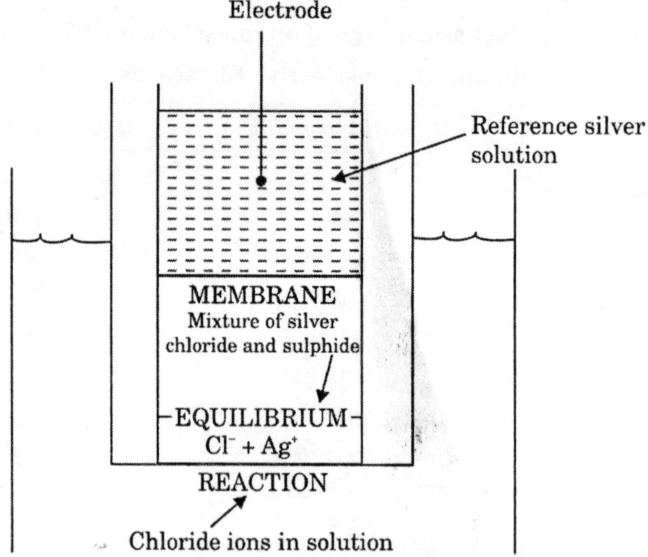

Fig. 6.12. A solid-state electrode showing a second-order response. The electrode shown in Figu. 6.11 can be modified by the incorporation of silver chloride into the membrane to enable the activity of chloride ions in a sample to be measured. A surface reaction between the test chloride ions and the membrane silver ions alters the activity of the latter, resulting in a change in the potential difference across the membrane.

Test anion $\rightleftharpoons$ Membrane cation

In solid-state electrodes the membrane is a solid disc of a relatively insoluble, crystalline material which shows a high specificity for a particular ion. The membrane permits movement of ions within the lattice structure of the crystal and those ions which disrupt the lattice structure the least are the most mobile. These usually have the smallest charge and diameter. Hence, only those ions that are very similar to the internal mobile ions can gain access to the membrane from the outside, a feature that gives crystal membranes their high specificity.

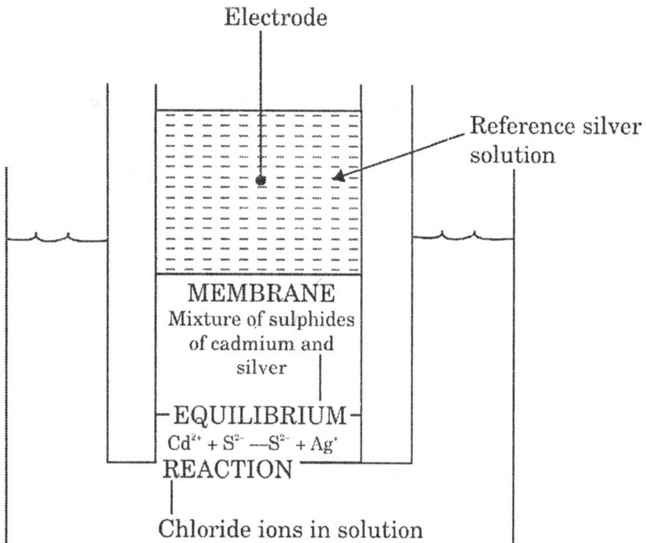

Fig. 6.13. A solid-state electrode showing a third-order response. An alternative modification to the electrode will permit the measurement of cadmium ions in solution. The membrane is composed of a mixture of silver and cadmium sulphides. The surface reaction between the cadmium ions in the test solution and the sulphide ions in the membrane will affect the equilibrium between the sulphide ions and the silver ions in the membrane.

Test cation $\rightleftharpoons$ Membrane anion

$\rightleftharpoons$ Membrane cation

When the electrode is immersed in the sample solution, an equilibrium is established between the mobile ions in the crystal and similar ions in the solution and the resulting potential created across the membrane can be measured in the usual manner. The Simplest solid-state membranes are designed to measure test ions, which are also the mobile ions of the crystal (first-order response) and are usually single-substance crystals. Alternatively, the test substance may be involved in one or two chemical reactions on the surface of the elec-trode which alter the activity of the mobile ion in the membrane.

Such membranes, which are often mixtures of substances, are said to show second- and third-order responses. While only a limited number of ions can gain access to a particular membrane, a greater number of substances will be able to react at the surface of the membrane. As a result, the selectivity of electrodes showing second-and third-order responses is reduced. Liquid-membrane electrodes consist of an ion-selective material dissolved in a solvent that is not miscible with water. The liquid is held in a porous, inert membrane (often plastic) which allows contact between the test solution on one side and the reference electrolyte on the other. The ions from the reference solution will partition themselves between the two immiscible solvents (the aqueous and the organic phases), giving the electrode a particular potential.

The presence of the test ions in the sample affects the activity of the reference ions in the membrane resulting in a change in the potential difference across the membrane. While the solvent is chosen to give the best partition effects for a particular ion, the specificity and sensitivity of the process can be greatly improved by the incorporation of a specific chelating or ion-selective substance. These are often large counter-ions to the test ions and have a low solubility in water, tending to form undissociated complexes with the test ion. Ion-selective electrodes are often incorporated in automated analysers, particularly for the measurement of sodium and potassium. The electrodes may be sited within the instrument or be available in a miniature disposable form.

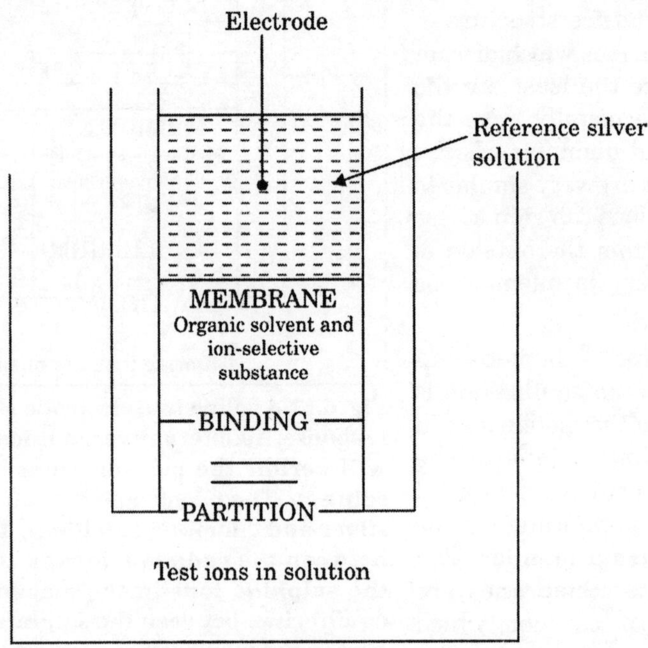

Fig. 6.14. The basic design of a liquid-membrane electrode. The partition of the ions from the sample in the immiscible solvent of the membrane will cause a change in the potential across the membrane. The effect can be enhanced by the binding of the test ions by an ion-selective substance which can be incorporated in the membrane.

Use of Ion-Selective Electrodes

Because potentiometric measurements reflect the activity of an ion rather than its concentration, it is usually necessary to calibrate the system with standard solutions of known activity. It is possible to calculate the concentration of an ion in a sample if the value of the activity coefficient for the ion is known:

$$\log \frac{a}{c} = \log \gamma$$

where a is the activity of the ion,

c is the concentration of the ion,

γ is the activity coefficient.

The equation is probably more useful in the form

$$\log c = \log a - \log \gamma$$

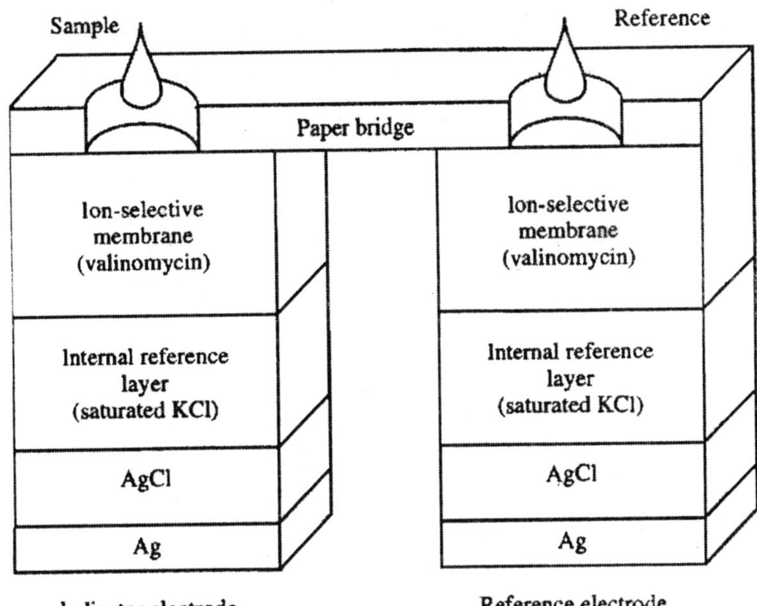

Fig. 6.15. Representation of the disposable slide for the measurement of potassium by the Vitros Chemistry System. The difference in potential between the two half-cells, one receiving the sample and the other a reference solution of known potassium ion concentration, is mathematically converted to give the concentration of potassium ions in the sample.

It is often more convenient to relate the potentiometer reading directly to concentration by adjusting the ionic strength and hence the activity of both the standards and samples to the same value with a large excess of an electrolyte solution which is inert as far as the electrode in use is concerned. Under these conditions the electrode potential is proportional to the concentration of the test ions.

The use of such solutions, which are known as TISABs (total ionic strength adjustment buffers), also allows the control of pH and their composition has to be designed for each particular assay and the proportion of buffer to sample must be constant. Ion-selective electrodes are particularly useful for monitoring the disappearance of an ion during a titration. In many cases it is not necessary to calibrate the instrument because there is often a significant change in the potential at the end-point of a titration. However, some electrodes have a slow response time and care must be taken to ensure that titration is not performed too quickly.

Apart from interference the greatest problem in the use of ion-selective electrodes is that of contamination. Any insoluble material deposited on the surface of the electrode will significantly reduce its sensitivity and oil films or protein deposits must be removed by frequent and thorough washing. It is possible to wipe membranes with soft tissue but they can be easily damaged. Solid-state membranes are more robust but they must not be used in any solu-tion which might react with the membrane material.

CONDUCTIMETRY

The simplest application of an electrolytic cell is the measurement of conductance. If a fixed voltage is applied to two electrodes which dip into a test solution, depending upon the conductivity of the solution, a current will flow between them. Although conductivity measurements do not give any information about the nature of the ions in the solution they can be used quantitatively. They are, however, more frequently used to monitor the changing composition of a solution during a titration. The current passing between the two electrodes is carried by the ions in the solution and is related to the number present, which is a result of both the molar concentration of the compound and its extent of ionization.

The anions donate electrons to the anode while the cations accept electrons from the cathode. It is this transfer of electrons that determines the amount of current flowing and the contribution of each ionic species is determined by its ionic mobility (velocity). Because of the nature of the reaction by which molecules dissociate into ions there is not always a simple relationship between the increase in conductivity with increasing molar concentration and often a limiting value is reached.

For instance, increasing concentrations of sulphuric acid result in increased conductivity up to a concentration of about 35% (v/v), above which the conductivity decreases.

The current flowing through a conductor is defined by Ohm's law as:

$$\text{Current } (I) = \frac{\text{voltage applied } (V)}{\text{resistance } (R)}$$

or in terms of conductance (C)

$$I = C \times V$$

The specific conductance (κ) of a solution is defined as the conductance (S) per centimetre of a solution that has a cross-sectional area of 1 cm^2, and is measured in S cm^{-1} (or in non-SI units as Ω^{-1} cm^{-1}). The molar conductance (Λ) is the specific conductance of a solution corrected for the concentration of ions in the solution. $\Lambda = \kappa \times$ volume of solution which contains 1 gram mole. The value for κ will normally decrease as the concentration of the solution decreases but the value for will increase because of the increased dissociation of molecules in dilute solution.

A value for the molar conductance at infinite dilution (Λ_0) can be determined by plotting the calculated values for Λ against them polar concentration of the solution used and determining the plateau value for Λ. From such investigations it is possible to determine the ionic mobilities of ions and calculate the molar conductance of an electrolyte at infinite dilution, *e.g.* the value for potassium sulphate (234) is the sum of the ionic mobilities of potassium (74) and sulphate (160).

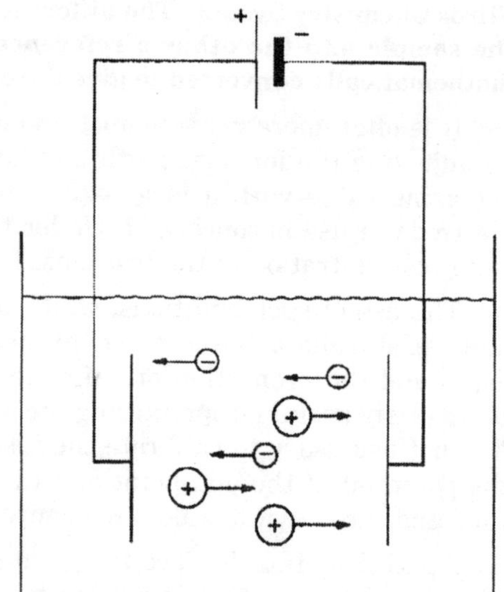

Fig. 6.16. The conductivity of a solution. The total transfer of electricity is not necessarily shared equally because different ions move at different speeds. The Kohlrausch law says that the total conductivity of an electrolyte is the sum of the con-ductivities of the anions and the cations.

Table 6.3. Molar Ionic Conductivity

Cation	*Molar conductance (Ω^{-1} cm^2 mol^{-1})*	*Anion*	*Molar conductance (Ω^{-1} cm^2 mol^{-1})*
Na^+	50	HCO^-	45
Ag^+	62	HSO_4^-	52
K^+	74	MnO_4^-	61
NH^{4+}	74	NO_3^-	71
Zn^{2+}	106	Cl^-	76
Mg^{2+}	106	I^-	77
Cu^{2+}	107	Br^-	78
Fe^{2+}	108	CO_3^{2-}	119
Ca^{2+}	119	SO_4^{2-}	160
Fe^{3+}	205	OH^-	197
H^+	350	PO_4^{3-}	240

Conductimetric Measurements

The basic instrumentation for conductimetric measurements is a Wheatstone bridge arrangement. An alternating rather than direct current is used in order to prevent polarization of the solution (*i.e.*, anions moving to the anode and cations moving to the cathode) and any electrolysis that might result from this polarization.

With fixed resistances, R_1 and R_2, the variable resistance R_v is adjusted until no current flows through the galvanometer. Under these conditions:

$$R_1 \times R_V = R_{cell} \times R_2$$

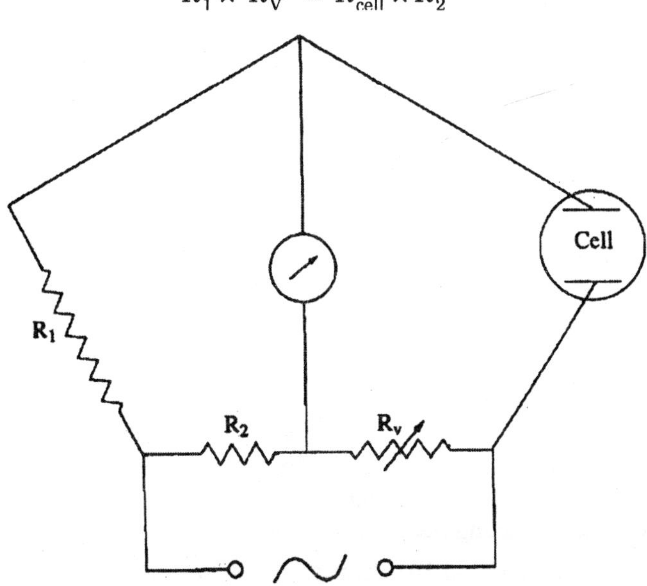

Fig. 6.17. A wheatstone bridge.

This gives a value for resistance of the cell which can be converted to conductance by calculating the reciprocal. The electrodes of conductivity cellsy are usually made of platinum coated with platinum black with a known area.

Although in many cells the distance between the electrodes is adjustable, for any series of experiments it must be held constant and for many calculations the precise value is required. The cells must be thermostatically controlled because any changes in temperature will cause significant alteration of conductivity values.

Applications of Conductimetric Measurements

Although conductimetry is useful in determining various physical constants, *e.g.* dissociation and solubility constants, its major analytical application is for monitoring titrations. When two electrolyte solutions that do not react with each other are mixed, provided that there is no appreciable change in volume, the conductance of the solution will rise because of the increase in the numbers of ions.

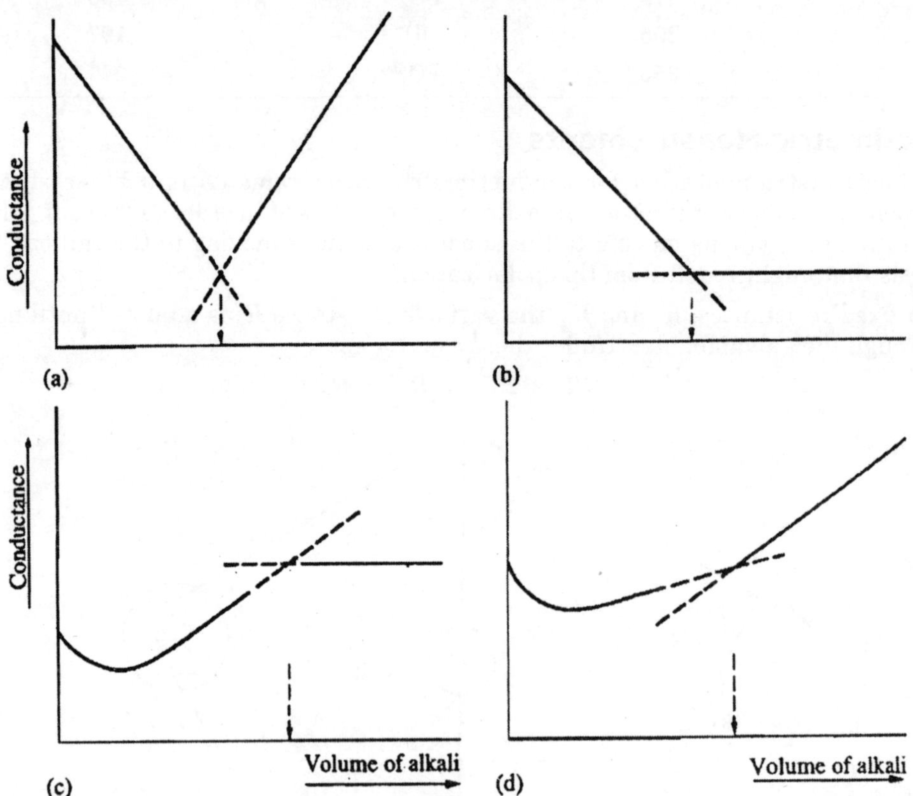

Fig. 6.18. Conductimetric titration curves. As an acid is titrated with an alkali, so the ionic composition of the mixture changes and is reflected in the conductivity of the solution, *(a)* A strong acid and a strong base, *(b)* A strong acid and a weak base, *(c)* A weak acid and a weak base, *(d)* A weak acid and a strong base.

However, if there is a reaction between the ions, one ion being replaced by another, the conductance of the solution will alter depending upon the relative mobilities of the ions involved. In the reaction in which CD is the titrant, the effect of the reaction is to replace

$$A^+B^- + C^+D^- = AD + C^+B^-$$

A^+ by C^+ in the solution. If these two ions have different ionic mobilities, the conductance of the solution will change accordingly. This is illustrated in the titration of a strong acid by a strong base: the hydrogen ions are replaced by the cation from the base, which will have a smaller ionic mobility and hence a lower conductivity and will result, in a gradual fall in the conductance of the solution. However, at the end-point of the reaction, the increasing concentration of hydroxyl ions will result in an increasing conductance.

An extrapolation of the lines before and after the end-point will give an intercept at the end-point. Conductimetric titrations, particularly when they are automated, are useful if either of the reactants is deeply coloured, so preventing visual monitoring of the titration using indicators, or if both reactants are very dilute. Flow through conductance cells are useful detectors in ion-exchange chromatographic separations where the analytes are ionic when they enter the detector cell. At the low concentrations encountered, conductivity is proportional to the mobility of the ions involved as well as their concentration, and this, together with the background conductivity of the solvent, demands that standards are used. Conductivity increases with an increase in temperature and it is important in such measurements that temperature is monitored and appropriate corrections made when calculating the results.

COULOMETRY

Coulometry is an electrolytic method of analysis. In general, electrolytic methods have limited applications in analytical biochemistry but they are useful in the analysis of substances which, while not strictly biochemical, are often important in biological and physiological chemistry.

Faraday's laws of electrolysis form the basis of quantitative coulometric analysis. They are:

1. The weight of substance liberated during electrolysis is directly proportional to the quantity of electricity passed.
2. The weights of substances liberated by the same quantity of electricity are in direct proportion to their equivalent weights.

There are two types of *coulometric analysis* using either constant current or constant voltage. In the latter (known as cgulometric methods), a fixed voltage is applied which causes the test substance to react at an electrode, and the current that flows decreases as the number of test ions remaining falls. The total amount of current that flows during the reaction can be measured and related to the amount of test substance originally present. This type of measurement has only limited biochemical applications. In the past it was used for the measurement of oxygen, but devices known as oxygen electrodes are more commonly used nowadays.

The alternative technique of constant current coulometry (known as coulometric titration) is more useful in analytical biochemistry because it permits the generation of specific, often very labile, reagents at electrodes.

Table 6.4. Colulometic Titrations

Analyte	*Titrant*	*Generating generated electrolyte*	*Electrode*
calcium	EDTA	Mercuric EDTA in ammonia buffer	Mercury cathode
chloride	Silver ions	—	Silver anode
$Iron^{2+}$	Ce^{4+}	Cerous sulphate in H^2SO_4	Anode
Oxygen	Cr^{2+}	Chromium sulphate in H_2SO_4	Anode
(Acids)	Hydroxyl ions	Sodium sulphate solution	Cathode
(Bases)	Hydrogen ions	Sodium sulphate solution	Anode

Coulometric Titrations

When a constant current is passed through the cell, the time taken for the reaction to go to completion can be used as a measure of the amount of test substance originally present. The electrode reaction may directly involve the test ions (as in constant voltage techniques) or alternatively may generate another substance which will react with the test substance. Reagents can be generated at an electrode which cannot be produced in any other way because they are so unstable.

The end-point can be detected using a visual system, *e.g.*, indicator dyes, but more usually by means of an additional electrode system, *e.g.* potentiometry. Coulometric titrations offer methods of analysis that are very precise and readily lend themselves to automation and the analysis of very small samples. While such methods do not strictly require standardization, the quantity of electricity being the standard, it is usual to use control samples to check or adjust the analysis conditions. The electrolytic system consists of a generating or working electrode and an auxiliary electrode. The auxiliary electrode is usually held in a separate glass tube with a fine porous glass disc to prevent the titrating reagent produced at the working electrode from reaching it and so reducing the efficiency of the reaction.

In addition a pair of indicator electrodes is necessary in order to detect the end-point of the reaction. The electrodes vary depending upon the nature of the test reaction taking place in the cell. In order for these electrodes to function effectively, there must be ions present at all stages of the reaction and it is therefore usual to incorporate a non-reacting carrier electrolyte in the reaction mixture.

The amount of substance titrated can be calculated from the following equation:

$$W = \frac{ItM}{Fn}$$

where W is the weight of substance,

M is the relative molecular mass,

F is the Faraday constant (96 493),

n is the number of equivalents,

t is the time in seconds,

I is the current in amperes.

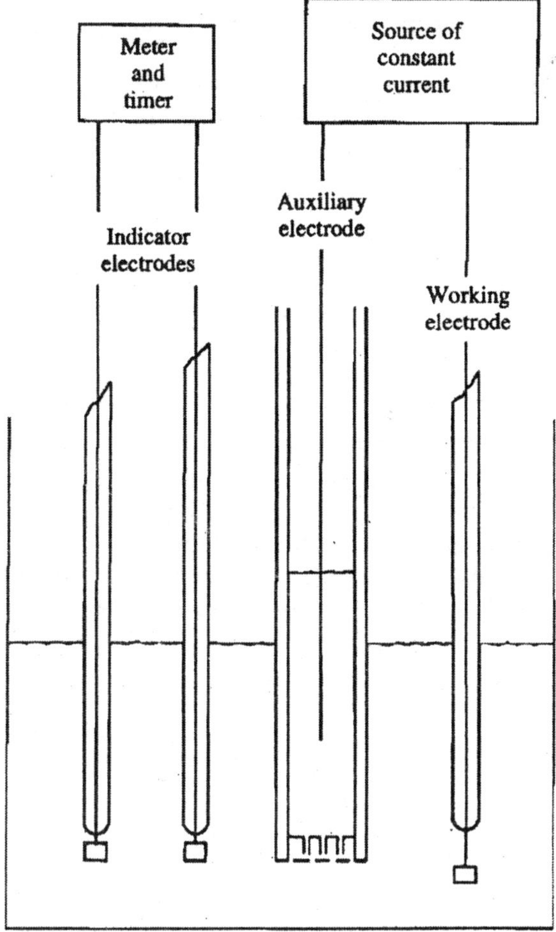

Fig. 6.19. A constant current coulometry titration cell. The reagent is produced at the working electrode and reacts with the sample. The indicator electrodes detect the changing potential or conductivity of the solution and the amount of change that takes place is measured and related to the concentration of the reactant in the sample.

However, most instruments designed for coulometric titration are at least semi-automated and directly convert the time into a measure of the sample concentration.

The determination of chloride using an instrument known as *chloride meter* is probably the most common application of coulometry in biochemistry. The instrument is designed to generate silver ions electrolytically from a silver anode. These ions are removed from the solution as undissociated silver chloride, which is either deposited on the anode or precipitated in the solution.

A low concentration of carrier electrolyte (nitrate ions) permits a small current to flow between the indicating electrodes during the titration but this rises rapidly at the end-point, when silver ions appear in the solution. This increase in current can be used to trigger a relay circuit which switches the timer off and records the total time for the reaction.

VOLTAMMETRY

Voltammetry is the term given to electrochemical techniques which monitor the relationship between the voltage applied to an electrode system and the current that flows as a result of the reaction. It covers a wide range of different electrode techniques, many of which are specifically designed to monitor a particular chemical reaction. Voltammetry is generally divided into two main subdivisions of polarography and amperometry. During electrolysis, ions are discharged at the appropriate electrodes and a continuous supply of the ions is necessary for the current to continue flowing.

An ion is said to be discharged when it either accepts or donates an at an electrode and as a result loses its charge. If there is only a low ncentration of reactive ions they will be quickly discharged around the electrode and the current will fall to a level that depends upon the rate at which more ions can defuse from the bulk of the solution to the electrode. As the voltage is increased, so the rate of diffusion will increase up to a maximum level, which is determined not only by the nature of the ion but also by its concentration. This results in a maximum current, which is known as the diffusion current.

The level of the diffusion current is used in quantitative polarographic techniques but the voltage at which it occurs forms the basis for qualitative polarography. Variations of voltammetry in which the voltage is held constant and the variations in current measured are known as amperometric techniques and are frequently used to monitor the end-point of titration or the presence of a particular analyte in the effluent from a chromato-graphic column.

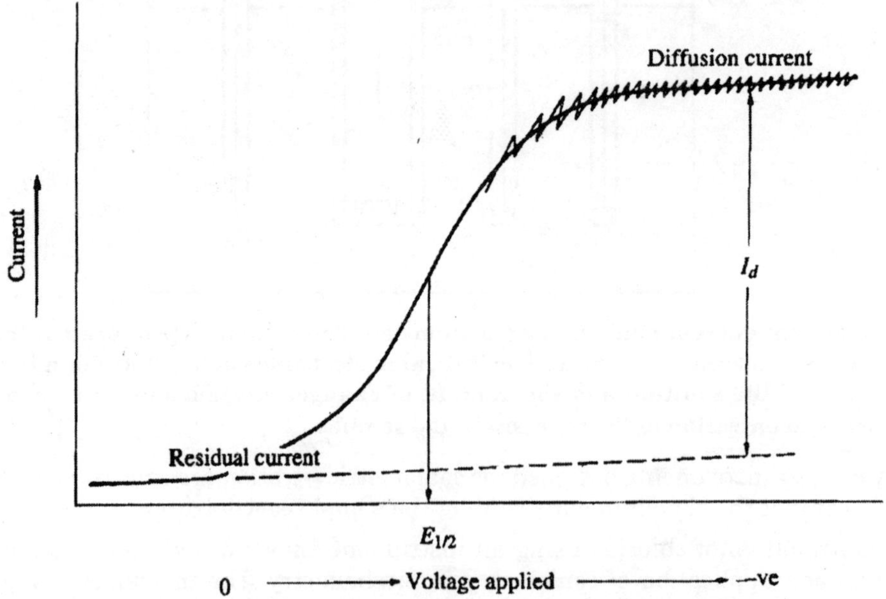

Fig. 6.20. A polarographic titraiion curve. A polarogram has oscillations about a mean current-voltage curve owing to the regularly changing area of the mercury drop.

Polarography

In polarographic methods the voltage that is applied to an electrode system is gradually increased and the resulting current is continuously measured. The information is usually presented in a graphical form known as a polarogram, which is a plot of current against voltage. If there is a very high concentration of non-reacting ions present, *i.e.* ions that are not discharged at the voltage applied, they will be responsible for carrying most of the current.

The amount of current carried by these ions will be almost independent of the voltage applied and is known as the residual current. However, as the voltage is increased and reaches the discharge potential of other ions, the current will rise rapidly owing to their discharge but it will stabilize at a new maximum level (the diffusion current) dependent upon their subsequent rate of diffusion to the electrode. These changes are reflected in the shape of the polarogram. The potential or voltage at which a particular ion will be discharged is characteristic of the ion although it is slightly affected by its concentration. However, the half-wave potential ($E_{1/2}$) is independent of concentration and can be used to identify a particular ion, giving polarography a qualitative application. The difference between the residual current and the diffusion current is dependent on concentrations and is used for quantitive measurements. It is necessary in quantitative analysis to standardize the instrument using solutions of known concentration and to prepare a calibration graph using the difference between residual current and the diffusion current as the variable parameter. In *classical polarographic techniques,* a dropping mercury electrode is used. This is a complex device in which continuously produced small droplets of mercury are used as the active electrode in order to prevent poisoning of the electrode and to provide constant conditions throughout the analysis.

For many applications, specifically designed electrodes are available which are simpler to use. One specific variant of the technique is known as *direct current cyclic voltammetry* (DCCV), in which the voltage sweep is over a limited range and a short tune and is immediately reversed. The cycle is repeated many times and the pattern of current change is monitored. The technique uses relatively simple electrodes and is used to study redox reactions and there are a range of sophisticated variants of the technique.

Electrochemical detectors for HPLC

Amperometric devices are gaining popularity as HPLC detectors. A fixed voltage is applied and the resulting current is measured. The system consists of a flow cell containing a working and a reference electrode. There is a range of electrodes available for use in the oxidation mode including glassy carbon, which is hard and impervious to organic solvents, platinum and gold, which tend to become poisoned by absorbed solutes, and others with improved properties which are being commercially developed. In designing a method it is necessary to know the operating voltage of the electrode for a particular analyte.

The classical polarogram (Fig. 6.20) shows a half-wave potential ($E_{1/2}$) and the diffusion current. As the potential applied is increased beyond the diffusion current level, eventually the mobile phase will react at the electrode and there will be a further significant increase in the current flowing. The potential used should be between the half-wave potential for the analyte and the discharge potential of the mobile phase. Most frequently the voltage at the beginning of the diffusion current is selected. By using more than one electrode it is possible to measure different compounds simultaneously by selecting a different operating voltage for each electrode.

Oxygen Electrode

The dissolved oxygen content of a solution can be determined by measuring the diffusion current that results at a selected voltage. The Clark electrode was developed for this purpose and various modifications have subsequently been introduced. It consists basically of a platinum electrode separated from the sample by a membrane which is permeable to oxygen, *e.g.* Teflon or polyethylene. A reference electrode of silver/silver chloride in potassium chloride is used to complete the system. When a voltage that is sufficient to give the diffusion current for oxygen (often –0.6 to –0.8 V with respect to the reference electrode) is applied, the amount of current that flows is related to the partial pressure of oxygen (PO_2) in the sample. It is necessary, however, to calibrate the electrode using solutions of known oxygen content. Other gases as well as oxygen can be reduced at the operating voltage of the electrode, e.g. halogens, halogenated hydrocarbons and sulphur dioxide, and will interfere with quantitative measurements, while hydrogen sulphide will poison the electrode.

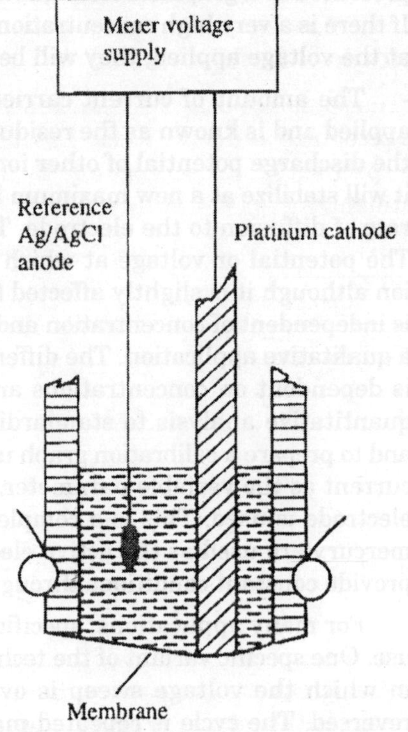

Fig. 6.21. An oxygen electrode.

Anodic Stripping Voltammetry

This is a technique for trace metal analysis and is a reversal of the usual polarographic method. Using a mercury electrode, either as a hanging drop or a graphite rod coated with mercury, a negative (cathodic) potential is applied relative to a reference silver/silver chloride electrode. Any metal ions, with discharge potentials less than that applied, will be deposited on the electrode and will form an amalgam with the mercury.

It is important to control the time of electrolysis and the method of stirring the sample because the amount of metal deposited will be taken as a measure of its concentration in the sample. The analysis stage involves scanning the voltage applied to the electrode towards a more positive (anodic) potential. As the potential becomes more positive, so the metal atoms are re-oxidized, giving an anodic wave of current. The current will show a maximum value as all the atoms of a particular element are stripped off the electrode as ions.

The height of the peak that results is proportional to the amount of metal and the voltage at which stripping occurs will be characteristic of the element. The technique is useful for the quantitation of many metals including lead, copper, mercury, cadmium and zinc with detection limits as low as 10pg. Its sensitivity makes it a very suitable method for trace metal analysis in biological samples.

BLOSENSORS

Biosensors are analytical devices that incorporate a biological component and a transducer. These must be in close proximity with one another and preferably in intimate contact, *i.e.*, the biological component immobilized on to the transducer. Such devices are available in disposable forms, *e.g.*, for measurement of blood glucose in diabetic patients, evaluation of the freshness of uncooked meat. Other designs are suitable for continuous use, *e.g.*, on-line monitoring of fermentation processes, the detection of toxic substances.

BIOSENSOR COMPONENTS

Biological Components

The possible biological components fall into two main categories (Table 6.5) and their function is to recognize and bind a specific analyte. It is the properties of the biological component that give specificity to biosensors and make them suitable for the analysis of samples without pre-treatment.

Table 6.5. Biological Components of Biosensors

Component	*Examples*
Biocatalysts	Enzymes
	Microbial, plant, animal cells
	Sub-cellular organelles
Bioreceptors	Antibodies
	Lectins
	Cell membrane receptors
	Nucleic acids
	Synthetic molecules

For *biocatalysts,* binding is followed by a chemical reaction and release of products. Enzymes were the first catalysts used in biosensors and a large number of these natural proteins are available. Although they remain the most commonly employed, the use of purified enzymes is not always satisfactory and in some situations cell preparations containing the required enzymes in their natural environment may be preferable.

This approach reduces specificity but can be used to advantage when the analysis of the range of related substances is required as in the case of pollution monitoring. In devices incorporating *bioreceptors* binding is non-catalytic and essentially irreversible. Of major significance to the development of this type of biosensor was the commercial availability of monoclonal antibodies. This, together with the availability of other bioreceptors and improvements in methods for monitoring binding, *e.g.*, optical, piezoelectric, has resulted in the introduction of devices capable of detecting trace amounts of a wide range of substances such as drugs, hormones, viruses and pathogenic bacteria.

Transducers

The transducer converts the reaction between the analyte and the biological component into an electrical signal which is a measure of the concentration of the analyte. A wide range of transducers are available and are broadly divided into electrochemical, optical, thermal and piezoelectrical.

Electrochemical Biosensors

Electrochemical biosensors are the most common especially when the biological component is an enzyme. Many enzyme reactions involve electroactive species being either consumed or generated and can be monitored by amperometric, potentiometric or conductimetric techniques, although the latter are the least developed and will not be discussed further.

Amperometric Biosensor

These devices measure the current that is produced when a constant potential difference is applied between the transducing and reference electrodes. A current is produced when the species of interest is either oxidized at the anode or reduced at the cathode. The amount of current flowing is a function of the concentration of that species at the electrode surface. This type of transducer is applicable to substances that are easily electrochemically oxidized or reduced at an electrode.

Biosensors for the measurement of glucose which are based on glucose oxidase are examples of devices which use amperometric detection. The over-all reaction can be monitored in several ways.

$$\text{glucose} + O_2 = \text{gluconic acid} + H_2O_2$$

Originally, a Clark oxygen electrode was used to measure a reduction in current due to the consumption of oxygen. Anodic detection of the hydrogen peroxide by oxidation at a platinum or carbon electrode was then introduced but, owing to the high electrode potential required, suffered from interference from other electroactive compounds in the sample.

Attempts to reduce interference and minimize the effect of variations in oxygen tension have resulted in the development of biosensors with improved linear ranges which operate at lower electrode potentials. They incorporate artificial electron acceptors, called mediators, to transfer electrons from the flavoenzyme (*e.g.* glucose oxidase) to the electrode and thus are not dependent on oxygen. Ferrocene (bis(η^5-cyclopentadienyl)iron) and its derivatives are examples of redox mediators for flavoenzymes. The reaction now becomes

$$\text{glucose} + \underset{\text{(ferricinium)}}{2Fe^+} = \text{gluconic acid} + \underset{\text{(ferrocene)}}{2Fe} + 2H^+$$

The ferrocene is detected at the anode, which also regenerates the ferricinium ion for further use.

Construction of enzyme-based amperometric biosensors usually involves coating the electrode surface with both the enzymes and the mediator. The method adopted and concentrations used are important factors in the performance of the device. For extended use, covalent immobilization is the most satisfactory method.

Potentiometric Biosensors

These measure the potential difference between the transducing electrode and a reference electrode under conditions of zero current. Three types of potentiometric detectors are commonly employed: ion-selective electrodes (ISE), gas-sensing electrodes and field effect transistors (FET).

Gas-sensing, electrodes are modified ISEs and consist of either a hydrophobic gas-permeable membrane or an air gap between the sample solution and a pH electrode. Only gases that can permeate through the membrane and produce a change in pH, *e.g.*, carbon dioxide, ammonia, will be detected by the electrode. Gas-sensing electrodes have been used in devices incorporating entrapped microorganisms as the biological component which metabolize the analyte of interest and produce a detectable gas.

Field effect transistors are miniature, solid-state, potentiometric transducers which can be readily mass produced. This makes them ideal for use as components in inexpensive, disposable biosensors and various types are being developed. The function of these semiconductor devices is based on the fact that when an ion is absorbed at the surface of the gate insulator (oxide) a corresponding charge will add at the semiconductor surface and the current flowing between the source and drain electrodes will change. The resulting voltage shift is a measure of the absorbed ionic species. Variation of the nature of the gate electrode results in the different types of FET. For example, in the metal oxide semiconductor FET (MOS-FET) palladium/ palladium oxide is used as the gate electrode.

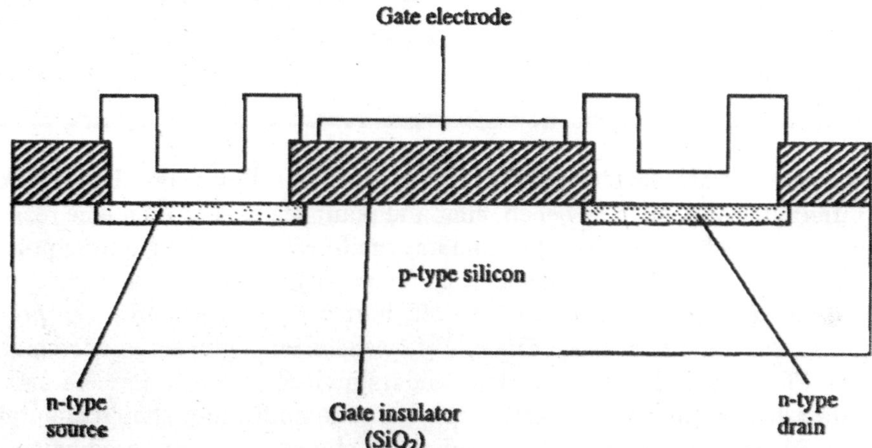

Fig. 6.2. Schematic diagram of a field effect transistor. The silicon-silicon dioxide system exhibits good semiconductor characteristics for use in FETs. The free charge carrier concentration, and hence the conductivity, of silicon can be increased by doping with impurities such as boron. This results in *p*-type silicon, the '*p*' describing the presence of excess positive mobile charges present. Silicon can also be doped with other impurities to form *n*-type silicon with an excess of negative mobile charges.

This catalytically decomposes gases such as hydrogen sulphide or ammonia with the production of hydrogen ions, which pass into the semiconductor layer. An enzyme may be coated on the palladium, *e.g.*, urease, which catalyses the production of ammonia from urea and thus provides a device for the measurement of this substrate. In ion-selective FETs (ISFETs), an ion-selective membrane replaces the gate electrode. When an enzyme-loaded gel is combined with the membrane, the device can be used to measure substrates which enzymically generate charged species.

CHAPTER 7

Scaling Bioprocess

Microbial processes are developed and implemented in many ways. These processes are operated at different scales, such as bench, pilot and plant scales. For economic reasons, large-scale bioprocesses should not be disturbed and any modification to the running processes may not be considered. For research, many small-scale processes are developed to screen and develop better strains, improve existing culture, use efficient media and handle the process more efficiently with maximum productivity. The role of biochemical engineers in microbial processes is well defined. First, they should have clear understanding of the bioprocess and know the action of biocatalysts to stimulate biochemical reactions. Secondly, they should maintain optimal growth and the environmental conditions required to obtain maximum production yield. The microbial environment involves substrate and product concentrations, media composition, pH and temperature.

There are many physical and chemical parameters involved in the growth environment once the organisms are propagated in a well-advanced bioreactor with instrumentation control. The process parameters are the flow dynamics and chemical dynamics of the biochemical system such as mass transfer, mixing, shear generated by agitation, power input, dilution rate, nutrients and growth stimulants. The chemical parameters of the process are well controlled and optimised for growth kinetics; physical factors are based on process design. The process design, geometric shape and the size of the bioreactor are affected by dilution rate and substrate consumption rate. The operating parameters are well controlled. Changing bioreactor size must follow certain rules to meet special criteria. Most bioprocess designers have attempted to follow only vessel geometric similarities and some dimensionless numbers. However, only increasing volume and dimensions by keeping some dimensionless number constant may not satisfy the needs of bioprocess design.

PROCEDURE

The usual procedure for scale-up of a fermenter based on concepts is well known. The following criteria are translated between two scales of operation. They are selected as a procedure for scale-up.

1. Similar Reynolds number or momentum factors.
2. Constant power consumption per unit volume of liquid, P_g/V.
3. Constant impeller tip velocity, ND_i.

4. Equal liquid mixing and recirculation times, t_m.
5. Constant volumetric of mass transfer coefficient, K_La.
6. Keep all environmental factors for the microorganism constant.

Based on the practical history of scale-up, most fermentation processes for alcohol and organic acid production have followed the concepts of geometric similarity and constant power per unit volume. From the above concept, and as a strong basis for translation of process criteria, only physical properties of the process were considered in the scale-up calculation. For power consumption in an agitated vessel, there is a fixed relation between impeller speed, N, and impeller diameter, D_i. The constant power per unit volume, for a mechanical agitated vessel is given by.

$$\frac{P}{V} = \frac{\rho N^3 D_i^5}{D_i^3} \text{ for Re} \geq 10^4 \quad ...(7.1)$$

The power per unit volume is constant. From power consumptions in a bench-scale bioreactor, the necessary agitation rate is calculated for the scale up ratio, using Equation (7.1). The choice of criterion is dependent on what type of fermentation process has been studied. The following equation expresses relations for the impeller size and agitation rate in small and large bioreactors.

$$N_1^3 D^2_{i,1} = N_2^3 D^2_{i,2} \quad ...(7.2)$$

The agitation rate is proportional to impeller diameter to the power of 2/3.

$$N_2 = N_1 \left(\frac{D_{i,1}}{D_{i,2}}\right)^{2/3} \quad ...(7.3)$$

Similarly, that is true for the rotational speed of large tank, which is related to a small tank with the ratio of impeller diameter of large and small tanks to the power of 2/3.

$$rpm_2 = rpm_1 \left(\frac{D_{i,1}}{D_{i,2}}\right)^{2/3} \quad ...(7.4)$$

The characteristics of an agitated vessel with impeller spacing are illustrated in Fig. 7.1

Constant K_La

The mass transfer coefficient K_La is constant; the general correlation is considered by many as proportional to the power per unit volume with constant exponent, and gas superficial velocity to another constant power as shown below:

$$K_La = \alpha\left[\left(\frac{P_g}{V_L}\right)^a (V_s)^b\right] \quad ...(7.5)$$

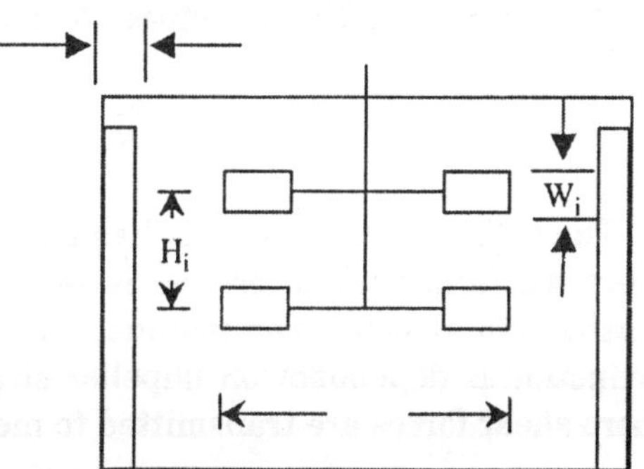

Fig. 7.1. Characteristics of agitated vessel with impeller spacing.

There has to be a relation between k_La with aeration rate and agitation speed, and scale-up factor has to be determined. To eliminate the effect of viscous forces, the rheology of the media and broth for a large vessel have to be similar to that of a bench-scale vessel. For scale-up based on geometric similarity, the constant values *a* and *b* are proposed for the mass-transfer correlation in Table 7.1

The oxygen transfer rate is calculated based on oxygen concentration gradient, by determination of the oxygen level in the liquid phase and the equilibrium value.

$$\text{OTR} = K_L{}^a(C^*_{AL} - C_{AL}) - XQ0_2 \quad ...(7.6)$$

At steady-state condition, change of oxygen concentration with time approaches zero. The value of $K_L a$ should be estimated by a correlation developed for various sizes of fermentation vessel.

$$K_L{}^G = \frac{XQo_2}{C^*_{AL} - C_{AL}} \quad ...(7.7)$$

where Qo_2 is oxygen transfer rate and X is biomass concentration.

Table 7.1. Constants in Mass Transfer Correlation for Various Fermenter Size

Fermentation vessel size, 1	*a*	*b*
5	0.95	0.67
500	0.60–0.7	0.67
50,000	0.40–0.5	0.50

Scale-up Based on Shear Forces

Scale-up calculation is based on constant shear forces, where shear is directly related to impeller tip velocity. Shear forces are defined as

$$\tau = \mu \frac{du}{dx} \quad ...(7.8)$$

where, τ is the shear stress, du/dx is the shear rate, and μ is the fluid viscosity. Since the shear is defined as *S*, which is proportional to πND_i, the concept of constant impeller tip velocity and constant power per unit volume were applied. Now introduce variable *S* for impeller tip velocity ND_i into Equation (7.8) then multiply both sides and divide by impeller diameter, so that the power equation is reduced to shear forces related to impeller diameter :

$$S_L = S_s \left(\frac{D_{i,L}}{D_{i,s}}\right)^{1/3} \quad ...(7.9)$$

The constant shear concept has been applied for bioreactor scale-up that utilises mycelia, where the fermentation process is shear sensitive and the broth is affected by shear rate of impeller tip velocity. For instance, in the production of novobicin, the yield of antibiotic production is dependent on impeller size and impeller tip velocity. The impeller is a device where shear forces are transmitted to me fermentation broth.

Since the process is sensitive to high agitation rate, it is necessary that the special bioreactor be scaled up, based on shear forces for maximising product yield such as antibiotics. This is the main reason for keeping impeller tip speed constant; in practice, it is recommended that this is in the range 0.25–0.5 ms^{-1}. If the above conditions hold, it is not necessary to follow the geometric similarity constraint; however, it may violate the rule of power consumption per unit volume. Therefore keeping the impeller at a constant speed and keeping K_La constant may be required for special case studies for bioreactor design in antibiotic fermentation.

Constant Mixing Time

The problem associated with poor mixing in a large vessel was identified as low dissolved oxygen in the aerated vessel. The mixing time has been correlated with turbulent flow. In a large-scale operation, the corrected function for mixing time incorporates the actual mixing time with impeller speed, tank diameter and liquid height. The function of mixing, $f(t)$ is defined as constant mixing time:

$$f(t) = t_m (ND_i^2)^{2/3} g^{1/6} D_i^{1/2} Y^{1/2} D_t^{3/2} \quad ...(7.10)$$

where, t_m is the mixing time, g is gravity, D_t is the tank diameter, and Y the depth of the liquid. Substitute (7.8) into (7.9) as power per unit volume is constant; the resulting equation is the mixing time :

$$\frac{t_{mL}}{t_{mS}} = \left[\left(\frac{N_S}{N_L}\right)^4 \left(\frac{D_{iL}}{D_{iS}}\right)\right]^{1/6} \quad ...(7.11)$$

That is proportional to impeller diameter to the power of 0.65. Then, solving for mixing time :

$$t_{mL} = t_{mS}\left(\frac{D_{iL}}{D_{iS}}\right)^{11/18} = t_{ms}\left(\frac{D_{iL}}{D_{iS}}\right)^{0.65} \quad ...(7.12)$$

Equation (7.12) is used for geometric similarity, where t_{mS} and t_{mL} represent the mixing time for small and large fermentation vessels. Mixing and mixing time have been major concerns for several investigators, because large-scale vessels may not have uniform mixing whereas in small vessels mixing is not a problem. The mixing time in large vessels has been correlated based on dimensional analysis. In 1953, Rushton proposed a dimensionless number that is used for scale-up calculation. The dimensionless group is proportional to N_{Re} as shown by the following equation:

$$\psi = k(N_{Re})^a \quad ...(7.13)$$

where the Reynolds number for an agitated vessel was defined as :

$$Re = \frac{\rho N D_i^2}{\mu} \quad ...(7.14)$$

Since the process is more complex, the proposed method may not be valid for scale-up calculation. The combination of power and Reynolds number was the next step for correlating power and fluid-flow dimensionless number, which was to define power number as a function of the Reynolds number. In fact, the study by Rushton summarised various geometrics of impellers, as his findings were plotted as dimensionless power input versus impeller Reynolds number; the plot as known as a power graph. Equation (7.15) correlates the power number as a function of Reynolds number.

$$N_\phi = k(Re)^{-m} \quad ...(7.15)$$

In fact, the power number is a dimensionless number that is the ratio of ungassed power to gassed power in a normal bioreactor.

$$N_p = \frac{Pg_c}{\rho N^3 D_i^5} \quad ...(7.16)$$

where P is ungassed power, in W or hp.

In a mixed agitated vessel with high agitation rate, at the centre of the vessel a vortex often forms. To prevent a central vortex in tanks less than 3 m in diameter, four baffles each with a baffle width of 15–20 cm are necessary. A basic assumption is to select a ratio of liq-uid height to tank diameter from 2:1 to 6:1.

$$\frac{H_L}{D_t} = 2:1-6:1 \quad ...(7.17)$$

It is common for an agitated vessel to have the tank diameter equal to the liquid working volume.

$$\frac{H_L}{D_t} = 1 \quad ...(7.18)$$

Also, select a ratio of tank diameter to baffle width from 10 to 12 :

$$\frac{D_t}{D_b} = 10-12 \quad ...(7.19)$$

A suitable impeller interspacing (H_i) is preferred for good mixing that has to be in distance of less than $2D_i$:

$$D_i < H_i < 2D_i \quad ...(7.20)$$

It is well understood that mixing and mass transfer are affected by agitation and aeration rates. Typically, in an agitated vessel, the working volume is about 75% of nominal CSTR volume.

BIOREACTOR

The Bioreactor is the major equipment used in biochemical processes. It differs totally from a simple chemical reaction vessel. To control physical operating parameters and microbial environmental conditions, there are several influential variables:

- Biomass concentration
- Sterile condition
- Effect of agitation and mass transfer
- Heat removal for temperature control
- pH control of the media
- Correct shear conditions

Bioreactors in terms of operation and specific criteria are well defined. The most useful and applied types of bioreactor are:

- CSTR

- Air lift: air is used for fluid circulation by pressurised air
- Loop reactor: modified air lift, pump transport the air and fluid through the vessel
- Immobilised system: air and fluid circulate over a film of microorganisms grown on a solid support
- Tower fermenter
- Membrane bioreactor

General Cases

If the height of liquid, H_L is equal to D_t, one set of agitators is enough to obtain sufficient mixing. For H_L equal to $2D_t$ or $3D_t$ additional sets of agitators separated by distance H_i are required, mounted on the same shaft. Spargers are placed at a distance $D_i/2$ from the bottom of the tank. Power input is greater than 100 Wm^{-3} and the tip impeller speed is πND_i greater than 1.5 ms^{-1}. Also, Froude number is $\frac{N^2 D_i}{g}$ greater than 1.0. H_i is impeller inter spacing.

Foam in the bioreactor is troublesome: it can reduce the oxygen transfer rate (OTR). Antifoam is use to prevent foam formation. However, excess antifoam may cause growth inhibition in the course of fermentation. The simplest device is known as a foam breaker, which is mounted on the stirrer shaft located on the surface of liquid. It is a flat blade.

Bubble Column

For production of baker's yeast, beer, vinegar, and in wastewater treatment, bubble columns are used. Generally these columns should be long enough so that bubbles are utilised as they rise while passing through the column. Column diameters range from 10 cm to 7.5 m. The height of column is three to six times greater than the diameter. The common height and diameter ratio *(H/D)* is in the range 2–6. The H:D ratio that is commonly used is 3:1; the ratio especially used for baker's yeast production is 6:1.

Hydrodynamics and mass transfer in bubble columns are dependent on the bubble size and the bubble velocity. As the bubble is released from the sparger, it comes into contact with media and microorganisms in the column. In sugar fermentation, glucose is converted to ethanol and carbon dioxide :

$$C_6H_{12}O_6 \rightarrow 2C_2H_5OH + 2CO_2 \quad ...(7.21)$$

The superficial gas velocity is the air flow rate per unit cross sectional area of the column. The superficial gas velocity is less than 0.4 ms^{-1} and media flow rate is based on liquid velocity. Then, the column may have a hydraulic retention time that is a function of fluid velocity and column diameter. The liquid velocity is given by:[4]

$$U_L = 0.9(gDU_g)^{0.33} \quad ...(7.22)$$

where, U_g is the upward gas bubble velocity at the centre of column, the gas superficial velocity is in the range of $0 < Ug < 0.4$ ms^{-1} and the column diameter is in the range $0.1 < D < 7.5$ m. The hydraulic retention time is obtained by division of column height and the actual liquid velocity :

$$U_g = \frac{\text{Air flow rate}}{\text{Cross-sectional area}} = \frac{Q_g}{A} \quad ...(7.23)$$

The mixing time, t_m is given :

$$t_m = 11\left(\frac{H_L}{D}\right)\left(\frac{gU_g}{D^2}\right)^{-0.33} \quad ...(7.24)$$

where, H_L is the height of bubble column.

Example 1. *Let us have a bubble column with an HID ratio of 3, diameter 0.5 m and a gas flow rate of 0.1 $m^3 \cdot h^{-1}$ which gives a superficial gas velocity of 0.25 m × s^{-1}. What would be the liquid media flow rate?*

Solution : Once superficial gas velocity is defined, then liquid velocity is given by (7.22)

$$U_g = \frac{Q_g}{A} = \frac{4(0.1)}{0.5\pi} = 0.255 \text{ m} \cdot \text{s}^{-1}$$

$$U_L = 0.9\,[(9.81)(0.5)(0.255)]^{0.33} = 0.96 \text{ m} \cdot \text{s}^{-1}$$

Equation (7.23) is used to calculate mixing time.

$$\tau_m = 11\left(\frac{1.5}{0.5}\right)\left(\frac{(9.81)(0.255)}{(0.5)^2}\right)^{-0.33}$$

Based on mixing time, the liquid media flow rate is:

$$Q_L = \frac{V}{\tau} = \frac{(1.5)(0.25\pi)/4}{31} = 9.5 \times 10^{-3} \text{m}^3 \cdot \text{h}^{-1}$$

Example 2. *Calculate the speed of an impeller and the power requirements of a production-scale bioreactor with 60 m^3 using two different methods. Also, match the volumetric mass transfer coefficient. The following optimum conditions were obtained with a small-scale fermenter of volume 0.3 m^3 and 60% of the vessel working. The density of broth, r_{broth}, 1200 kg $\cdot m^{-3}$, working volume 0.18 m^3, aeration rate of one volume of gas per volume of liquid (1 vvm), oxygen transfer rate 0.25 kmol $\cdot m^{-3} \cdot h^{-1}$, liquid height inside the vessel, H_L, 1.2D_t. Two sets of standard, flat-blade turbine impellers were installed.*

Solution :

$$V_1 = \frac{\pi}{4} \times D_t^2(1.2D_t) = 0.3\pi D_t^3$$

Diameter of the vessel :

$$D_t = \left(\frac{V_1}{0.3\pi}\right)^{1/3} = \left(\frac{0.18}{0.3\pi}\right) = 0.576\text{m}$$

Diameter of the impeller :

$$D_{i,1} = \frac{1}{3} \times D_t = \frac{0.576}{3} = 0.192 \text{ m}$$

Height of liquid media was assumed to be 1.2 times the diameter of the fermentation vessel.

$$H_{L,1} = 1.2 \times 0.576 = 0.691 \text{ m}$$

Diameter of the larger vessel :

$$D_{t,2} = \left(\frac{V_1}{0.3\pi}\right)^{1/3} = \left(\frac{60}{0.3\pi}\right)^{1/3} = 3.36\text{m}$$

The impeller size for the larger vessel is :

$$D_{i,2} = \frac{3.36}{3} = 1.12 \text{ m}$$

and the liquid media height in the second fermenter is :

$$H_{L,2} = 1.2 \times 3.36 = 4.03 \text{ m}$$

Assume the fermentation broth has the same viscosity as water :

$$\bar{\mu}_1 = 1 \text{ cp}$$

Aeration rate for 1 vvm is :

$$F_1 = 1 \times 0.18 = 0.018 \text{ m}^3 \cdot \text{min}^{-1} = 3 \times 10^{-3} \text{ m}^3 \cdot \text{s}^{-1}$$

Gas superficial velocity is :

$$Us = \frac{0.18 \times 60}{\frac{\pi}{4}(0.576)^2} = 41.45 \text{ m} \cdot \text{h}^{-1}$$

Let us take average values for the partial pressure of oxygen

$$\bar{P}_{O_2} = \frac{1 \text{ atm} + \left(1 + \frac{H_L m}{10.3 \text{ m}} \cdot \text{atm}\right)}{2} \times 0.21 = 0.213 \text{ atm}$$

The oxygen transfer rate is

$$(Kv\, \bar{P}_{O_2}) = 0.25 \text{ kmol} \cdot \text{m}^{-3} \cdot \text{h}^{-1}$$

The mass transfer coefficient is

$$Kv_1 = \frac{0.25}{0.213} = 1.174 \text{ kmol} \cdot \text{m}^{-1} \cdot \text{h}^{-1} \cdot \text{atm}^{-1}$$

Use imperial correction based on the following equation for mass transfer in the bioreactors.[1, 2] The general equation for evaluation of $K_L a$ is

$$K_L a = x\left(\frac{P_g}{V}\right)^Y (U_s)^Z \qquad ...(7.25)$$

where x, y and z are empirical constants. For Newtonian fluids, non-coalescing broth and gas bubbles, the following correlation is valid for a working volume of less than 4 m^3 and a power per unit volume of 500–10,000 $W \cdot m^{-3}$.

$$Kv = 0.002\left(\frac{Pg}{V}\right)^{0.7} Us^{0.5} \qquad ...(7.26)$$

$$1.174 = 0.002\left(\frac{P_g}{V}\right)^{0.7} (41.45)^{0.5} \qquad ...(7.27)$$

For the gassed power per unit volume $\left(\frac{P_g}{V}\right)$ is 630.7 *W*; the passed power, P_g, was 0.15 *hp*.

Since the flow regime is turbulent, Power number versus Reynolds number, *Re,* reads $P_{no} = 6$. For two sets of impellers, $N_p = 2(6) = 12$.

$$N_p = \frac{P_1 g_c}{\rho N_1^3 D_i^5} \quad ...(7.28)$$

$$P_1 = \frac{12 \times 1200 \times N_1^3 \times (0.192)^5}{9.81} = 0.383 N_1^3 W = 5 \times 10^{-4} N_1^3 hp$$

Using Michel and Miller's correction factor for power calculation:

$$P_{g1} = 0.5\left\{\frac{P_1^2 \eta_1 D_{i1}^3}{\mu_1^{0.56}}\right\}^{0.45} \quad ...(7.29)$$

Knowing the power input, we can calculate the rotational speed:

$$0.15 = 0.5\left\{\frac{(5 \times 10^{-4})^2 (N_1^3)^2 (0.192)^3}{(3 \times 10^{-3})^{0.56}}\right\}^{0.45}$$

$$N_1 = 10 \text{ rps}$$

$$N_1 = 600 \text{ rpm}$$

The power input for ungassed system is

$$P_1 = 5 \times 10^{-4} N_1^3 = 5 \times 10^{-4} (10)^3 = 0.5 \text{ hp}$$

$$\frac{P_g}{P} = \frac{0.15}{0.5} = 0.3$$

For constant power input based on geometric similarity of the vessels, agitation rate is calculated.

$$\frac{\rho N_1^3 D_{i1}^5}{V_1} = \frac{\rho N_2^3 D_{i2}^5}{V_2} \quad ...(7.30)$$

$$\left(\frac{N_2}{N_1}\right)^3 = \left(\frac{V_2}{V_1}\right)\left(\frac{D_{i1}}{D_{i2}}\right)^5 \quad ...(7.31)$$

$$N_2 = N_1\left(\frac{V_2}{V_1}\right)^{1/3}\left(\frac{D_{i1}}{D_{i2}}\right)^{5/3} = 600\left(\frac{60}{0.3}\right)^{1/3}\left(\frac{0.192}{0.12}\right)^{5/3} = 185 \text{ rpm} \leftarrow$$

For constant input velocity for a large system :

$$N_2 = N_1\left(\frac{D_{i1}}{D_{i2}}\right) = 600\left(\frac{0.192}{1.12}\right) = 103 \text{ rpm} \leftarrow$$

TUBULAR PLUG FLOW

The performance of a biochemical reactor is designed and evaluated based the reaction rate equation. The rate of biomass generation is based on the Monod rate model :

$$qx = \frac{kC_s^{\,x}}{K_m + C_s} = \frac{kxS}{a+S} \qquad ...(7.32)$$

where $$S = \frac{C_s}{C_{sf}} \text{ and } a = \frac{K_m}{C_{sf}}$$

dimensionless, K_m is the Monod rate constant, $g \cdot l^{-1}$, and kx represents the maximum specific growth rate, $g \cdot l^{-1} \cdot h^{-1}$. The fresh media dilution rate for cell production is D, h^{-1}

$$D = \frac{qx}{x} = \frac{q_s}{C_{sf} - C_s} \qquad ...(7.33)$$

The substrate concentration is defined as :

$$S = (1-S)\, Y_{x/s}\, C_{sf} \qquad ...(7.34)$$

The yield is incorporated into (7.34).

$$qx = \frac{kS}{a+S}(1-S)Y_{x/s}C_{sf} \qquad ...(7.35)$$

Example 1. *A special CSTR fermentation known as a chemostat bioreactor is used for microbial cell growth. The rate of biomass generation is given by :*

$$qx = \frac{4}{3}\frac{V_{sx}}{C_s + 4}\ \frac{g \text{ cells}}{m^3 \cdot h} \qquad ...(7.36)$$

We wish to compare the performance of two reactor types: plug flow versus CSTR with a substrate concentration of $Cs_f = 60\, g \cdot m^{-3}$ and a biomass yield of $Y_{x/s} = 0.1$. In a plug flow bioreactor with volume of 1 m^3 and volumetric flow rate of 2.5 $m^3 \cdot h^{-1}$, what would be the recycle ratio for maximum q_x compared with corresponding results and rate models proposed for the chemostat?

Solution : To maximise q_x, we need to take the derivative of the following function, if $\frac{S(1-S}{a+S}$ is maximum, then the derivative must be set to zero

$$\frac{d}{dS}\left[\frac{S(1-S)}{a+S}\right] = \frac{(1-2S)-(a+S)-S(1-S)}{(a+S)^2} = 0$$

$$(1-2S)(a+S) - S(1-S) = 0$$

$$S^2 + 2a\,S - a = 0$$

$$S = -a + \sqrt{a + a^2}$$

The quadratic equation is solved for a

$$a = 5 = \frac{K_m}{C_{sf}} = \frac{4}{60}$$

$$Y_{x/s} = 0.1$$

Maximum cell production is obtained in a chemostat at :

$$S = -a + \sqrt{a + a^2} = \frac{1}{15} + \sqrt{\frac{1}{15} + \left(\frac{1}{15}\right)^2} = \frac{1}{5}$$

$$\frac{1}{q_x} = \frac{4+C_s}{C_s \times x} \times \frac{3}{4}$$

Now, we can generate data by plug-in numbers in the above equation. We need to illustrate inverse rate versus substrate concentration. After the data are plotted we can justify a suitable type of bioreactor

$$x = Y_{X/S}(C_f - C_s) = 0.1\,(60-3) = 5.7$$

$$x = 0.1\,(55) = 5.5$$

The performance data for plug versus mix reactor were obtained. The data were collected as the inverse of q_x vs inverse of substrate concentration. Table 7.2 shows the data based on obtained kinetic data. From the data plotted in Fig. 7.2, we can minimise the volume of the chemostat. A CSTR works better than a plug flow reactor for the production of biomass. Maximum q_x is obtained with a substrate concentration in the leaving stream of 12 $g \cdot m^{-3}$.

$$S = \frac{1}{5} = \frac{C_S}{60}$$

Then, the substrate concentration is:

$$C_s = 12 \text{ g.m}^{-3}$$

Table 7.2. Substrate Concentration Versus Inverse Biomass Concentration

Substrate concentration Cs, $g \cdot m^{-3}$	*Inverse of biomass rate l/q_x, $m^3 \cdot g^{-1} \cdot h^{-1}$*
3	0.310
5	0.250
10	0.210
15	0.210
20	0.225
25	0.220
30	0.280
35	0.330
40	0.410
45	0.540
50	0.810
55	1.610

A tubular bioreactor design with operational may lead to a CSTR, having sufficient recycle ratio for plug flow that behave like chemostat. The recirculation plug flow reactor is better than a chemostat, with maximum productivity at $C_s = 3\ g \cdot m^{-3}$. Combination of plug flow with CSTR which behave like chemostat was obtained from the illustration minimised curve with maximum rate at $C_{Sf} = 3\ g \cdot m^{-3}$.

In plug flow reactor the value for C_s is reduced from 12 to 3 $g \cdot m^{-3}$, then the retention time and rate model with recycle ratio in a plug flow reactor can be written as :

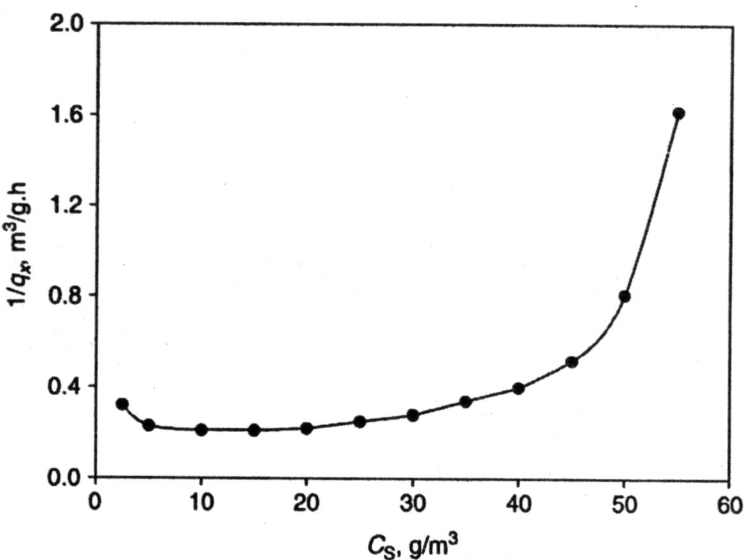

Fig. 7.2. Performance plot for plug versus mix reactor.

$$\tau = \frac{C_{sf}V}{v_o} = -(R+1)\int_{\frac{C_{sf}+RC_s}{R+1}}^{Cs} \frac{dC_s}{\dfrac{4v_{sx}C_s}{3(C_s+4)}}$$

$$\frac{Km}{S_f}\ln\left(\frac{C_{sf}+RC_s}{RC_s}\right)+\ln\left(\frac{R+1}{R}\right)=\frac{R+1}{R}\times\frac{K_m}{f+R}+\frac{1}{R}$$

where S_f is the inlet to outlet substrate ratio, R is the recycle ratio, and f is the ratio of inlet to outlet substrate concentration. Solve R by trial and error, with calculation for right-and left-hand sides with the assumption of a value for R :

$$\frac{4/3}{60/3}\ln\left(\frac{60+2.5(3)}{2.5(3)}\right)+\ln\left(\frac{2.5+1}{2.5}\right)=\frac{2.5}{2.5}\times\frac{4/3}{20+2.5}+\frac{1}{2.5}$$

Let us plug in f = 20, k_m = 4/3 and R = 2.5 into (7.36), then calculate $k\tau$

$$k\tau = R\left[\ln\left(\frac{R+1}{R}\right)+\frac{K_m}{f}\ln\left(\frac{C_{sf}+RC_s}{R}\right)\right]$$

$$k\tau = (2.5)\left[\ln\left(\frac{3.5}{2.5}\right)+\frac{4/3}{20}\ln\left(\frac{60+2.5(3)}{2.5}\right)\right]$$

$$k\tau = 1$$

Example 2. *The bioreactor will be scaled up by a factor of 125. If is necessary to discuss the effect of operating variables resulting from constant power per unit volume, agitation rate, and speed tip velocity, N_{Re}. The data for a small-scale bioreactor are a 100 litres fermenter with 187.5 rpm and*

$$V_1 \times 125 = V_2 = 12.5 \text{ m}^3$$

$$\frac{P}{V} = \text{constant}$$

$$N_i D_i = \text{constant}$$

Factoring for scale-up, we make some assumptions and may use a few known correlations to progress our calculations. Let us say $P \alpha V = D_i^3$, $H/D = 1$ and $D/D_i = 3$. Also, use constant power per unit volume for the small-and large-scale bioreactor.

$$\frac{P}{V} \approx (N_i^3 D_i^3)_{\text{Small}} = (N_i^3 D_i^2)_{\text{Large}}$$

Since we assumed power per unit volume is constant, it is possible $(N_i D_i)_1$ may not be equal with $(N_i D_i)_2$. Now calculate the agitation rate for a large vessel.

$$N_{i2} = N_{i1}\left(\frac{D_{i1}}{D_{i2}}\right)^{2/3} = 600\left(\frac{1}{5}\right)^{2/3} = 204 \text{ rpm}$$

$$\frac{H}{V} \propto \frac{N_i D_i^3}{D_i^3}$$

$$V = HD^2\frac{\pi}{4} = \frac{\pi D^3}{4} = \frac{9}{4}\pi D_i^3$$

Now assume $H \approx D$, $D = 3D_i$ and bioreactor volume is proportional to impeller diameter to the power of $V \propto D_i^3$. The Reynolds number is written as:

$$Re_i = \frac{\rho N D_i}{\mu}$$

The power for laminar flow is proportional to agitation rate, N^2, and the flow is turbulent the power is proportional to ~ $N^3 D_i^2$. Let us assume the mass transfer coefficients remain constant ($K_L a$ unchanged):

$$\frac{D_i}{D_t} = 0.33$$

$$ND_i = 0.5 \text{ m/s}$$

Also assume the height of liquid media is 1.2 times the tank diameter.

$$\frac{H}{D_t} = 1.2$$

$$V_s = 1001\frac{1.2}{4}\pi D_i^3$$

The diameter of small tank is 47.4 cm. Since the impeller is one third of the tank diameter, $D_i = 16$ cm

$$Re_i = \frac{(1 \text{ g/cm}^3)(50 \text{ cm/s})(16)}{0.01} = 80{,}000 \text{ Tubulent flow}$$

Power is defined for a well-agitated vessel with a mechanical stirrer; then read power number for turbulent flow.

$$P_{no.} = \frac{P \cdot g_c}{\rho N^3 D_i^5} = 6$$

$$ND_i = 0.5$$

$$N = \frac{0.5}{0.16}$$

Calculate power for a small-scale vessel:

$$P_{small} = 6 \times \frac{(1000 \text{ kg/m}^3)\ (3.125)^3 (0.16)^5}{9.81} = 1.96 \text{ W}$$

The power for large vessel is based on scale factor, that is calculated as following:

$$P_{Large} = 125 \times 1.96 = 245 \text{ W}$$

For mass transfer calculation

$$K_L a = \text{constant}$$

$$V_s = 1001 \quad D_i = 16 \text{ cm} \quad D_t = 47.43 \text{ cm}$$

$$ND_i = 0.5$$

$$N_i = \frac{0.5}{0.16} = 3.125 \text{ rps} = 187.5 \text{ rpm}$$

When Re = 80,000 for turbulent flow, the power number versus Re, we can read

$$P_{no} = 6$$

$$P_{NO} = \frac{P \cdot g}{\rho N_i^3 D_i^5}$$

$$P_{small} = \frac{6 \times 1000 \times 3.12 - 5^3 \times 0.16^5}{9.81} = 1.96 \text{ W}$$

$$\frac{1.96}{0.1} = 19.6 \frac{\text{W}}{\text{m}^3}$$

Scale-up factor = 125

$$\frac{P_s}{P_L} = \frac{(N_i^3 D_i^5)s}{(N_i^3 D_i^5)_L} = \frac{1}{125}$$

The power required for a large vessel is:

$$P_L = 19.6 \times 125 = 2450 \text{ W}$$

The power per unit volume of small bioreactor is

$$P/V = 19.6/0.1 = 196 \text{ W.m}^{-3}$$

The volume of a large vessel is

$$V_L = 2450/196 = 12.5 \text{ m}^3$$

$$\left(\frac{D_{iL}}{D_{is}}\right)^2 = 25$$

$$D_{iL} = (0.84\ \text{m})$$

$$\frac{\pi}{4} D_t^3 = \frac{\pi}{4}(2.515)^3 = 12.5\ \text{m}^3$$

We have understood the following concepts are true; our calculations also support the fact that power per unit volume is constant.

$$ND_i \neq \text{constant} \quad N^3D_i^2 = \text{constant}$$

$$V_L = 12.5\ V_s$$

$$D_{i,L} = \frac{4}{\pi}\sqrt[3]{12.5}\ D_{tL} = 2.515\ \text{m},\ D_{iL} = 0.84\ \text{m}$$

$$N_{i,L} = N_{i,S}\left(\frac{D_{i,S}}{D_{i,L}}\right)^{2/3} = (3.125)\left(\frac{0.16}{0.8}\right)^{2/3} = 1\ \text{rps}$$

$$Re_{i,L} = \frac{\rho N_i D_i^2}{\mu} = \frac{(1000)(0.8)^2(1)}{0.001\frac{\text{kg}}{\text{m.s}}} = 6.2\times 10^5$$

$$P_{no} = 6$$

Power = 0.3 hp

Example 3. *A 20 m^3 working volume of bioreactor is used for production of penicillin. What is the initial substrate concentration, S_0, that you choose when there is a limitation in sufficient oxygen transfer rate and there are no limiting reactants?*

Characteristics

D_{tank}	= 2.4 m	*Turbine*	
D_i	= 0.8 m	μ_{broth}	= 1 mPa · s
N_i	= 2.3 rps	*P*	= 1.2 ρH_2O
Number of blades	= 8	*Aeration rate*	= 1 *vvm*
P_g/P	= 0.4		

If you aerate a bioreactor, power consumption is much less than a non-aerated bioreactor.

Oxygen transfer rate, OTR: 6 × 10^{-3} kg · m^{-3}

Oxygen uptake rate, OUR: 0.65 mmol $O_2 kg^{-1}$ cell

Specific growth rate, v_{max}: 0.5 h^{-1}

Specific sugar consumption rate = 1.0 kg · kg^{-1} cell · h^{-1}

The mass transfer coefficient is calculated by the following correlation

$$K_La(s^{-1}) = 2\times 10^{-3}\left(\frac{P_g}{V}\right)^{0.6} V_s^{\ 0.667}$$

where P_g/V is the power per unit volume (hp · m^{-3}) and V_s is the gas superficial velocity (cmmin^{-1}).

Solution : The flow regime is turbulent as the Reynolds number is large.

$$N_{Re} = \frac{\rho D v}{\mu} = \frac{\rho D_i (N_i D_i)}{\mu} = \frac{1200(0.8)^2(2.3)}{0.1} = 1.92\times 10^4$$

Read power number N_p from the graph P_{no} versus $R_e : N_p = 6$

$$N_P = \frac{P g_c}{\rho N^3 D_i^5} \Rightarrow P = N_p \rho N^3 D_i^5 / g_c$$

$$= 6 \times 1200 \times (2.5)^3 \times (0.8)^5 / 9.81$$

$$= \frac{3758 \text{ kg}\cdot\text{m/s hp}}{745.7 \text{ kg}\cdot\text{m/s}} = 5.04 \text{ hp}$$

Correction factor for non-geometrical similarity,

$$f_c = \sqrt{\frac{(D_t / D_i)^* (H_L/D_i)^*}{(D_t / D_i)(H_L/D_i)}} = \sqrt{\frac{3\times 4.42}{3\times 3}} = 1.25$$

$$(D_t/D_i)^* = \frac{2.4}{0.8} = 3$$

$$(H_L/D_i)^* = \frac{20 \text{ m}^3}{(2.4)^2 \frac{\pi}{4} / 0.8} = 4.42$$

The power input for three sets of impellers in a single shaft is:

$$P = (5.04 \text{ hp})(1.25)(\text{three sets of impellers}) = 19 \text{ hp}$$

$$\text{Aeration rate} = N_a = \frac{F_g}{N_i D_i^3}$$

$$F_g = 20 \text{ m}^3\cdot\text{min}^{-1} = 1 \text{ vvm} \times V_L = 0.333 \text{ m}^3\text{s}^{-1}$$

Given gassed power is 40% of ungassed power,

$$P_g/P = 0.4 \; P_g = 0.4 \times 19 = 7.6 \text{ hp}$$

$$N_a \frac{F_g}{N_i D_i^3} = \frac{0.333}{0.5\times(0.8)^3} = 0.26$$

$$V_s = \frac{20 \text{ m}^3/\text{min}}{\frac{\pi}{4}\times(2.4)^2 \text{ m}^2} = 4.42 \text{ m/min} = 7.4\times 10^{-2} \text{ m/s}$$

The mass transfer coefficient is calculated.

$$K_L a = 2\times 10^{-3}\left(\frac{7.6\times 754.7 \text{ W/hp}}{20}\right)^{0.6} (4.42)^{0.667} = 3.57 \; s^{-1}$$

With an assumption of oxygen concentration at the interface, equilibrium with liquid phase is 6 ppm, the oxygen transfer rate is calculated.

$$OTR = K_La(C^*o_2 - Co_2) = (3.57)(6 \times 10^{-3}) = 0.02 \text{ kg/m}^3\text{ s}$$
$$OTR = xqo_2 = (0.65 \times 10^{-3})(32 \times 10^{-3})x$$
$$OURqo_2 = 2.08 \times 10^{-3} \text{ kg O}_2\text{/kg cells}$$
$$x = \frac{0.02}{2.08 \times 10^{-3}} = 10.3 \text{ kg/m}^3$$

Cell Balance

$$19.375 = x_S = x_0 + \overbrace{\frac{\mu_{max}}{q_S}}^{\text{Sugar to cell}} Cs = 0 + \frac{0.5}{1.0}\, Cs$$

$$C_S = \frac{10.3}{0.5} = 20.6 \text{ kg/m}^3$$

Example 4. *Calculate mass transfer, gas hold up, gassed and ungassed power for the fermenter with the given data :*

$\rho_{borth} = 1200 \text{ kg} \cdot m^{-3}$ $\mu = 0.002 \text{ N} \cdot s \cdot m^{-2}$ $N_i = 90 \text{ rpm}$

$D_t = 4 \text{ m}$ $D_i = 2 \text{ m}$ $w_b = 0.4 \text{ m}$ $H_L = 6.5 \text{ m}$

Air is sparged with 0.4 vvm, it is equipped with two sets of impellers and a flat-blade turbine with four baffles.

Calculate: (a) power, P; (b) Power, P_g; (c) K_La; (d) gas holdup.

Solution : $$(D_t/D_i)^* = 21 \text{ cm } (H_L/D_i)^* = \frac{6.5}{2} = 3.25$$

$$N = 90 \text{ rpm}/60 = 1.5 \text{ rps}$$

$$N_{Re} = \frac{ND_i^2\rho}{\mu} = \frac{1.5 \times 2^2 \times 1200}{0.002} = 3.6 \times 10^6$$

Read N_p power versus *Re*, for a turbulent regime, $N_p = 6$

$$6 = \frac{P \times 9.81}{(1.5)^3(2)^5(1200)} \Rightarrow P = 105 \text{ hp}$$

Correction factor for non-geometrical similarity,

$$f_C = \sqrt{\frac{(D_t/D_i)^*(H_L/D_i)^*}{(D_t/D_i)(H_L/D_i)}} = \sqrt{\frac{2 \times 3.25}{3 \times 3}} = 0.85$$

$$P = 2 \times 0.8 \times 105 = 168 \text{ hp}$$

The gas flow rate is

$$F_g = 0.4(4)^2\left(\frac{\pi}{4}\right)(6.5) = 32.67 \text{ m}^3\text{/min} = 0.5445 \text{ m}^3\text{/s.}$$

The aeration number is

$$N_a \frac{F_g}{N_i\, D_i^3} = \frac{0.5445}{1.5 \times (2)^3} = 0.045$$

Reading gassed power to ungassed power ratio from the plot for the ratio of P_g/P, defined

$$\frac{P_g}{P} = \frac{P_a}{P} = 0.74$$

The gassed power is always less than the ungassed power.

$$P_g = 0.74P = 0.74(168) = 124.3 \text{ hp}$$

The mass transfer coefficient for turbulent flow is:

$$K_L a = 2 \times 10^{-3} \left(\frac{P_g}{V_L}\right)^{0.6} Vs^{0.667}$$

$$K_L a = 2 \times 10^{-3} \left(\frac{124.3 \times 754.7 \text{ W/hp}}{26\pi}\right)^{0.6} \cdot \left(\frac{0.54}{4\pi}\right)^{0.677} = 58.62 \text{ s}^{-1}$$

where the gas and liquid velocities are

$$V_s = \frac{0.5445 \text{ m}^3/\text{s}}{4\pi \text{ m}^2} \quad V_L = 4\pi \times 6.5 = 26\pi \text{ m}^3$$

Example 5. *Calculate mass transfer coefficient in a 60 m^3 fermenter with a gas and liquid interfacial area of $a = 0.3$ $m^2 \cdot m^{-3}$, given $\rho_{broth} = 1200$ $kg \cdot m^{-3}$. The small reactor has working volume of $0.18 m^3$, 1 vvm aeration rate. Oxygen transfer rate (OTR) is 0.25 $kmol \cdot m^{-3} \cdot h^{-1}$. There are two sets of impellers, and flat-blade turbine types of impeller were used, $H_L = 1.2D_t$. Find the exact specifications of a large fermenter.*

$$V_1 = 0.18 = \frac{\pi}{4} D_t^2 (1.2D_t) = 0.3\pi D_t^3$$

The diameter of the tank is

$$D_t = \left(\frac{0.18}{0.3\pi}\right)^{1/3} = 0.576 \text{ m}$$

Also, the impeller diameter is

$$D_i = \frac{1}{3} D_t = 0.576/3 = 19.2 \text{ cm}$$

The liquid media height is

$$H_L = 1.2 \times 0.576 = 0.691 \text{ m}$$

The large tank diameter is

$$D_{t2} = \left(\frac{V_2}{0.3\pi}\right)^{1/3} = \left(\frac{60}{0.3\pi}\right)^{1/3} = 3.36 \text{ m}$$

and the impeller for the large tank is

$$D_{i2} = 3.36/3 = 1.12 \text{ m}$$

$$H_{L2} = 1.2 \times 3.36 = 4.03 \text{ m}$$

The air flow rate is

$$F_1 = 1 \times 0.18 \text{ m}^3/\text{min} = 3 \times 10^{-3} \text{ m}^3 \text{ s}^{-1}$$

Gas superficial velocity,

$$U_s \frac{0.18 \times 60}{(\pi/4)(0.576)^2} = 41.45 \text{ m/h}$$

Oxygen partial pressure:

$$\overline{P_{O_2}} = 1 \text{ atm} + \left(1 + \frac{H_L m}{10.3 \text{ in}} \cdot \text{atm}\right) \times 0.21 = 0.213 \text{ atm}$$

$$OTR = K_L \cdot \overline{Po_2} \Rightarrow K_{L1} \frac{0.25 \text{ kmol/m}^3 \cdot \text{hr}}{0.213 \text{ atm}} = 1.174 \text{ kmol/m}^3 \cdot \text{h} \cdot \text{atm}$$

$$\text{H:D} = 3:1 = 6:1$$

The column diameter is in the $0.1 < D < 7.5$ m.

The mass transfer for freely rising gas bubbles is governed by conservation of mass and momentum balance. When the distance, velocity and concentration of substances provide the major driving forces for fluid motion, three dimensionless numbers explain the mass transfer. The dimensionless numbers are the Schmidt number, Grashof number and Sherwood number. Schmidt number relates viscous forces with mass diffusivity.

Grash of number deals with free convection and buoyant forces. The Rayleigh number results from the multiplication of the Schmidt number and the Grashof number. Finally, Sherwood number is the ratio of conductive forces and mass diffusivity. Based on analogy in mass transfer, the Sherwood number is similar to the Nusselt number in heat transfer. The Sherwood number can be correlated with Schmidt and Grashof numbers for calculating mass transfer coefficients in a mathematical expression.

$$S_C = \frac{\nu}{Do_2}$$

$$Sh = \frac{k_1 D}{D_{O2}}$$

$$Gr = \frac{D^3 \Delta\rho g}{18\mu_1{}^2}$$

And Rayleigh number

$$Gr.Sc = R_a = \frac{D^3 \Delta\rho g}{\mu_1 D_{O_2}}$$

Calderbank's correlation for turbulent flow aeration shows mass transfer is proportional to D_i^2 and $Re^{3/4}$

$$Sh = 0.13\, Sc^{1/3}\, Re^{3/4}$$

$$U_g = \frac{Q_g}{A}$$

OXYGEN TRANSFER RATE

The oxygen requirements for an activated sludge system with aeration and agitation for complete mixing are shown by mathematical model as the rate of oxygen transferred is equal to the difference of inlet and outlet concentration plus the oxygen utilised for generation of biomass:

$$V\frac{DCo_2}{dt} = a \cdot \bar{\mu}\ (Cso^- Cs) + b\ VX$$

where V is aeration tank volume in m^3, dCo_2/dt is oxygen accumulation rate in kg O_2. m^{-3}. day^{-1}, a and b are empirical constants (kg $O_2 \cdot kg^{-1}$ BOD) and (kg $O_2 \cdot kg^{-1}$ MLSS-day^{-1}), respectively. Also $\bar{\mu}$ is the flow rate of fresh wastewater ($m^3.day^{-1}$), MLSS stands for mixed liquor suspended solids, C_s is the BOD in effluent from aeration tank (kg BOD $\cdot$ m^{-3}), C_{So} is BOD in fresh wastewater and X is the sludge concentration in the aeration tank (kg MLSS $\cdot$ m^{-3}).

Example 1. *In a specific activated sludge plant, the organic load is carried out at 0.8 kg BOD per kg MLSS $\cdot day^{-1}$ with an 80% BOD removal efficiency. Values for the above mathematical model are as follows.*

Fresh feed flow rate 1 $m^3\ h^{-1}$, initial BOD concentration is 20,000 ppm and V is 10 m^3. The yield of biomass on substrate is 0.5 g $\cdot g^{-1}$ BOD.

$$a = 0.55\ kg\ O_2 \cdot kg^{-1} BOD.$$

$$b = 0.35\ kg\ O_2 \cdot kg^{-1} MLSS \cdot day^{-1}$$

What is the oxygen requirement per kg BOD removal?

Solution : Making oxygen balance for the aeration tank, the dynamic, transfer rate of oxygen is obtained :

$$V\frac{dCo_2}{dt} = a \cdot \bar{\mu}\ (Cso - Cs) + b\ VX$$

Dividing the oxygen balance by the volume of the aeration tank and concentration of biomass, XV, the above equation will be simplified as follows :

$$\frac{1}{X}\frac{dCo_2}{dt} = a\frac{\bar{\mu}}{VX}(Cso - Cs) + b$$

$$\frac{\bar{\mu}\ Cso}{VX} = \frac{\left(\frac{m^3}{day}\right)\left(\frac{kg\ BOD}{m^3}\right)}{(m^3)\left(\frac{kg\ MLSS}{m^3}\right)}$$

or

$$\frac{kg\ BOD}{kg\ MLSS \cdot day} = 0.8 \leftarrow \text{BOD loading}$$

$$\frac{1}{X}\frac{dCo_2}{dt} = \frac{a \cdot \bar{\mu} \cdot Cso}{VX}\left(1 - \frac{Cs}{Cso}\right)$$

The BOD removal is 80%

$$\frac{80}{100} = \frac{Cso - Cs}{Cso} = \left(1 - \frac{Cs}{Cso}\right)$$

$$0.8 = 1 - \frac{Cs}{Cso}$$

$$Cs = 0.2Cso$$

$$\frac{1}{X}\frac{dCO_2}{dt} = 0.4 \times 0.55(0.8) + 0.35 = 0.526 \ \frac{\text{kg } O_2}{\text{kg MLSS} \cdot \text{day}}$$

Now we can increase the BOD removal efficiency with decreasing organic loading. What would be the amount of oxygen required per kg BOD removal, with 90% removal efficiency?

Given data for constant coefficient $a = 0.5$ and $b = 0.3$

$$\frac{1}{X}\frac{dCO_2}{dt} = a\left(\frac{F}{V}\right) \cdot \frac{1}{X}(Cso - Cs) + b$$

Given *BOD loading*

$$\frac{FCso}{VX} = \frac{(\text{m}^3/\text{day})(\text{kg BOD/m}^3)}{(\text{m}^3)(\text{kg MLSS/m}^3)} = 0.4$$

$$\frac{Cso - Cs}{Cso} = 0.9 \Rightarrow 1 - \frac{Cs}{Cso} = 0.9$$

$$Cs = 0.1\ Cso$$

$$\frac{1}{X}\frac{dCO_2}{dt} = 0.5 \times 0.4 \times 0.9 + 0.3 = 0.48 \text{ kg of } O_2\text{/kg MLSS} \cdot \text{day}$$

In fact, the more organic load there is, the more oxygen is needed in the treatment process.

Example 2. *A bioreactor of $V = 20\ m^3$ working volume produces penicillin. What sugar concentration will you choose if OTR is not limited?*

Hint : if you aerate a bioreactor, the power consumption is less than a non-aerated bioreactor and the specific sugar consumption rate is 1 kg $\cdot kg^{-1}$ cell $\cdot h^{-1}$.

Data Given

D_t	$= 2.4$ m	*Three sets of impellers*	
D_i	$= 0.8$ m	*M*	$= 1$ mPa $\cdot$ s
N_i	$= 2.5$ rps	*P*	$= 1200$ kg $\cdot m^{-3}$
Number of blades	$= 8$, turbine	*Aeration rate*	$= 1$ vvm
P_g/P	$= 0.4$	*OTR*	$= 6 \times 10^{-3}$ kg $\cdot m^{-3}$
Specific O_2 uptake	$= 0.63$ mmol $O_2 \cdot kg^{-1}$ cell		$v_m = 0.5\ h^{-1}$

Solution : Let us find the Reynolds number

$$Re = \frac{\rho N D_i^2}{\mu} = \frac{1200 \times 0.8^2 \times 2.5}{0.1} = 1.9 \times 10^4$$

When flow is turbulent, the power number is using *Re* from power curves.

$$N_p = P_{no} = 6$$

$$P_{no} = 6 = \frac{P \cdot g}{\rho N_i^3 D_i^5} \Rightarrow P = \frac{6\rho N_i^3 D_i^5}{g} = \frac{6 \times 1200 \times 2.5^3 \times 0.8^5}{9.81} = 5 \text{ hp}$$

Ungassed power = 5 × 3 sets of impellers = 15 hp

Correction factor for scale-up using non-geometry similarity,

$$f_c = \sqrt{\frac{(D_t / D_i)^* (H_L / D_i)^*}{(D_t / D_i)(H_L / D_i)}} = \sqrt{\frac{3 \times 4.42}{3 \times 3}} = 1.25$$

$$D_t/D_i = 3$$

$$H_L/D = 3$$

$$(H_L/D_i)^* = (20)/(2.4)^2 (\pi/4) = 4.42$$

Gassed power, $P_g = 0.4\ (15) = 6$ hp ⇒ for three sets of impellers using aeration

$$Vs = \frac{20\text{m}^3 / \text{min}}{(\pi / 4)(2.4)^2 \text{m}^2}\ 4.4 \text{ m/min} = 7 \times 10^{-2} \text{ m} \cdot \text{s}^{-1}$$

The mass transfer coefficient for a turbulent regime is

$$K_L a = 2 \times 10^{-3} \left(\frac{6}{20}\right)^{0.6} (4.4)^{0.667} = 6 \times 10^{-2} \text{s}^{-1}$$

The oxygen transfer rate is

$$OTR = K_L a(C_1^* - C_1) = 6 \times 10^{-2} (6 \times 10^{-3} - 0)$$

$$OTR = xqo_2 = 4 \times 10^{-4} \text{ kg } O_2 \cdot \text{m}^{-3} \cdot \text{s}^{-1}$$

$$qo_2 = \left(6 \times 10^{-3} \frac{kg}{m^3} \times \frac{1}{32}\right) \text{kg } O_2 \cdot \text{kg}^{-1} \text{ cell} \cdot \text{m}^{-1} \cdot \text{s}^{-1}$$

$$2.08 \times 10^{-3} = 0.65 \times 10^{-3} \frac{\text{mole } O_2}{\text{kg cell}} \times \frac{32 \text{ kg}}{\text{kmol}}$$

Biomass concentration is calculated.

$$x = \frac{6 \times 10^{-3}}{2.08 \times 10^{-4}} = 30 \text{ kg cells} \cdot \text{m}^{-3}$$

What is your desired sugar concentration?

To answer the above question, we truly need to know the yield of biomass on substrate. Based on availability of oxygen, there is a good chance of sufficient biomass being generated, therefore nutrients and carbon sources must be available to carry on the process.

Example 3. *Let us scale-up a small fermenter with volume (V) of 0.3 m³. The scale-up factor is 200 fold. The large fermenter has a volume (V_2) of 60 m³. The working volume is 60% of nominal volume, that is*

$$V_L = 0.18\ m^3$$

Aeration rate = 1 vvm

Density, ρ = 1200 kg $\cdot m^{-3}$

Oxygen transfer rate, OTR = 0.25 kmol $\cdot m^{-1} \cdot h^{-1}$

$H_L/D_t = 1.2$

Two sets of impellers, turbine and flat blade.

The usual procedures for scaling up are summarised as following:

1. ***Reynolds number.***
2. ***Power consumption per unit volume of liquid.***
3. ***Tip velocity of an impeller, $N_i D_r$***
4. ***Liquid circulation time, mixing time.***
5. ***Volumetric oxygen transfer coefficient.***

Solution :

$$V_1 = \frac{\pi}{4} D_t^2 (1.2D_t) = 0.18 \text{ m}^3$$

$$D_t = \left(\frac{0.18}{0.3\pi}\right)^{1/3} = 0.576 \text{ m}$$

$$D_i = 1/3D_t = 0.192 \text{ m}$$

$$H_{L1} = 1.2 \times 0.576 = 0.691 \text{ m}$$

Large reactor :

$$D_{t2} = \left(\frac{V_2}{0.3\pi}\right)^{1/3} = \left(\frac{60}{0.3\pi}\right)^{1/3} = 3.36 \text{ m}$$

and

$$D_i = 3.36/3 = 1.12 \text{ m}$$

$$H_{L2} = 1.2 \times 3.36 = 4.03 \text{ m}$$

Air flow rate

$$F_1 = 1 \times 0.18 \text{ m}^3/\text{min} = 3 \times 10^{-3} \text{ m}^3/\text{s}$$

Gas supply velocity,

$$Us \frac{0.18 \times 60}{(\pi/4)(0.576)^2} = 41.45 \text{ m/h}$$

$$\overline{Po_2} = \frac{1 \text{ atm} + \left(1 + \frac{H_L}{10.3}\right)}{2} = 0.213 \text{ atm}$$

$$OTR = K_v \cdot \overline{Po_2}$$

$$K_V = \frac{0.25 \text{ kmol/m}^3 \cdot \text{h}}{0.213 \text{ atm}} = 1.174 \text{ kmol} \cdot \text{m}^{-3} \cdot \text{h}^{-1} \cdot \text{atm}^{-1}$$

$$K_v = 0.0318 \left(\frac{P_g}{V}\right)^{0.95} Us^{0.67} = 1.174$$

$$P_g = 3.22V = 3.22 \times 0.18 = 0.58 \text{ hp}$$

For turbulent flow the power number is about 6 and for two sets of impellers is going to be doubled.

$$Re \to N_p = 6 \times 2 = 12$$

$$N_p = \frac{P_1 g}{\rho N_1^3 D_i^5} \Rightarrow P_1 = 12 \times 1200 \times N_1^3 (0.192)^5 / 9.81 = 0.383 N_1^3 \text{ hp}$$

If you have access to the graph Pg/P that is good, read gassed power to ungassed power ratio off of the graph.

Read aeration number (Na) from Fig. 7.7

Now we make use of Miller's correlation for gassed power calculations:[5]

$$P_g = 0.5\left(\frac{P_1^2 N_1 D_{i1}^3}{F^{0.56}}\right)^{0.43}$$

$$0.58 = 0.5\left(\frac{(0.383)^2 (N_1^3)^2 N_1 D_{i1}^3}{(3 \times 10^{-3})^{0.56}}\right)^{0.43} \Rightarrow N_1 = 116 \text{ rpm}$$

$$P_1 = 0.383 N_1^3 = 0.383(116)^3 = 2.75 \text{ hp}$$

$$P_g/P = 0.58/2.75 = 0.21$$

Power input constant, geometric similarities

$$\frac{\rho N_1^3 D_{i1}^5}{V_1} = \frac{\rho N_2^3 D_{i2}^5}{V_2} \Rightarrow \left(\frac{N_2}{N_1}\right)^3 = \left(\frac{V_2}{V_1}\right)\left(\frac{D_{i1}}{D_{i2}}\right)^5$$

$$N_2 = N_1\left(\frac{V_2}{V_1}\right)^{1/3}\left(\frac{D_{i1}}{D_{i2}}\right)^{5/3} = 116\left(\frac{60}{0.3}\right)^{1/3}\left(\frac{0.192}{1.12}\right)^{5/3} = 35 \text{ rpm}$$

Constant impeller tip velocity :

$$N_1 D_{i2} = N_2 D_{i2}$$

$$N_2 = N_1\left(\frac{D_{i1}}{D_{i2}}\right) = 116\left(\frac{0.192}{1.12}\right) = 20 \text{ rpm}$$

$$P_g/V = 0.58/0.18 = 3.22 \text{ hp}$$

$$P_g = \left(\frac{0.58}{0.18}\right) \times (36 \text{ m}^3) = 116 \text{ hp}$$

$$P = 2.75\left(\frac{36}{0.18}\right) = 550 \text{ hp}$$

Example 4. *A batch production of penicillin of 40 m^3 capacity is required to supply sterile air through bioreactor at 1 volume of air per volume of culture per min (1 vvm). Incoming air contains 3000 bacteria per cubic metre of air, for 100 hours' operation. Calculate the filter depth, if the penetration of bacteria is 1 in 1 million.*

Solution : Assuming :

$$D_{filter} = 60 \text{ cm based on availability}$$

Air flow rate,

$$F_{air} = (40 \text{ m}^3)(1 \text{ vvm})(60 \text{ min/h}) = 2400 \text{ m}^3\cdot\text{h}^{-1}$$

Microbial load = 3000 cells m^{-3}

$$= 24000 \times 3000 = 7.2 \times 10^8 \text{ cells}$$

Filter cross-sectional area = $(\pi/4)(0.6)^2 = 0.0283 \text{ m}^2$

$$\text{Air velocity } \frac{\text{Vol. flow rate}}{A} = \frac{2400}{0.0283} = 8488 \text{ m}\cdot\text{h}^{-1}$$

Removal of organisms may follow the following exponential format of equation

$$N_t = N_0 e^{-kt}$$

$$\ln\frac{N_1}{N_2} = kL$$

$$\text{In}\frac{7.2\times10^8}{10^{-6}} = kL$$

$$N_1 = 7.2 \times 10^8 \text{ cells}$$

$$N_2 = 1 \text{ cell in } 10^{-6} \text{ cells} = 10^{-6} \text{ cells}$$

where L is the length of filter and k is rate constant for bed materials 40 m^{-1} and/or 84 m^{-1}. The length of filter is based on filter materials: for large k, a shorter length of filter was obtained.

$$L = \frac{1}{k}\ln(7.2\times10^{14}) = \frac{1}{40}\ln(7.2\times10^{14}) = 0.85 \text{ m}$$

$$L = \frac{1}{84}\ln(7.2\times10^{14}) = 0.41 \text{ m}$$

Example 5. *In a batch fermentation of ethanol, kinetic data were collected as product formed. The data are shown in Table E.7.1. The data will be used to design a continuous bioreactor (CSTR) with a 1001 working volume.*

(*a*) If the feed rate, F, is 5×10^{-3} $m^3\cdot h^{-1}$, calculate the ethanol concentration in the product stream. Keep in mind there was no ethanol in the feed stream.

(*b*) If the product yield is 45%, what would the sugar concentration in the feed stream be?

Experimental data are shown in Table E. 3.3.

The data are plotted in Fig. E. 7.3.

Use material balance for ethanol production :

$$\frac{dC_p}{dt} = D(C_{p0} - C_p) + r_p$$

At steady-state condition the product concentration remained constant, Initially there was no ethanol in feed stream, $C_{po} = 0$; When there is no accumulation, the balance equation is reduced to:

Table 7.3. Ethanol Production in Batch Fermentation with Respect to Inoculation Time

Time, h	*Ethanol,*
0	0.0
10	2.0
15	4.0
18	6.2
21	10.0
27	16.0
33	30.0
42	56.5
48	75.7
51	58.0
54	96.0
57	103.5
60	107.3
66	110.5
72	113.0

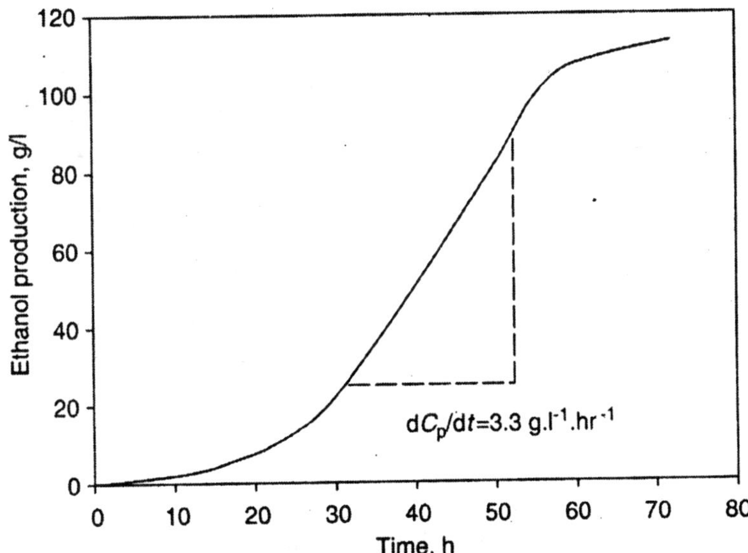

Fig. 7.3. Ethanol concentration profile with respect to fermentation time in batch mode of operation.

$$D(-C_p + r_p)$$

$$r_P \doteq DC_p = \frac{F}{V}C_p$$

$$C_P = \frac{V}{F} r_P$$

There is no wall temperature effect,

$$\left(\frac{\mu}{\mu_w}\right)^{0.14} = 1$$

From plotted data

$$\frac{dC_p}{dt} = 3.3\, g \cdot l^{-1} \cdot h$$

$$\frac{V}{F} r_P = \left(\frac{dC_p}{dt}\right)\frac{V}{F} = \frac{F}{V} C_p$$

$$C_P = \left(\frac{0.1\ m^3}{5\times10^{-3}}\right)(3.3) = \frac{330}{5} = 66\ kg \cdot m^{-3}$$

Ethanol concentration in the product stream is 66 g · l⁻¹

$$Y_{P/S} = 0.45$$

$$S_0 = \frac{Cp}{Y_{p/s}} = \frac{66}{0.45} = 147\ g \cdot l^{-1}$$

Example 6. *Ethanol is oxidised to acetic acid in the production of vinegar using Acetobacter. The reaction is exothermic:*

$$C_2H_5OH + O_2 \xrightarrow{\text{Acetobacter}} CH_3COOH + H_2O \quad \Delta H_{rxn} = -119\,\text{kcal.mol}^{-1}$$

The biochemical reaction rate followed the Monod rate model with a Monod rate constant of $k_s = 6.2 \times 10^{-6}\, g \cdot cm^{-3}$ and a specific growth rate of $v_{max} = 6.67 \times 10^{-7}\, g \cdot cm^{-3} \cdot s^{-1}$. Design the bioreactor with a suitable heat transfer area.

Given data

$N = 500\ rph = 30000\ rph$ $\qquad$ $T = 32\ °C = 90\ °F$

$D_i = 0.75\ ft$ $\qquad$ $k = 0.356\ Btu \cdot h^{-1} \cdot ft^{-1} \cdot °F^{-1}$

$P = 62.4\ lbm \cdot ft^{-3}$ $\qquad$ $C_p = 1\ Btu \cdot lb^{-1} \cdot °F^{-1}$

$M = 1\ cp = 2.42\ lbm \cdot ft^{-1} \cdot h^{-1}$

Solution :

$$-r_S = \frac{v_{max}\, S}{K_S + S}$$

$K_s = 6.2 \times 10^{-6}\ g \cdot cm^{-3}$ and $v_{max} = 6.67 \times 10^{-7}\ g \cdot cm^{-3} \cdot s^{-1}$

Fig. 7.4. Shows the jacketed bioreactor.

From the graph, the overall heat transfer coefficient for jacketed vessel[7] is

$$U \approx 100\ \text{Btu.h}^{-1}.\ \text{ft}^{2}\text{°F!}$$

$$Re = \frac{\rho N D_i^2}{\mu}$$

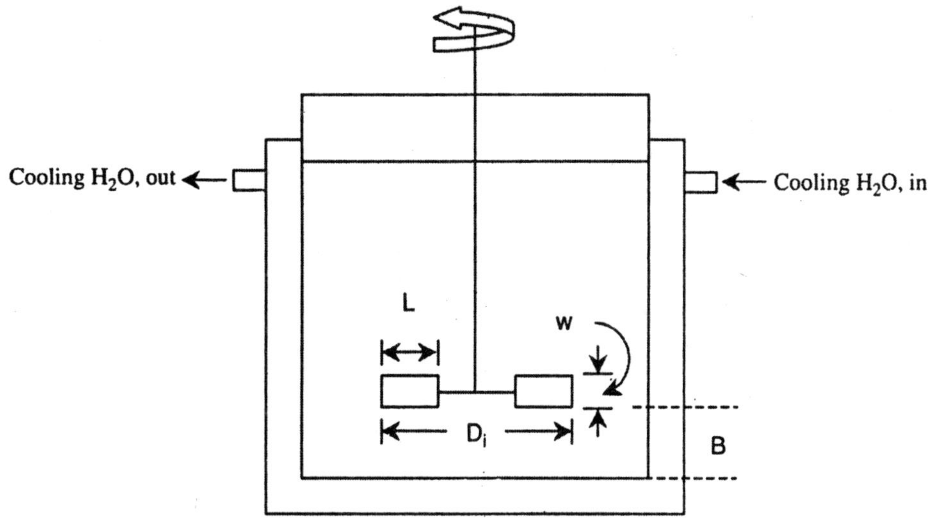

Fig. 7.4. Diagram of jacketed vessel with impeller location.

$$Nu = \frac{hDt}{k} = 0.36\ Re^{2/3}\ Pr^{1/3}$$

$$\frac{h_1}{h_2} = \left(\frac{D_1}{D_2}\right)^{1/2}\left(\frac{N_1}{N_2}\right)$$

The power for the fermentation vessel was projected by the following general equation

$$\text{Power(hp)} = 1.29 \times 10^{-4} D^{1.1}\ L^{2.72} N^{2.86}\ w^{0.3} H^{0.6}\ \mu^{0.14}\ \rho^{0.86}$$

$$Re = \frac{\rho N D_i^2}{\mu} = \frac{(62.4)(30{,}000)(0.75\ \text{ft})^2}{2.42} = 4.3 \times 10^3$$

The given Re, read $hD/k \times 2000$

$$\frac{D_i}{D_t} = \frac{1}{3}$$

$$h = \frac{2000 \times 0.356}{3 \times 0.75} = 600\ \text{Btu/hr} \cdot \text{ft}^2 \cdot {}^\circ\text{F}$$

Assuming the thickness of jacket = 1 in

$$h_{\text{cold water}} \approx 550\ \text{Btu/hr} \cdot \text{ft}^2 \cdot {}^\circ\text{F}$$

$$U_c = \frac{h_i\ h_o}{h_i + h_o} = \frac{550 \times 600}{1150} = 300\ \text{Btu/hr} \cdot \text{ft}^2 \cdot {}^\circ\text{F}$$

Dirt factor, $R_d = 0.005$

$$\frac{1}{U_D} = \frac{1}{U_e} + R_d = \frac{1}{300} + 0.005 + 120\ \text{Btu/hr} \cdot \text{ft}^2 \cdot {}^\circ\text{F}$$

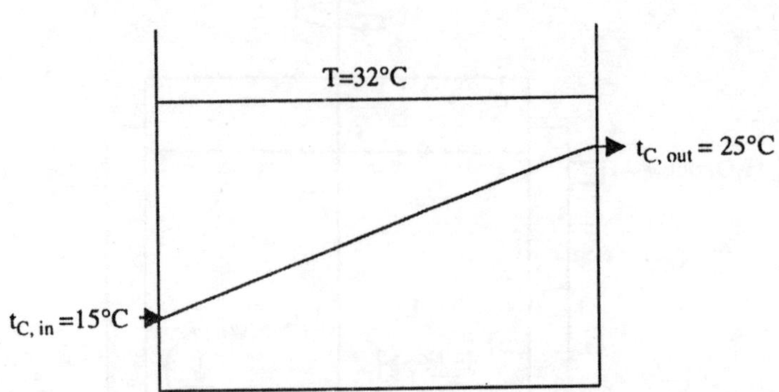

Fig. 7.5. Temperature pattern in a jacketed bioreactor for log mean temperature.

The inlet and outlet temperature of the jacketed bioreactor is shown in Fig. 7.5. The log mean temperature is :

$$\Delta T_{\ln} = \frac{(32-15)-(32-25)}{\ln\left(\frac{17}{7}\right)} = 11.27°\text{C}$$

Heat transfer based resistance theory for composite wall. The heat transfer overall coefficient is calculated.

$$\frac{1}{U} = \frac{1}{h_i} + \frac{\Delta x}{k} + \frac{1}{h_o}$$

Heat transfer area,

$$A = \pi[(D^2/4) + (DH_L)] = \pi[(2.25)^2/4 + 3(2.25)] = 24.25 \text{ ft}^2$$

$$Q = (24.45)(120)(1.8)(11.27) = 350{,}000 \text{ Btu} \cdot \text{h}^{-1}$$

$$Q = AU_D \Delta T_{\ln}$$

Mass flow rate of cooling water = 35000 $\text{lb}_m \cdot \text{h}^{-1}$

AEROBIC WASTEWETER TREATMENT

The well-known aerobic downflow process a trickled bed filter. Attached growth is used in the biological treatment of wastewater. Air passes through the bed while the liquid is forced to down by gravity. The liquid gas system for the mass transfer process in a biological filter is known as attached growth. Modes of bioreactor operation are :

- Batch
- Fed batch
- Continuous

$$\frac{dm}{dt} = m_i - m_o + r_p - r_s$$

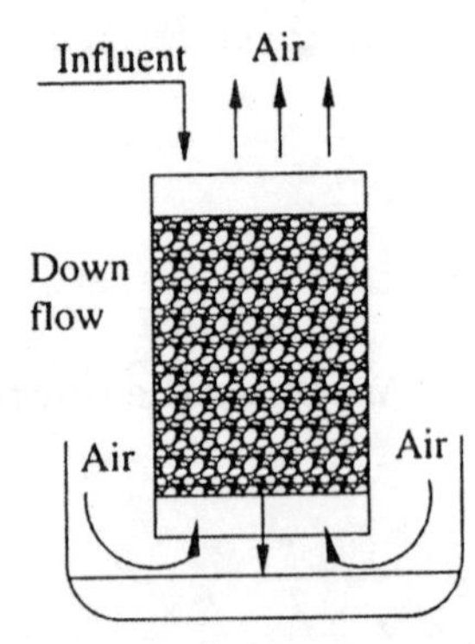

Fig. 7.7. Liquid gas mass transfer process in biological filter, attached growth system.

For batch operation

$$m_i - m_o = 0$$

$$-r_s = -r_A = -\frac{1}{V}\frac{d(SV)}{dt} = \frac{n_{\max} S}{K_m + S}$$

$$-r_A = -\frac{dC_A}{dt}$$

$$V = \text{constant}$$

$$\frac{dS}{dt} = -\frac{\nu_{\max} S}{K_{M^8} + S}$$

$$-\int_0^t dt = \int_{s_o}^{s}\left(\frac{K_M + S}{\nu_{\max} S}\right) dS$$

$$-t_{\text{batch}} = \frac{K_M}{n_{\max}} \ln\frac{S_f}{S_0} + \frac{1}{n_{\max}}(S_f - S_0)$$

where, S_0 and S_f are initial and final substrate concentration in mol · l^{-1}

Substrate Balance in a Continuous System

$$\frac{dS}{dt} = \frac{F}{V}(S_i - S) - \underbrace{\frac{\mu P_{\text{cell}}}{Y_{x/s}}}_{\text{Yield of cell}} - mP_{\text{cell}} - \underbrace{\frac{q_p P_{\text{cell}}}{Y_{p/s}}}_{\text{Nutrient maintenance}}$$

$$Glucose \xrightarrow[\text{ATP}+P_i \rightarrow \text{ADP}+7.8 \text{ kcal}]{} G6P$$

At steady-state condition

$$\frac{dS}{dt} = 0$$

$$m\ P_{\text{cell}} \le \frac{\mu\ P_{\text{cell}}}{Y_{s/x}}$$

$$D(S_i - S) = \frac{\mu\ P_{\text{cell}}}{Y_{x/s}} \text{ for SCP}$$

$$= \frac{\mu\ P_{\text{cell}}}{Y_{x/s}} + \frac{q_p\ P_{\text{cell}}}{Y_{p/s}} \text{ for ethanol}$$

when

$$\mu = D$$

$$\rho_{\text{cell}} = Y_{x/s}(S_i - S)$$

$$\frac{d_{\text{tank}}}{dt} = \frac{E}{RT^2}$$

$$k = ke^{-\frac{E}{RT}}$$

Example 1. *For production of an enzyme used for synthesis of a sun protection lotion, the required kinetic data and constants are:*

$$v_{max} = 2.3\ mmol.m^{-3}.s^{-1}$$
$$K_m = 8.9\ mmol.l^{-1}$$
$$So = 12\ mmol.l^{-1}$$

Find the reaction time for 95% conversion of raw materials.

$$t_{batch} = ?\ for\ x_A = 0.95$$

Solution :

$$\upsilon_{\max} = \left(\frac{2.5\,\text{mmol}}{\text{m}^3\cdot\text{s}}\right)\left(\frac{3600\,\text{s}}{\text{h}}\right)\left(\frac{1\,\text{m}^3}{1000\,\text{L}}\right) = \frac{9\,\text{mmol}}{\text{L}\cdot\text{h}}$$

$$t_{\text{batch}} = \frac{K_m}{\upsilon_{\max}}\ln\frac{S_o}{S_f} + \frac{S_o - S_f}{\upsilon_{\max}}$$

$$= \frac{8.9}{9}\ln\frac{12}{(1\text{–}0.95)(12)} + \frac{0.95\times 12}{9}$$

$$= 4.2 \text{ hours}$$

Material Balance in Fed Batch

In fact, fed batch is a batch system operating without any outlet stream. The differences in substrate in and out are equal to the rate of product generation.

$$\frac{F}{V}(S_i - S) = \left(\frac{\upsilon_{\max}\ S}{K_m + S}\right)\frac{V}{V}$$

The retention time:

$$\tau = \frac{1}{D} = \frac{V}{F}$$

$$D(Si - S) = \frac{\upsilon_{\max}\ S}{K_m + S}$$

$$\underbrace{\frac{d\rho_{\text{cell}}}{dt}}_{\text{stead state=0}} = \frac{F}{V}\left(\rho_1 - \underbrace{\rho_0}_{\text{sterile system}}\right) + \left(\underbrace{\mu}_{\text{growth}} - \underbrace{\alpha}_{\text{death}}\right)\rho_{\text{cells}}$$

The simplified material balance would lead to growth rate

$$\mu = \frac{F}{V}$$

Example 1. *A 10 m^3 bioreactor with $H/d = 2.5D$, 75% working volume, with two sets of standard flat-blade impellers was used for a baker's yeast production unit. The operating dilution rate was set at 0.5 h^{-1}. The rate equation satisfying the reactor design was a Monod model with μ_{max}= 0.65 h^{-1} and $k_s = 3\ kg \cdot m^{-3}$. The initial substrate concentration was 65 g $\cdot I^{-1}$ with Ivvm aeration, broth density 1200 kg.,$^{-m}$ and viscosity $\mu = 0.02N \cdot s \cdot m^{-2}$. The yield of biomass* *on glucose was 0.5 g cell · g*$^{-1}$ *glucose. Calculate power consumption for a 90 rpm agitation rate and OTR of 0.05 g* · *l*$^{-1}$. *State the controlling resistance in the mass transfer process.*

Solution :

$$10 = \frac{\pi}{4}(2.5D)D^2$$

$$D_t = 1.7 \text{ m}$$

$$D_i = 0.57 \text{ m}$$

$$Re = \frac{(1.5\,\text{rps})(0.57)^2(1200)}{0.02\frac{\text{N}\cdot\text{s}}{\text{m}^2}} = 2.9 \times 10^4$$

The flow regime is turbulent flow.

$$N_p = 6$$

$$N_a = \frac{F_g}{\mu D_i^3} = \frac{7.5\,\text{m}^3/\text{min}}{(90\text{ rpm})(0.57)^3} = 0.45$$

where N_a, is known as the aeration number,

$$\frac{P_g}{\rho} = 0.95$$

$$N_p = \frac{Pg_c}{\rho n^3 D_i^3}$$

$$P = \frac{6 \times 1200(1.5)^3(0.57)^5}{9.81} = 149\frac{\text{kg}\cdot\text{m}}{s}$$

$$= \frac{149}{745.7} = 0.2 \text{ hp}$$

Conversion factor:

$$1 \text{ hp} = 745.7 \text{ kg} \cdot \text{m} \cdot \text{s}^{-1}$$

$$P = 2 \times 0.2 = 0.4 \text{ hp}$$

$$P_g = 0.95(0.4) = 0.38 \text{ hp}$$

$$\frac{P_g}{V_L} = 0.051 \text{ w} \cdot \text{l}^{-1}$$

Oxygen Transfer Rate

$$OTR = K_L a(C - C_L) = 7.9 \times 10^{-4}(0.03)$$

$$= 4 \times 10^{-5} \text{ kg} \cdot \text{m}^{-3} \cdot \text{s}^{-1}$$

$$K_L = 2 \times 10^{-3}\, V_s^{0.7}\left(\frac{P_g}{V_L}\right)^{0.2}$$

$$V_s = \frac{7.5\text{m}^3/\text{min}}{(\pi/4)(1.72)^2} = 3.23 \text{ m/min} = 194 \text{ m.h}^{-1}$$

$$K_L = 2 \times 10^{-3} (323 \text{ cm/min})^{0.7} (0.051)^{0.2} = 7.9 \times 10^{-4} \text{ S}^{-1}$$

Liquid film was the controlling resistance.

NOMENCLATURE

N	Impeller speed, rpm
D_i	Impeller diameter, m
Q_{O2}	Oxygen transfer rate, mmol · min^{-1}
X	Biomass concentration, g · l^{-1}
P_g/V	Power per unit volume, hp · m^{-3}

CHAPTER

8

Electro-Magnetic Spectroscopy

Spectroscopy is the study of the absorption and emission of radiation by matter. The most easily appreciated aspect of the absorption of radiation is the colour shown by substances that absorb radiation from the visible region of the spectrum. If radiation is absorbed from the red region of the spectrum, the transmitted or unabsorbed radiation will be from the blue region and the substance will show a blue colour. Similarly substances that emit radiation show a particular colour if the radiation is in the visible region of the spectrum. Sodium lamps, for instance, owe their characteristic orange-yellow light to the specific emission of sodium atoms at a wavelength of 589 nm. Measurement of the intensity and wavelength of radiation that is either absorbed or emitted provide the basis for sensitive methods of detection and quantitation. Absorption spectroscopy is most frequently used in the quantitation of molecules but is also an important technique in the quantitation of some atoms. Emission spectroscopy covers several techniques that involve the emission of radiation by either atoms or molecules but vary in the manner in which the emission is induced. Photometry is the measurement of the intensity of radiation and is probably the most commonly used technique in biochemistry. In order to use photometric instruments correctly and to be able to develop and modify spectroscopic techniques it is necessary to understand the principles of the interaction of radiation with matter.

INTERACTION OF RADIATION WITH MATTER

Radiation is a form of energy and shows both electrical and magnetic characteristics, hence the term electromagnetic radiation (Fig. 8.1). It can be considered as being composed of a stream of separate groups of electromagnetic waves and the energy associated with the radiation can be mathematically related to the waveform.

Waveform can be defined in at least two different ways which are relevant to spectroscopic measurements. Wavelength (λ) is defined as the distance between successive peaks (Fig. 8.1) and is measured in subunits of a metre, of which the most frequently used is the nanometre (10^{-9} m). An angstrom unit (Å) is not acceptable in SI terminology but is still occasionally encountered and is 10^{-10} m (*i.e.*, 10 Å = 1 nm). The frequency of radiation (nu, υ) is defined as the number of successive peaks passing a given point in 1 second. Hence the relationship between these two units of measurement is :

$$\nu \propto \frac{1}{\lambda}$$

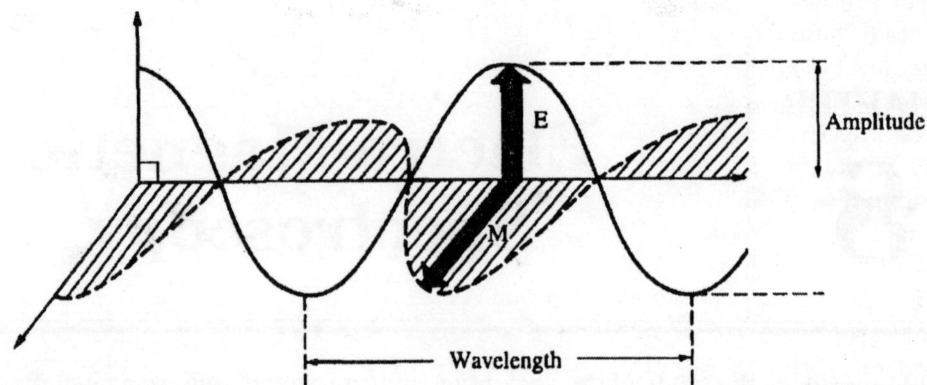

Fig. 8.1. Electromagnetic radiation. A representation of electromagnetic radiation with the electric field (*E*) and the magnetic field (*M*) at right angles to the direction of the wave movement. Both fields oscillate at the same frequency.

The visible region of the spectrum extends approximately over the wavelength range 400-700 nm, the shorter wavelengths being the blue end of the spectrum and the longer wavelengths the red end. Wavelengths between 400 and 200 nm make up the near ultraviolet region of the spectrum and wavelengths above 700 nm to approximately 2000 nm (2 μm) the near infrared region.

The energy associated with a particular waveform is directly related to the frequency of the radiation and therefore inversely related to the wavelength. It may be calculated using the equation:

$$E = h\nu = \frac{hc}{\lambda}$$

where h = Planck's constant (6.626×10^{-34} Js))

ν = frequency of the radiation

λ = wavelength of the radiation

c = speed of light ($3.0 \times 10^{8} \text{ms}^{-1}$)

While the energy associated with a waveform is related to its wavelength, the intensity of that radiation is related to its amplitude. The absorption or emission of radiation by matter involves the exchange of energy and in order to understand the principles of this exchange it is necessary to appreciate the distribution of energy within an atom or molecule. The internal energy of a molecule is due to at least three contributing sources:

1. The energy associated with the electrons.
2. The energy associated with the vibrations between the atoms.
3. The energy associated with the rotation of various groups of atoms within the molecule relative to the other groups.

These energy levels may be altered by the absorption or emission of energy as radiation and, because any given atom or molecule can only exist in a limited number of energy levels, these energy changes must be in definite packets or quanta.

The exact amount of energy required to produce a change in the molecule from one energy level to another will be given by the photons of one particular frequency which will be selectively absorbed or emitted. A study of the wavelength or frequency of radiation absorbed or emitted by an atom or a molecule will give information about its identity and this technique is known as *qualitative spectroscopy*. This information is usually reported as the wavelength of radiation involved and is most easily represent-ed as an absorption or emission spectrum (Fig 8.5). Measurement of the total amount of radiation will give information about the number of absorbing or emitting atoms or molecules and is called *quantitative spectroscopy.*

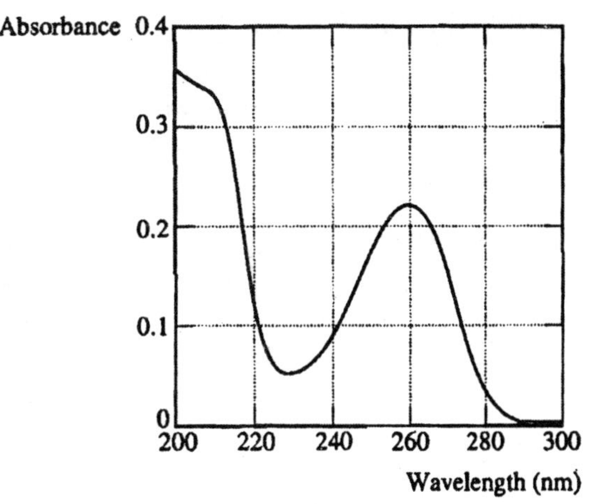

Fig 8.2. Absorption spectrum of adenosine diphosphate (ADP).

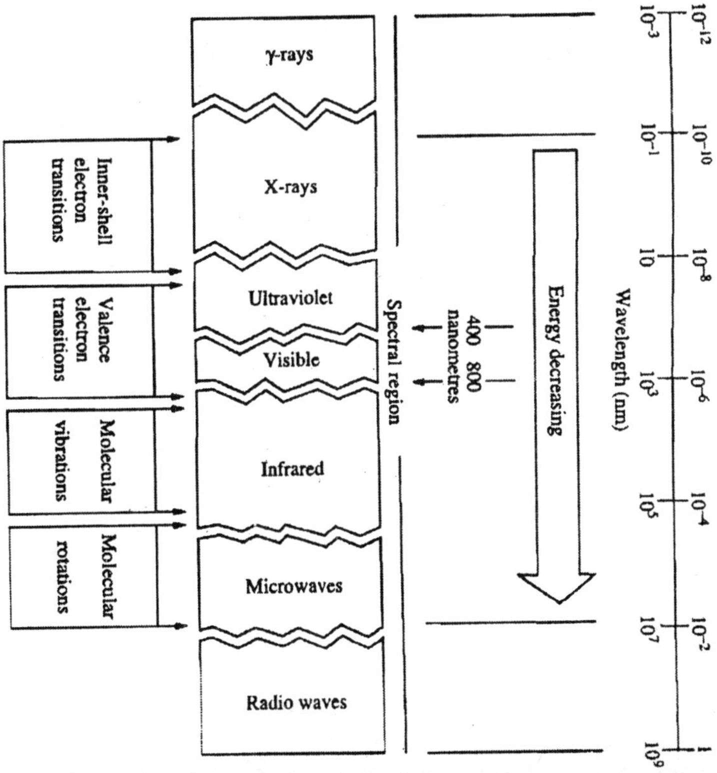

Fig 8.3. Range of electromagnetic radiation. All electromagnetic radiation travels at a constant speed of 3×10^8 but the energy associated with each waveform is inversely proportional to the wavelength. The energy required for different atomic and molecular transitions is provided by radiation of different wavelengths.

Absorption of Radiation

The increase in energy in a molecule that occurs when radiation is absorbed can be accommodated in the three ways already described. The range of energies involved is characteristic of each type of change and is associated with a well-defined region of the electromagnetic spectrum.

Atomic Absorption

Individual atoms cannot rotate or vibrate in the same manner as molecules and as a result the absorption of energy is only associated with electronic transitions that are limited in number and associated with very narrow ranges of radiation or absorption lines. The wavelength that is most strongly absorbed usually corresponds to an electronic transition from the ground state to the lowest excited state and is known as the resonance line. The limited number of electronic transitions for any atom results in the fact that excited atoms, when returning to the ground state, emit radiation of the same wavelength as that which was absorbed.

Molecular Absorption–Ultraviolet and Visible Region

This is the area of greatest interest to quantitative biochemistry and is dependent upon the electronic structure of carbon compounds. Absorption of radia-tion in this region of the spectrum causes transitions of electrons from molecular bonding orbitals to the higher energy molecular antibonding orbitals.

The electronic structure of the carbon atom is designated as $1s^2,2s^2,2p^1{}_x,2p_y{}^1$ and this means that in order to fill all the available orbitals four further electrons would be required,

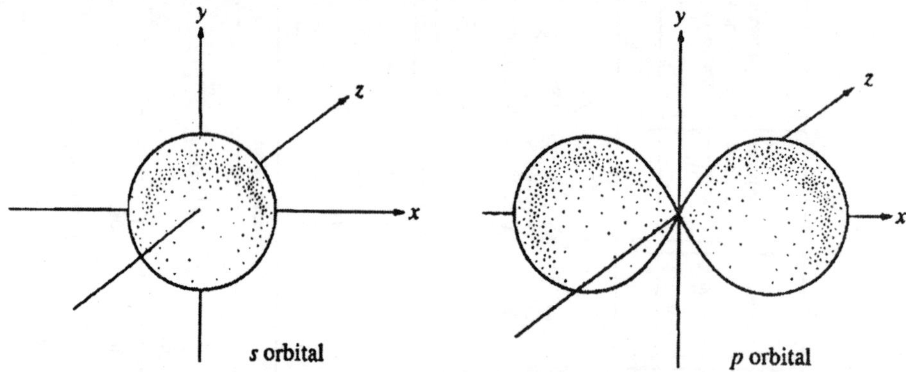

Fig. 8.4. Atomic orbitals. An *s* orbital is spherical and can only exist in one orientation but the *p* orbital is directional and can exist in either of three planes (*x*, *y* and *z*), all of equal energy status. All orbitals can accommodate two electrons provided that they are of opposite spin. The number of electrons in a particular orbital is indicated as a superscript to the orbital letter, *e.g.*, $1s^2$.

one in each of the $2p_x$ and $2p_y$ orbitals with two in the unoccupied $2p_z$ orbital. These four vacancies for electrons are not all equivalent in energy or in direction and yet the four valencies of the carbon atom as evidenced in organic compounds are equivalent and symmetrical. This symmetry is due to the combination of orbitals with the same basic energy level, *i.e.*, 2*s* and 2*p*, to give hybrid *sp* orbitals.

Three different combinations are possible depending upon the number of orbitals involved in the hybridization. The four existing outer electrons of the carbon atom (two in the 2*s* orbital and one in each of the $2p_x$ and $2p_y$ orbitals) are initially dispersed, one in each of the available orbitals (2*s*, $2p_x$, $2p_y$ and $2p_z$) and subsequent hybridization of these four orbitals results in the hybrid sp^3 orbital (*i.e.*, one *s* and three *p*) which has four symmetrical lobes.

Combinations that include only two of the *p* orbitals with the *s* orbital result in a hybrid sp^2 orbital, which has three symmetrical lobes and a residual 2*p* orbital while a combination of one *s* and one *p* orbital results in an *sp* hybrid. Atoms are linked to form a molecule by covalent bonds, which are a result of the sharing of electrons from two orbitals, one from each atom, and so effectively filling both orbitals with the maximum number of two electrons. The formation of a covalent bond between two carbon atoms which are in the sp^3 hybrid form results in a symmetrical (sigma, σ) bond, which is also known as a single or saturated bond. However, if the two carbon atoms are in the sp^2 hybrid form then not only is a bond similar to the sigma bond formed between the two lobes of the sp^2 orbitals but the residual *p* orbitals combine to produce what is known as a pi (π) bond. This bond is not symmetrical about the main sigma bond but is formed by a side-to-side overlap of the two *p* orbitals resulting in the pi bond electrons being displaced above and below the plane of the two carbon atoms.

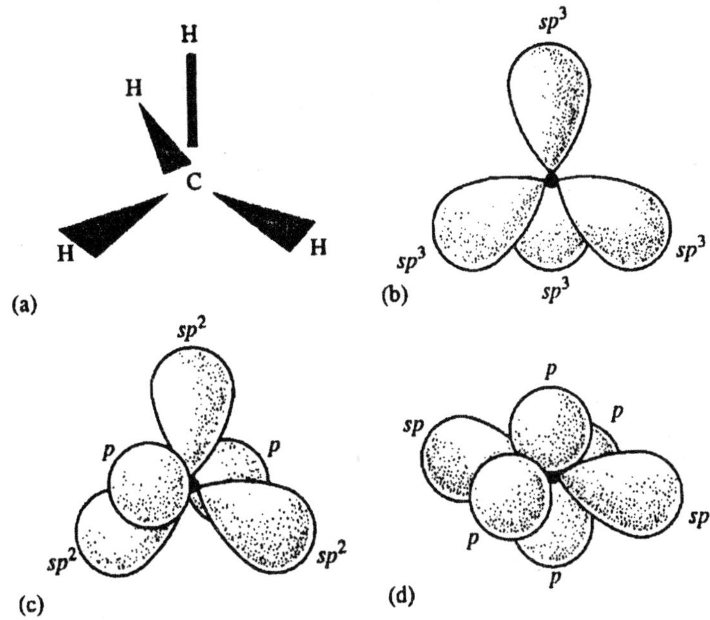

Fig. 8.6. Bonding in carbon compounds.

These are known as *delocalized electrons* and the overall bond is known as a double or unsaturated bond. If an alternating system of sigma and pi bonds is formed (a conjugated system), the delocalized electrons combine to form two molecular orbitals, one above and one below the entire length of the bond sequence (or ring in the case of cyclic compounds). As indicated earlier, all electrons are capable of existing in one of several energy levels and the change from one level to another involves the exchange of energy.

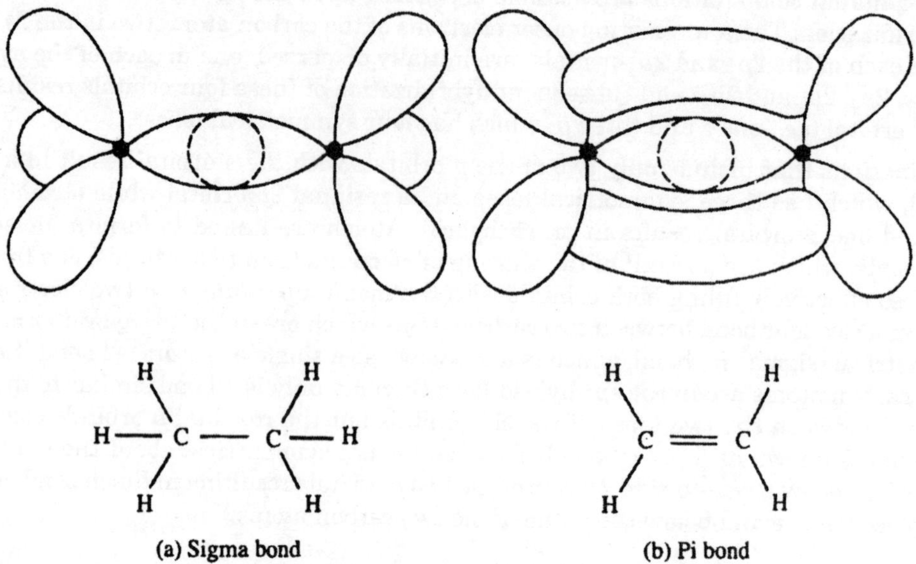

Fig. 8.6. Bonding in carbon compounds.

The ease with which an electron can be raised to its next higher energy level is related to the amount of energy required for such a change and, if the energy is provided by radiation, is related to the wavelength of the radiation. Electronic changes can be listed in order of the increasing ease with which the transition may take place, and this parallels the increasing wavelength causing the transition.

Table 8.1. Absorption Maxima of Hydrocarbons

Compound	*Structure*	*Type of transition*	*Absorption maximum (nm)*
Ethane	C—C	$\sigma \rightarrow \sigma^*$	< 180
Ethylene	C=C	$\pi \rightarrow \pi^*$	190
Benzene		Cyclic $\pi \rightarrow \pi^*$	256
Naphthalene		Cyclic $\pi \rightarrow \pi^*$	290
Anthracene		Cyclic $\pi \rightarrow \pi^*$	360

Molecules that show increasing degrees of conjugation require less energy for excitation and as a result absorb radiation of longer wavelengths. In addition, the introduction of saturated groups (methyl, hydroxyl, etc.) into a compound that already absorbs radiation may result in an increase in the wavelength of radiation absorbed. The effect of conjugation is most dramatically demonstrated by certain dyes which in the reduced forms are colourless but in the oxidized forms show a characteristic colour.

The absorption characteristics of the reduced form of methyl red are due to two separate conjugated systems which absorb radiation in the ultraviolet region of the spectrum and hence are not visible. The effect of oxidation is to join the two systems in a way that increases the extent of conjugation and shifts the wavelength of the absorbed radiation into the visible region of the spectrum, giving a colour to the compound.

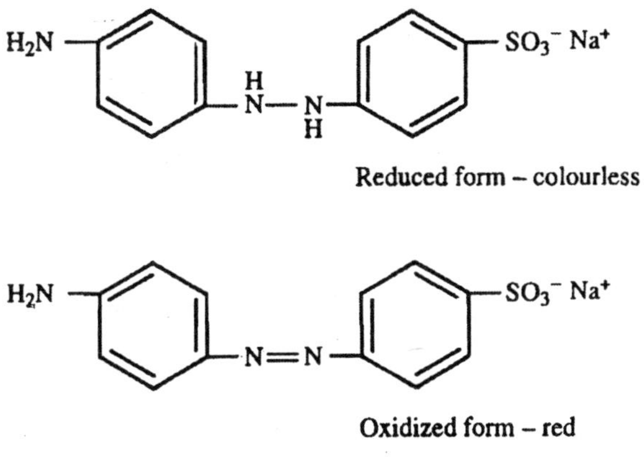

Reduced form – colourless

Oxidized form – red

Fig. 8.7. The effects of increased conjugation in the dye methyl red.

The absorption peaks found in ultraviolet and visible spectroscopy are much broader than those found in infrared spectroscopy. This is due to the additional effects of vibrational and rotational transitions being superimposed on the basic electronic transition. Molecules existing in an electronic ground state may be at different vibrational and rotational energy levels and as a result will require different total amounts of energy to undergo an electronic transition. Similarly the final excited molecule may exist at various vibrational and rotational energy levels.

The overall effect is that a range of wavelengths is absorbed and the spectrum shows the characteristic broad peak which is useful in quantitative studies but not so useful in qualitative work.

Not all the molecules in a sample show the same energy change (E) for a particular electronic transition because although they all exist in the same electronic energy level, they may have different rotational and vibrational energies. They will therefore require different amounts of energy (E_{alt}) in order to be raised to the next electronic energy level. Each different amount of energy will be provided by a different wavelength of radiation, resulting in the characteristic absorption peak.

Molecular absorption–infrared region

Not all organic compounds absorb radiation in the ultraviolet and visible regions of the spectrum but they do show the absorption of infrared radiation due to vibrational changes. Vibrational transitions within a molecule may be achieved with energy levels associated with the near infrared region of the spectrum while rotational changes in molecular energy are associated with the far infrared and microwave regions. The wavelength range of infrared radiation lies between 700 nm and 1×10^6 nm (1 mm) but because of the large numerical values involved in quoting wavelength, absorption band positions are usually quoted as the wavenumber, which is the reciprocal of the wavelength in centimetres:

$$\text{Wavenumber (cm}^{-1}) = \frac{1}{\text{wavelength (cm)}} = \frac{1 \times 10^7}{\text{wavelength (nm)}}$$

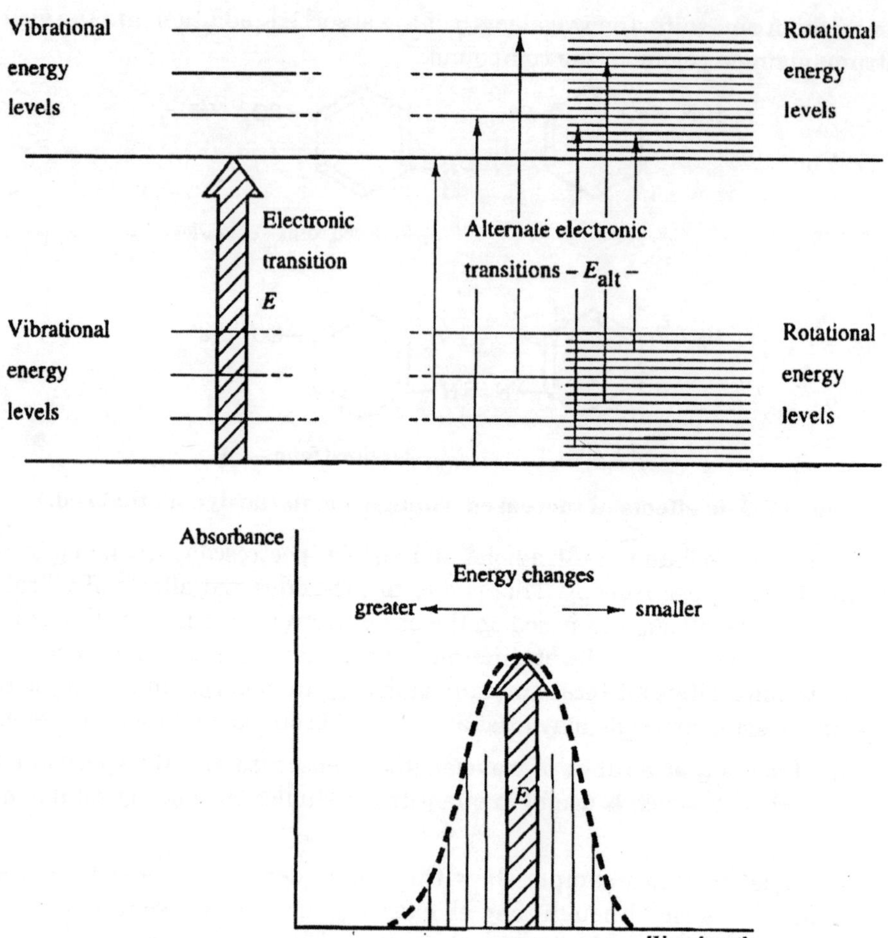

Fig. 8.8. Range of energy levels associated with electronic transitions.

The two main types of molecular vibrations associated with the absorption of infrared radiation involve the stretching and bending of bonds. Stretching is a vibrational movement in which the bond length alters and the two nuclei move harmonically relative to each other. For a compound involving two atoms there is only one bond that is capable of being stretched and in general terms it can be said that for a compound containing n atoms there are $(n-1)$ possible stretching effects. Bending vibrations present many more possible variations, the total number being defined as $(2n - 4)$ for linear molecules and $(2n - 5)$ for non-linear molecules. Although a large number or vibrational changes are possible for any given molecule, not all of them will result in the absorption of radiation.

An absorption maximum in the infrared spectrum of a compound can be demonstrated only when a vibration results in a change in the dipole of the molecule, a process for which energy is required. Carbon dioxide, for example, has a lin-ear structure and can undergo two stretching vibrations $(n-l)$ and two bend-ing vibrations $(2n- 4)$. An examination of the four vibrational changes possible will show that one stretching effect does not result in a change in the relative distribution of the charge associated with the molecule (dipole) and as a result will not show an absorption maximum. The other three vibrational changes do result in a change in the overall distribution of the charge and give absorption maxima in the infrared. However, the two bending vibrations are identical in energy level, although in a different plane, and therefore show the same absorption maximum.

Table 8.2. Fundamental Vibrations of Carbon Dioxide

Structure	*Vibrations*	*Absorption maximum (cm^{-1})*
O—C—O	None	Normal structure
O→C←O	Stretching	None
O→←C—O	Stretching	2349
C / \ O O	Bending (horizontal plane)	667
O←C→O	Bending (vertical plane)	

Infrared spectra show not only absorption bands due to fundamental vibrations but also additional and usually weaker bands due to multiples of the fundamental frequencies (overtones) and to the combination of fundamental vibrational effects and their multiples (combination bands). The interpretation of such spectra is further complicated by the fact that the spectrum for a given substance may vary depending upon the solvent used or the method of sample preparation. Infrared absorption involves a large number of quite different transitions, both vibrational and rotational, which result in a spectrum which is composed of a large number of separate, narrow peaks and is more useful in qualitative than in quantitative analysis.

Emission of Radiation

An atom or molecule is said to be in an excited or unstable state when it absorbs energy which results in either electronic or vibrational transitions. It will return to its ground state very rapidly and the energy may be lost in one of three ways:

1. As a result of a chemical reaction.
2. By dissipation as heat.
3. By emission as radiation.

If the atom or molecule loses all or part of this energy as radiation, photons of energy will be emitted which correspond to the difference between the energy levels involved. Since these levels are clearly defined for any given atom or molecule, the radiation emitted will be of specific frequencies and will show up as bright lines if the emitted light is dispersed as a spectrum.

Atomic emission is displayed by many elements, particularly the metals, which after being excited either thermally or electrically emit a discontinuous spectrum, the strongest lines of which are due to transitions ending in the ground state (resonance lines). Molecular emissions are more complex than atomic emissions, the radiation emitted consisting of broad bands of radiation rather than the narrow lines associated with atomic emission and resulting in a spectrum which is an approximate mirror image of the absorption spectrum of the compound. Molecular emissions are due to electronic transitions within the molecule but are modified by

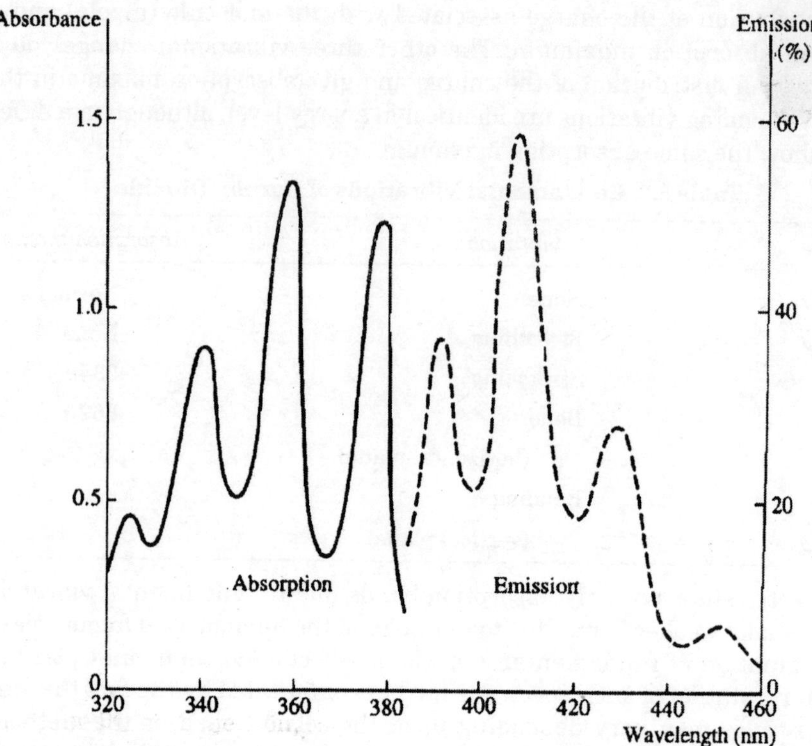

Fig. 8.9. Absorption and emission spectra of anthracene in solution in dioxane.

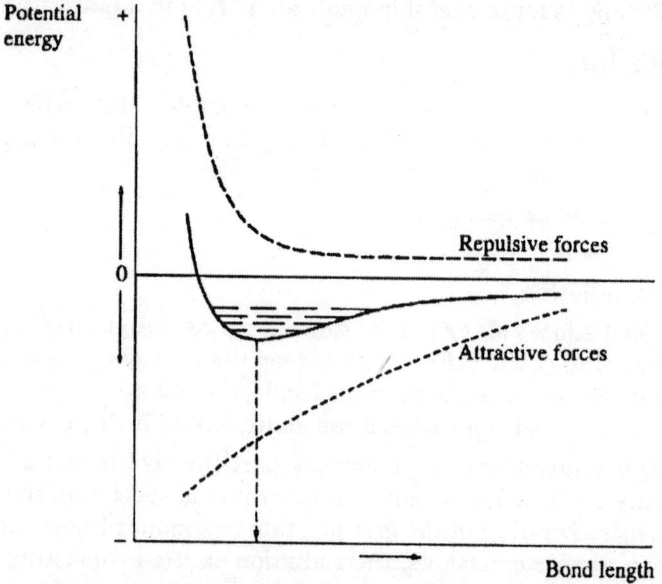

Fig. 8.10. Morse curve. Inter-atomic attractive and repulsive forces result in the formation of a bond length with a minimum energy level.

variations in bond length. The bond between two atoms assumes a particular length as a result of the various forces acting upon the atoms involved. The attractive forces between, the electrons of one atom and the nucleus of the other atom are balanced by the repulsive forces of the like-charges carried by both nuclei. The Morse curve describes the relationship between these forces and illustrates the fact that at a particular bond length the energy associated with the bond is minimal and the molecule is said to be in its ground state. However, this only describes the most stable and hence the most frequent form of the molecule.

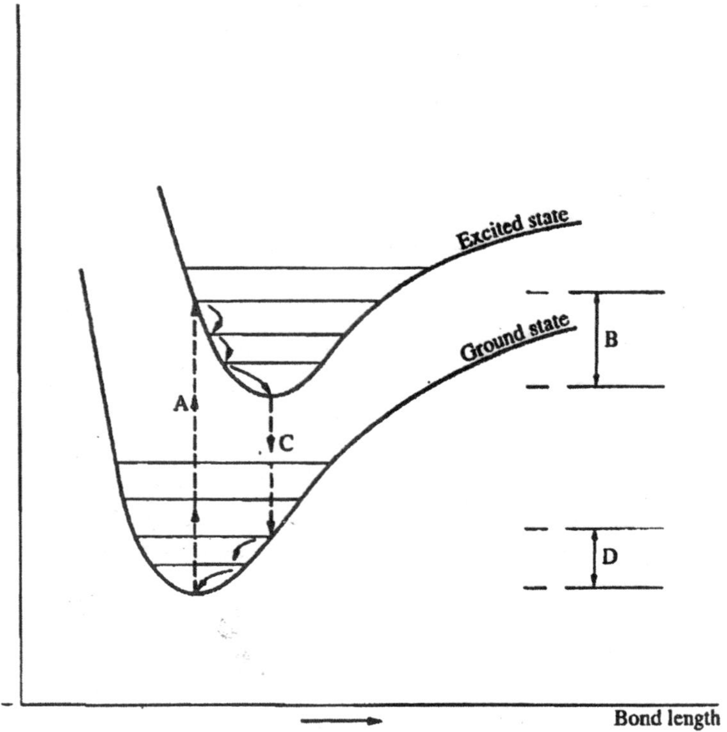

Fig. 8.11. Morse curve for an excited molecule. The energy required for excitation (A) is lost as the molecule returns to the ground state but only the energy lost between states (C) may be emitted as radiation. Energy losses due to internal rearrangements (B and D) are non-radiative.

More energetic forms will also exist and these are indicated by the higher energy levels in the Morse curve diagram. A new Morse curve is required to describe the energy levels associated with an excited molecule and this is displaced to the right and above the original curve. The excitation of a molecule with a particular bond length results in an excited molecule with the same bond length but with a higher internal energy as illustrated by the new Morse curve. An excited molecule initially will revert to the minimum energy state for the excited molecule before reverting to the ground state, with the result that the energy originally absorbed is lost in at least three stages (B, C + D) as the molecule returns to its ground state.

Some energy losses are due to internal rearrangements of each of the two molecular forms and as such involve a non-radiant loss of energy: it is only the energy lost during the interchange between the two molecular forms that may possibly be emitted as radiation. Any energy lost by radiation (C) must be less than the total energy absorbed (A) and hence the wavelength emitted will be greater than that which was absorbed in the first place. A molecule in the ground state has a pair of electrons with opposite spin in each molecular orbital.

On excitation, one electron is elevated to an anti-bonding orbital and because it is not restricted by the presence of another electron it can exist with either its original spin (a singlet transition) or reversed spin (a triplet transition). If the return to the ground state involves a fall from the singlet transition and radiation is emitted, the compound is said to fluoresce. However, in a return to the ground state from a triplet transition, fluorescence will not occur and the energy loss will probably be non-radiative. The intensity of fluorescence depends upon the proportion of the total molecules which undergo singlet transitions and it is mainly for this reason that different compounds show different degrees of fluorescence.

Some compounds do show radioactive emission during triplet transitions and these compounds are said to phosphoresce. The most demonstrable difference between these two types of emission is the fact that fluorescence occurs within about seconds and persists for only about 1 to nanoseconds and is much faster than phosphorescence, which persists for up to seconds due to the time taken for the spin change to occur. Chemiluminescence is another form of molecular emission in which the initial electronic transition is caused by an exergonic reaction rather than the absorption of radiant energy.

Most chemiluminescence reactions are of the oxidative type and those involving hydrogen peroxide are particularly useful biochem-cally. Luminol (5-amino-2, 3-dihydrophthalazine-1, 4-dione), for instance, will emit light when reacting with hydrogen peroxide and may be used in monitoring many oxidative enzymes such as glucose oxidase, amino acid oxidase, etc. Bioluminescence is a special type of chemiluminescence in which the process of light emission is catalysed by an enzyme. The enzyme luciferase, extracted from the firefly, uses ATP to oxidize the substrate luciferin with the emission of radiation and can be used to measure ATP concentrations. The enzyme extracted from bacterial sources can oxidize long-chain aliphatic aldehydes in the presence of oxygen, FMN and NADH with the emission of radiation. Luminescence methods are very sensitive, quantities as little as 1 femtomole of ATP being detectable and while measurements may be made using scintillation counters, much simpler equipment that requires neither a radiation source nor a monochromating system is satisfactory.

MOLECULAR ABSORPTIOMETRY

Photometric measurements provide the basis for the majority of quantitative methods in biochemistry and are related to the amount of radiation absorbed rather than the nature of such radiation. This relationship is expressed in two experimental laws, which provide the mathematical basis for such quantitative methods.

Beer-Lambert Relationship

Lambert's law states that the proportion of radiant energy absorbed by a substance is independent of the intensity of the incident radiation. Beer's law states that the absorption of radiant energy is proportional to the total number of molecules in the light path. Beer's law describes the basic relationship between the concentration of the absorbing substance and the measured value of absorbed radiation.

Lambert's law is of major significance in the manner in which measurements are made and the fact that a given sample always absorbs the same proportion of the incident radiation regardless of its intensity greatly simplifies the design of instruments. The amount of radiation absorbed by a substance cannot be measured directly and it is usually determined by measuring the difference in intensity between the radiation falling on the sample (incident radiation, I_0) and the residual radiation which finally emerges from the sample (transmitted radiation, I).

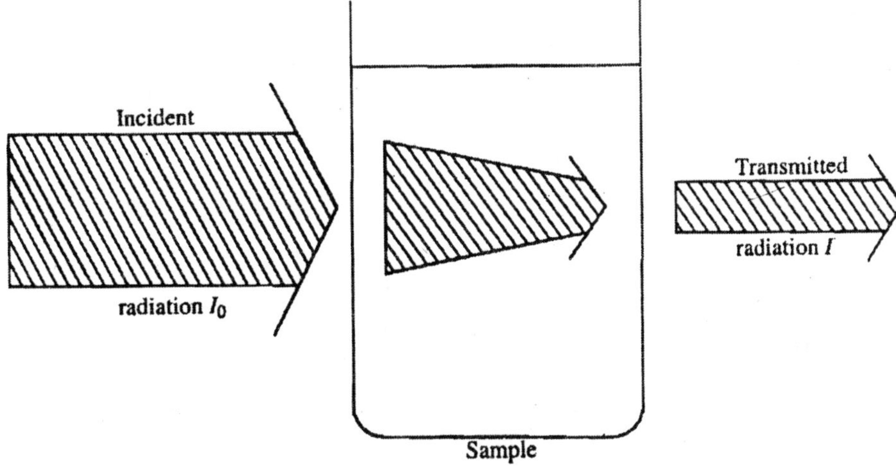

Fig. 8.12. Absorption of radiation.

Using such measurements, the Beer-Lambert law can be expressed as an equation:

$$\log_{10} \frac{I_0}{I} = \epsilon c l$$

where c is the concentration of the substance in gram, molecules per litre, l is the light path in centimetres and ϵ (epsilon) is known as the molar absorption coefficient for the substance and is expressed in litres per mole per centimetre (1 mol^{-1} cm^{-1}).

The values for I and I_0 cannot be measured in absolute terms and the measurements are most conveniently made by expressing I as a percentage of I_0. This value is known as the percentage transmittance (T) and only shows a linear relationship with the concentration of the test substance if the logarithm of its reciprocal is used. It is therefore more convenient to report the measurements initially as this logarithmic function of I and I_0, a parameter which is known as absorbance *(A)*:

Percentage transmittance $$(T) = \frac{I}{I_0} \times 100$$

$$A = \log_{10} \frac{100}{T}$$

In many instruments the meter read-out is calibrated in absorbance units using a logarithmic scale while other instruments retain the convenience of a linear scale but convert the signal from the detector to a logarithmic one by electronic or mechanical means. It is essential when using a photometric instrument to know if it is calibrated in absorbance or transmittance units.

Deviations from the Beer-Lambert Law

The assumption that the difference between the incident and the transmitted radiation is a measure of the radiation absorbed by the analyte is not completely true because this may not be the only reason why the incident radiation does not appear in the transmitted form. A certain amount of radiation will be reflected from the surface of the sample holder, usually a glass or plastic cell, or absorbed by the material of which the cell is composed.

The sample may also be dissolved in a solvent which itself may also absorb or reflect radiation:

incident (I_0) = absorbed + transmitted (I) + other losses

For this reason the radiation transmitted by a blank sample is measured and taken to be the effective incident radiation. This blank should be identical to the test sample in all aspects except the presence of the test substance:

blank reading = I_0 – other losses

Hence:

absorbed = blank–transmitted (I)

An important factor which influences the Beer-Lambert relationship is the wavelength range of the incident radiation. Ideally the radiation should only provide a specific unit of energy and not a range of energy levels. Such specific radiation is known as monochromatic radiation and can only be produced under very restricted conditions.

The best that can often be achieved is radiation with a very limited range of wavelengths. If the incident radiation contains wavelengths that will not be absorbed by the test substance, the difference between the intensities of incident and transmitted radiation will not be proportional to the concentration of the test substance. In such cases a non-linear relationship will exist as illustrated in Fig. 8.13. This problem is particularly relevant to instruments that use simple glass filters rather than monochromating systems such as prisms or diffraction gratings.

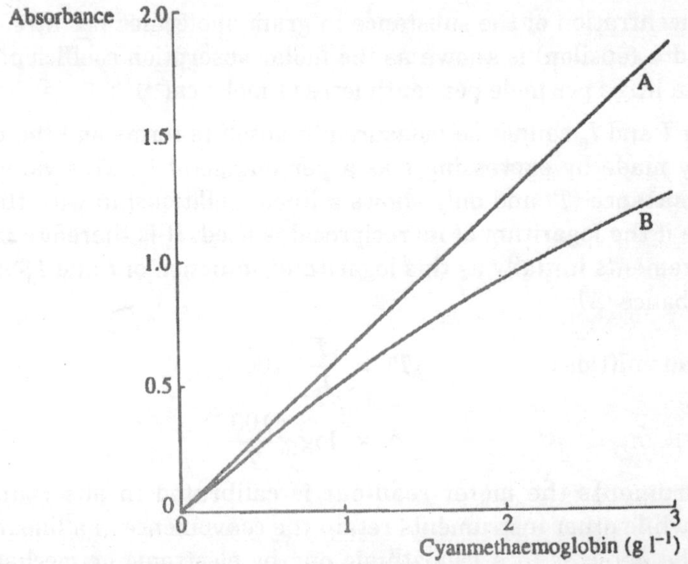

Fig. 8.13. Validity of the Beer-Lambert relationship for different monochromating Systems. The absorbance of varying concentrations of cyanmethaemoglobin was measured at 540 nm using a spectrophotometer (A) and a simple photometer (B) with a glass filter.

Alterations that may occur in the molecular nature of the sample due to changes in concentration may also result in deviation from the Beer-Lambert relationship. Molecules may tend to associate with one another when the concentration is high or, conversely, complexes may tend to dissociate in low concentrations. Both types of change may possibly affect the absorption characteristics the compound and result in non-linear graphs. Quantitative measurements rely on the assumption that the only radiation to reach the detector has passed through the sample but in practice it is very difficult to design instruments that are capable of effectively eliminating all extraneous radiation.

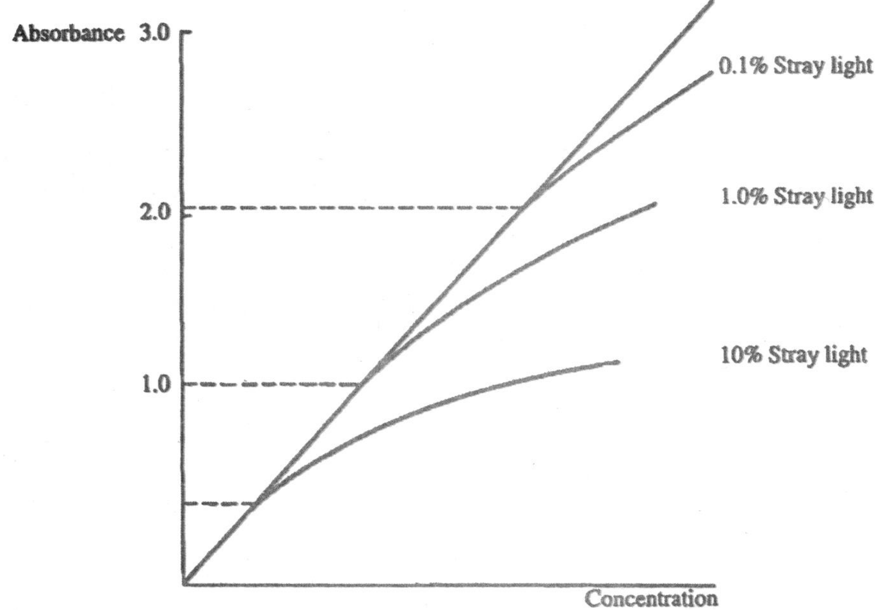

Fig. 8.14. Effect of stray light. In the absence of stray light samples of increasing concentration result in a linear plot against the measured absorbance value. Increasing proportions of stray light result in deviations from this linear relationship above the absorbance values indicated (——).

Much of this unwanted radiation arises from the scattering of the incident radiation by irregularities in surfaces caused by faults in manufacture or scratches, etc. Certain general precautions are usually taken to reduce this extraneous light in photometric instruments, such as the use of matt black interior surfaces to give maximal absorption and the insertion of baffles and windows to compartmentalize the instrument. Optical light paths should be designed to produce parallel beams of light and sharply focused spectra by the use of collimating lenses and mirrors. Imperfections in monochromators result in the presence of a small proportion of unwanted wavelengths in the incident radiation. Such **stray light** results in a deviation from a Beer-Lambert relationship and the effect is that absorbance measurements are lower than they should be. It is possible to assess the proportion or stray light by measuring the amount of radiation transmitted by samples that are optically opaque at the wavelength to be assessed but that transmit radiation of other wavelengths.

The instrument is set to zero and 100% transmittance in the normal way and the opaque substance introduced into the sample compartment. The amount of light transmitted by the sample, measured in percentage transmittance, is quoted as the stray light at a specified

Table 8.3. Stray Light Measurements for Various Instruments Measured at 370 nm

Optical system	*Stray light (%)*
Single beamprism monochromator	<0.02
Single holographic monochromator	<0.02
Double holographic monochromator	<0.0002

wavelength. At 220 nm a solution of sodium iodide (10 g l^{-1}) transmits less than 1×10^{-10} of incident radiation but transmits more than 50% at wavelengths greater than 265 nm. The stray light at 340 nm can be measured using a solution of sodium nitrite (50 g l^{-1}). Alternatively, various filters are available for use in a similar manner, a Vycor filter being opaque to radiation below 220 nm.

Examples of stray light measurements for various instruments are given in Table 8.3 Errors are most significant near the wavelength limit of an instrument aim may result in the recording of 'false' absorption peaks in addition to incorrect absorbance values.

Quantitative Measurements

The measurement of the intensity of radiation is indirect, involving the generation and measurement of an electric current, and it is necessary to standardize instruments before test readings are made. The method of standardization is in principle the same for all instruments but does vary in practice depending upon the design of a particular instrument. The basic procedure involves setting the minimum and maximum conditions of transmitted radiation and adjusting the metering system to give appropriate readings.

To set maximum transmittance a blank sample is used and the instrument is adjusted to give either a reading of 100% transmittance or zero absorbance. Zero transmittance is set when all light to the detector is cut off using an opaque shutter and the meter is adjusted to give a transmittance reading of zero. The corresponding value for absorbance is infinity and because it is technically difficult to set to this value, methods of adjustment vary from instrument to instrument but are always described in the instrument manual. Although the Beer-Lambert law describes the quantitative relationship between the absorption of radiation and the concentration of the absorbing substance, it does not specify the precise wavelength of radiation used, although we have noted that ideally the radiation should be monochromatic.

In order to achieve maximum sensitivity from the method, the absorbance measurements for any given concentration should be as great as possible and preferably should be made at the absorption maximum. Examination of the absorption spectrum of adenosine diphosphate (ADP) shows that at the absorption maximum at 258 nm an absorbance value of 0.22 is given while at 270 nm the absorbance value is only 0.14. In theory either wavelength would be suitable for the quantitation of ADP but obviously measurement made at 258 nm would offer greater sensitivity.

Absolute methods

It is possible to measure the absorbance of a sample of a known compound at its absorption maximum and to calculate the actual concentration of the compound in the sample using a known value for the molar absorption coefficient (often obtainable from published spectral tables). In Fig. 8.2 the absorption spectrum of ADP shows an absorbance of 0.22 at 258 nm. The quoted value for the molar absorption coefficient of ADP at this wavelength is 1.54×10^4 l mol^{-1} cm^{-1} and hence the concentration of ADP in the sample used can be calculated from the Beer-Lambert equation:

$$A = \epsilon c l$$

Hence:

$$c = \frac{A}{\epsilon} = 1.42 \times 10^{-5} \text{ mol l}^{-1}$$

In practice, however, this is not always possible owing to a variety of reasons. If the sample is a mixture of several organic compounds, the measured absorbance value will be the cumulative effect of all the substances that absorb at the selected wavelength. In addition the precise value of the molar absorption coefficient depends to some extent on the particular instrument used and ideally values for the constant should be determined rather than accepted from the literature.

The use of the molar absorption coefficient assumes that the Beer-Lambert relationship is valid over the range of absorbance values measured and, because this is not always true, it is essential that this relationship is always confirmed experimentally using the particular instrument in question.

Determination of the Molar Absorption Coefficient

In order to determine the molar absorption coefficient, a pure, dry sample of the compound must be available. Purity is often difficult to check in a routine analytical laboratory but dryness may be achieved by desiccation. Care must be exercised in the choice of desiccant and the temperature used, owing to the potential instability of the compound at even ambient temperatures or the effect of light.

It is advisable to determine the value for the coefficient using different samples of the compound and subsequently compare the results. A small amount of the substance is accurately weighed, dissolved and made up carefully in a volumetric flask to a definite volume, *e.g.* 100 ml. From the known relative molecular mass (RMM) of the compound it is possible to calculate the molar concentration of the solution:

$$\text{Molar concentration} = \frac{\text{weight}}{\text{RMM}} \times \frac{1000}{\text{volume(ml)}} \text{mol l}^{-1}$$

Concentrations in the region of 0.1 mol l^{-1} are often convenient but it obviously depends upon such factors as the amount of substance available, the cost, the solubility, etc. From this stock solution, a series of accurate dilutions are prepared using volumetric glassware and the absorbance of each dilution measured in a 1-cm cuvette at the wavelength of maximum absorbance for the compound. A plot of absorbance against concentration will give an indication of the validity of the Beer-Lambert relationship for the compound and a value for the molar absorption coefficient may be calculated from these individual measurements or from the slope of the linear portion of the graph:

$$\text{Molar absorption coefficient} = \frac{\text{absorbance}}{\text{molar concentration}}\ \text{l mol}^{-1}\ \text{cm}^{-1}$$

Comparative Methods

If the use of the molar absorption coefficient is inappropriate then it may be sufficient to use a single reference solution of known concentration (known as a standard or calibrator solution) and to compare the test absorbance with that of this standard. The principle of quantitation is exactly the same as before except that the molar absorption coefficient is eliminated from the calculation by measuring the absorbance of both the test and standard solutions at the same wavelength and comparing their absorbance values.

A standard solution of concentration c_s gives an absorbance value of A_s while the test solution gives an absorbance of A_t. Hence:

$$A_s = \epsilon c_2 l$$

and

$$A_t = \epsilon c_1 l$$

Therefore

$$\frac{A_s}{A_t} = \frac{\epsilon c_s l}{\epsilon c_t l} = \frac{c_s}{c_t}$$

and the concentration of the test

$$c_t = c_s \times \frac{A_t}{A_s}$$

The use of a single standard in this way assumes that the Beer-Lambert relationship is valid over the absorbance range measured and again it is necessary to confirm the relationship before using the method. The analysis of a range of known concentrations of the test substance is necessary to validate the Beer-Lambert relationship.

If the plot of absorbance values against concentration results in a straight line then either of the two previously outlined methods may be used. If, however, the resulting graph shows a curve instead of a straight line then the implication is that the actual value for the molar absorption coefficient is dependent to some extent upon the concentration of the compound and as a result invalidates both methods. In such circumstances a graphical plot (calibration curve) will be needed. Care must be taken in preparing the standard solutions because a pure preparation of the test compound will not be subject to the same interference effects of a crude or mixed test sample.

To compensate for these differences it may be necessary to prepare the standard by adding a known amount of the pure component to the actual sample and using the resulting increase in absorbance as the absorbance value for the standard.

Analysis of Mixtures

When there are several compounds which absorb radiation of the same wavelength, although not necessarily having the same absorption maxima, the absorbance value measured is the cumulative effect of all the absorbing compounds and no longer reflects the concentration of any one compound. The most frequent approach to such a problem involves the use of a chemical reagent to modify one of the compounds chemically and to produce a new compound with significantly different absorption characteristics.

This technique is particularly useful if it results in a shift of the absorption maximum from the ultraviolet to the visible region of the spectrum. Alternatively the absorbance of the mixture can be measured at different wavelengths (usually one for each compound present) and, provided that the molar absorption coefficients for each compound at each wavelength are known, the concentration of each compound can be calculated.

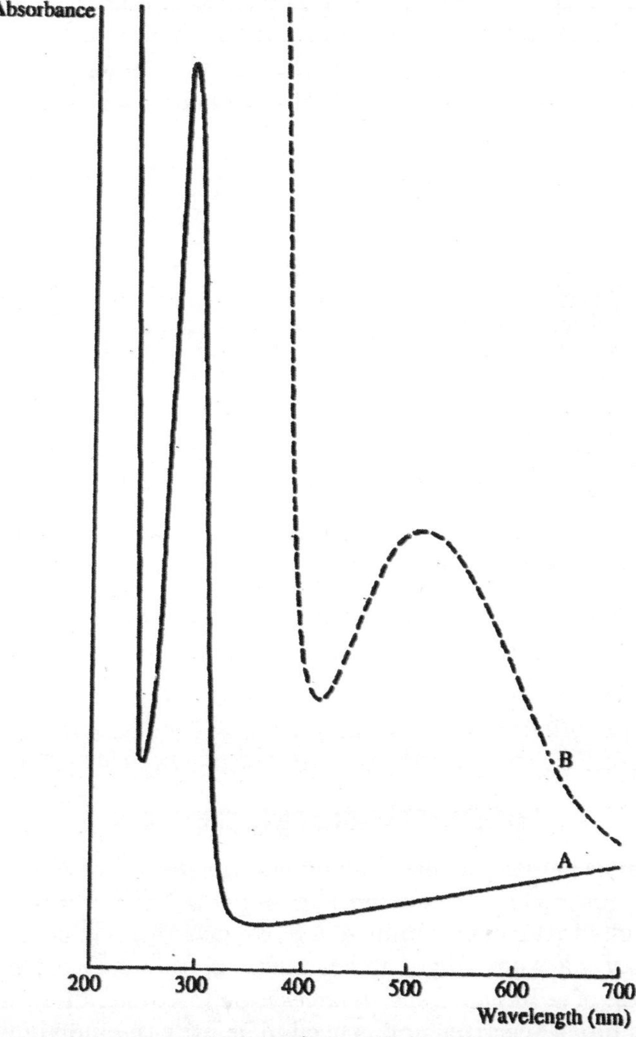

Fig. 8.15. Modification of the absorption characteristics of a compound. Salicylate has an absorption maximum at 290 nm (A). Complexing with ferric nitrate in an acid results in a shift of the absorption maximum into the visible region at 510 nm (B) and provides the basis for a method of quantitation.

Difference Spectroscopy

Some samples may contain different forms of the same substance which show very similar absorption spectra and hence make the detection or measurement of one rather difficult. This situation arises, for instance, in the measurement of the oxidized and reduced forms of cytochrome *c* and in the detection of various haemoglobin derivatives in blood. In these situations the use of difference spectroscopy may be useful. Difference spectroscopy requires a blank solution which includes the interfering substance.

This gives an absorption spectrum which is a plot of the difference in absorption between the blank and the sample at each wavelength. Compounds that are common to both blank and sample will give a zero difference in absorbance and any slight spectral differences in the sample will be shown as either peaks or troughs. The technique will demonstrate small differences in absorption characteristics but the sensitivity of the method is obviously significantly affected by the quality of the spectrophotometer.

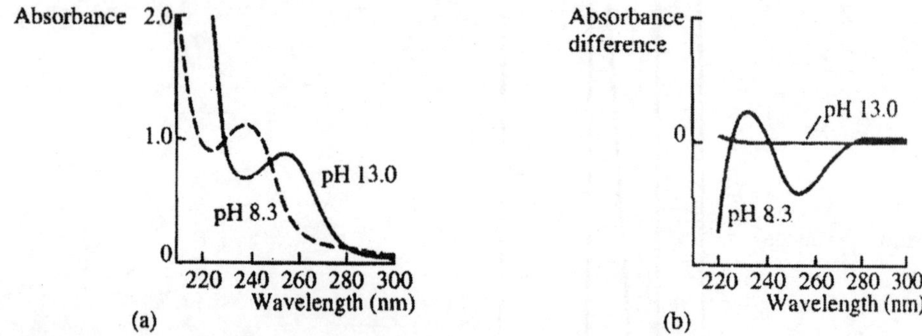

Fig. 8.16.Difference spectroscopy. Phenobarbitone (a) shows different absorption maxima at pH 13.0 (keto form) and pH 8.3 (enol form).

A difference spectrum of two solutions of phenobarbitone with identical concentrations but at different pH (b) (pH 13.0 as the reference and pH 8.3 as the sample) has a positive peak at 232 nm and a negative peak at 254 nm.

The difference in absorbance between the peak and the trough may be used as a direct measure of the concentration even in the presence of many interfering substances.

ABSORPTIOMETER DESIGN

There is a wide range of instruments available for the measurement of absorbed radiation and it is important to appreciate the characteristics of the different instruments and the nature and quality of the results that can be obtained. A *photometer* is an instrument that is designed to measure the intensity of a beam of light and usually does so by comparing it with the intensity of some reference source of radiation. A *spectrometer* is an instrument that is capable of splitting the incident radiation into a spectrum and is used to identify the individual wavelengths that are present. A *spectrophotometcr* is an instrument that combines both of these two functions. It is capable of producing radiation of defined wavelength characteristics from a mixed source of radiation and subsequently measuring the intensity of that radiation. The basic design of a spectrophotometer is illustrated in Fig. 8.17.

Light from the lamp passes through a monochromating device and the selected wavelength, after passing through the sample, is measured by a suitable photoelectric detector. The intensity of the initial radiation is controlled either by an attenuator, which is often a simple shutter system, or by varying the operating voltage of the lamp. The geometry of the light path is controlled by a series of lenses or focusing mirrors. The degree of sophistication of the instrument (and hence its price) and the quality of the results depend upon the individual components and the optical design but the basic method of operation is similar for all instruments. The simplest types of photometric instrument are designed for measurements in the visible region of the spectrum only and rely on coloured filters and simple photoelectric detectors.

The name colorimeter is often used to describe such instruments although this is not necessarily correct and the word should probably be reserved for visual comparators rather than photoelectric instruments.

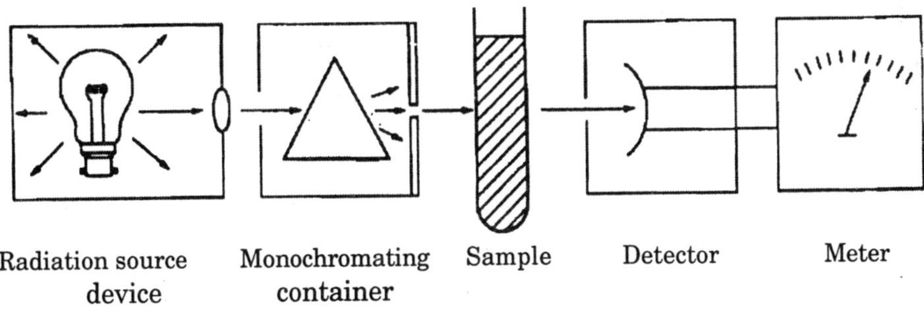

Fig. 8.17. Components of a spectrophotometer.

Radiation Sources

The tungsten lamp is a reliable and cheap source of radiant energy for use mainly in the visible region of the spectrum. It is not a satisfactory source of ultraviolet or infrared radiation owing to the nature of the emission and the absorbing characteristics of the glass envelope. Although it may be used in the near ultraviolet and infrared regions (330 nm to 3 μm) the large amount of visible light also emitted causes technical problems, particularly in the infrared. The usual sources of ultraviolet radiation are hydrogen or deuterium discharge lamps (the latter usually being preferred) or the mercury vapour lamp.

All ultraviolet sources must be fitted with quartz or silica glass windows and none of the lamps named emits any significant amounts of radiation above 400 nm. Black body radiators are used as sources of infrared radiation in the range 2-15 μm, *e.g.* the Nernst glower, which consists of a hollow rod made of the fused oxides of zirconium, yttrium and thorium. For use it is preheated and, when a voltage is applied, it emits intense continuous infrared radiation with very little visible radiation.

Monochromators

The ability to produce monochromatic radiation is a very desirable feature of photometric instruments because the Beer-Lambert relationship is only strictly true for monochromatic radiation. A good spectrophotometer may provide radiation of a specified wavelength with a range or bandwidth of as little as 0.1 nm, but it can be appreciated that even this is still not monochromatic when it is considered that the bandwidth of the sodium emission line is about 1×10^{-5} nm.

This fact provides one of the major problems in the design of photometric instrumentation, namely the production of so-called monochromatic radiation.

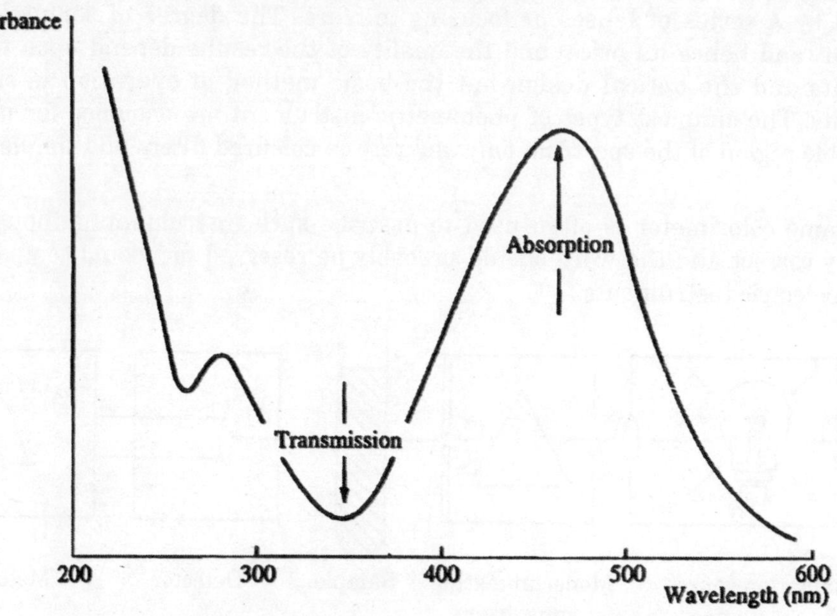

Fig. 8.18. Absorption spectrum of methyl orange.

There are various devices designed to produce radiation of limited wavelength range, some being extremely effective while others transmit a considerable range of wavelengths.

Coloured Filters

The absorption spectrum of a dye shows not only an absorption maximum in the visible region of the spectrum but also an absorption minimum which indicates the wavelengths of radiation that are transmitted by the dye. Such dyes, in the form of stained glass or gelatin filters, provide the simplest approach to the design of monochromating systems. Filters obviously do not provide monochromatic radiation and in many cases the bandwidth may be as great as 100-150 nm. Increasing the concentration of the dye may reduce the bandwidth to some extent but will also reduce the amount of radiation transmitted and this then will impose a strain on the detecting system and the sensitivity of the instrument.

Interference Filters

Interference filters modify the intensity of the radiation transmitted by the filter, enhancing the required wavelength and suppressing the unwanted wavelengths.

They give transmission spectra with considerably reduced bandwidths (10–30 nm) compared with coloured filters and can be used over the spectral range varying from the near ultraviolet (350 nm) to the near infrared (5 μm). Interference filters consist of two pieces of glass each mirrored on one side in such a manner that half of the incident radiation is reflected and the remainder transmitted. They are separated by a thin layer of transparent material, often calcium or magnesium fluoride, the thickness of which is the critical factor in determining the wavelength of the radiation transmitted by the filter.

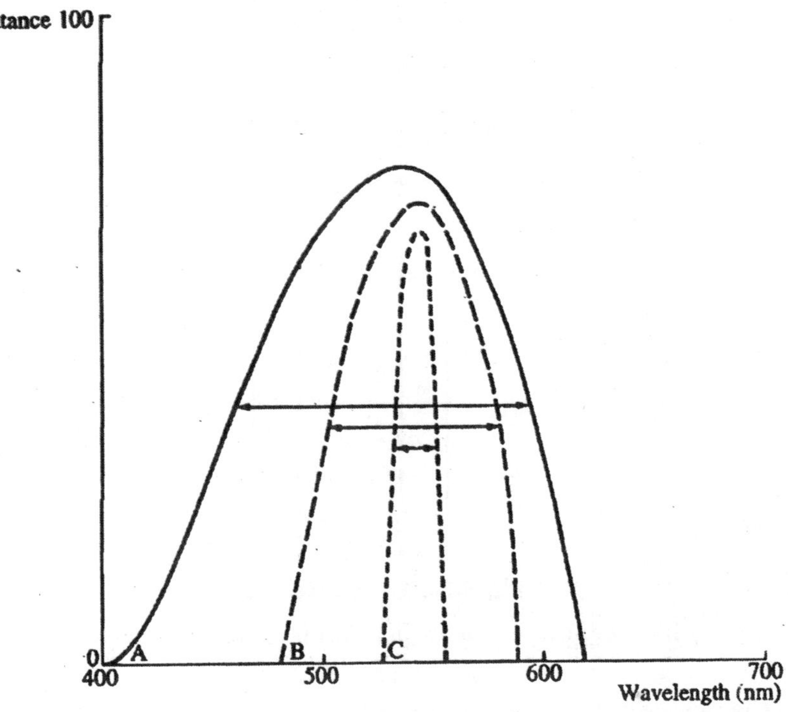

Fig. 8.19. Transmission characteristics of various filters.

Spectrum	*Filter*	*Colour or wavelength*	*Bandwidth (nm)*
A	Glass OGRI	Light green	135
B	Glass 625	Medium green	68
C	Interference	540 nm	14

Bandwidth is defined as the range of wavelengths transmitted at half maximum transmittance.

As a result of internal reflections, the beam of light that is incident upon the filter will emerge as a series of superimposed beams of light, each one retarded from the preceding beam by twice the thickness of the filter. The wavelength composition of the transmitted beam will be the same as the incident beam but those wavelengths for which the distance $2T$ is a whole multiple will be in phase in all transmitted beams. These wavelengths will be reinforced in the beam by constructive interference while all other wavelengths will be out of phase either completely or partially and will suffer destructive interference to a greater or lesser extent. The refractive index of the material used to separate the two pieces of glass also affects the process and the wavelength reinforced will be that radiation which in the material used has a wavelength of $2T$.

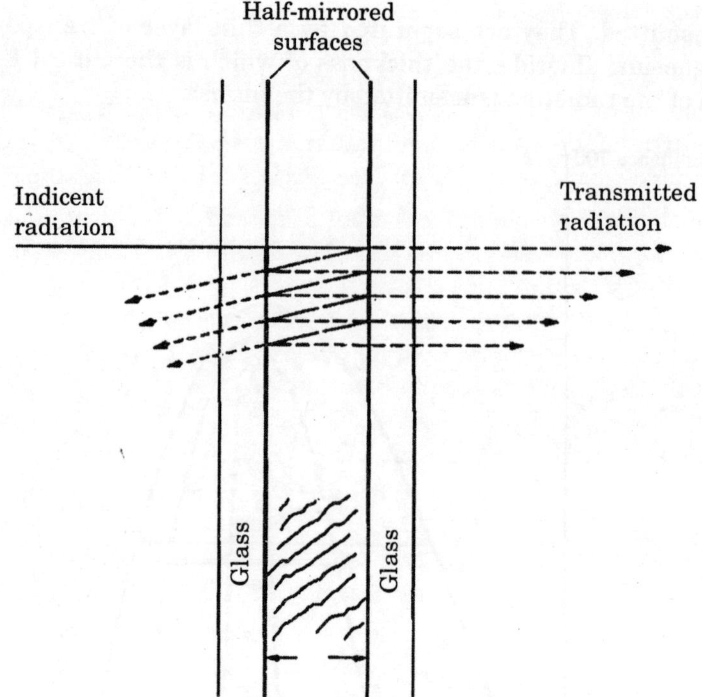

Fig. 8.20. An interference filter.

The wavelength of radiation in air which will be reinforced by such a filter will be:

$$\lambda = 2Tn$$

where n is the refractive index of the material and λ is the wavelength.

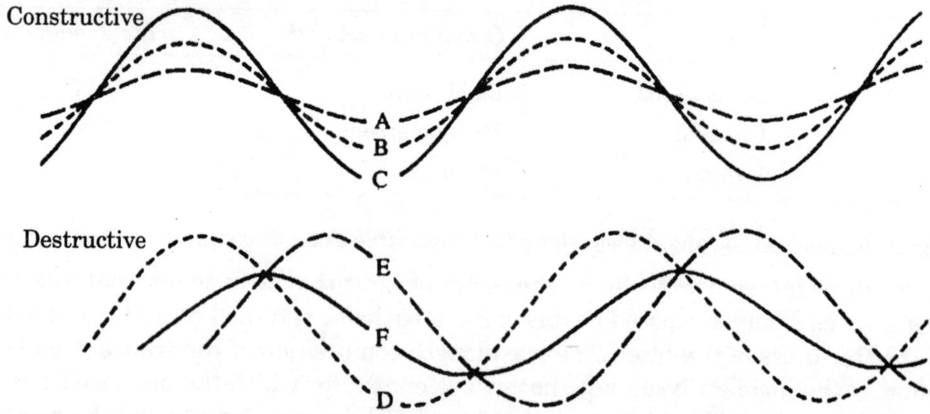

Fig. 8.21. Interference effects. Combination of the waveforms (A) and (B) which, despite having different amplitudes, are of the same wavelength and in phase with each other, results in a constructive effect and the final waveform (C) has an increased amplitude. Combination of waveforms (D) and (E), which are out of phase with each other, results in destructive interference and the final waveform (F) has a reduced amplitude.

Because any wavelength for which $2Tn$ is a whole multiple will also be reinforced, these filters will also transmit second-, third- and fourth-order bands in which the wavelength transmitted will be one-half, one-third and one-quarter of the original. In some cases second- and third-order bands are preferred to the first-order band despite a decrease in the intensity of the transmitted radiation, because the bandwidth of higher order transmissions becomes progressively narrower.

It is relatively simple to eliminate the unwanted bands of radiation using additional coloured filters because of the large differences in wavelength. Wedge-shaped interference filters which have an increasing distance between the two pieces of glass produce an apparent spectrum due to the range of wavelengths constructively enhanced. These are extremely useful in the design of variable wavelength photometers but the range of radiation produced is not a true spectrum and there is still considerable overlap between wavelengths.

Prisms

The refractive index of a material is different for radiation of different wavelengths and hence if a parallel beam of polychromatic white light is passed through a prism the various wavelengths of which it is composed will be bent through different angles to produce a spectrum. The introduction of a variable slit into the optical light path will permit the selection of a particular section of the spectrum.

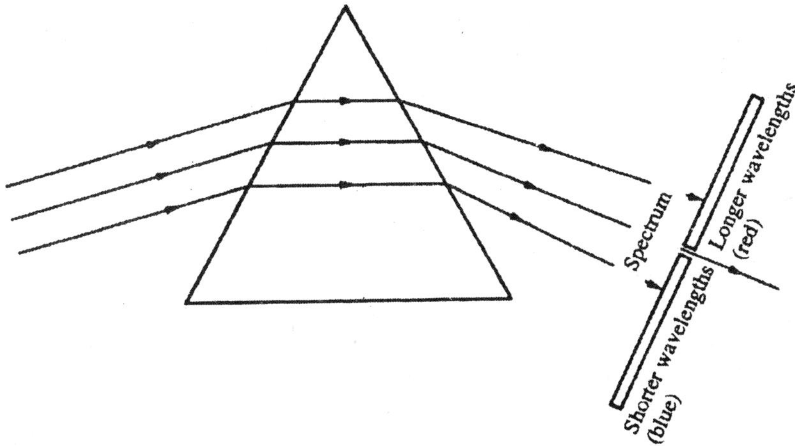

Fig. 8.22. A prism mono-chromator.

Silica or quartz prisms are necessary for the ultraviolet region of the spectrum. While glass may be used in the visible region, rock salt prisms are necessary for work in the infrared. Prisms are aligned so that the beam of light passing through the prism is parallel to its base with the result that the angle of deviation of the emerging beam is minimal and gives maximum resolution between the individual wavelengths. A Littrow mounting is designed so that the spectrum is focused on a plate which incorporates a slit and permits the position of the prism to be adjusted so that radiation from the selected section of the spectrum passes through the slit.

The width of the slit is obviously a major factor in determining the bandwidth of the transmitted radiation but the geometry and refractive index of the prism also have a significant effect. The dispersion of spectra produced by prisms varies depending upon the design of the prism and also on the region of the spectrum involved. Dispersion is less at shorter wavelengths and greater at longer wavelengths. Some typical values are:

Wavelength	*Linear dispersion*
250 nm	0.5 nm per mm
450 nm	5.5 nm per mm
600 nm	12.0 nm per mm

This means that if a fixed slit width is used, the radiation transmitted will have different bandwidths depending upon the region of the spectrum. The mechanical efficiency of the mobile slit is therefore extremely important in the ability to select specific wavelengths.

Diffraction Gratings

In its simplest form a diffraction grating consists of a series of parallel opaque lines drawn on a piece of glass. When the grating is held in a parallel beam of light each clear space acts as a source of radiation. When viewed from any angle (θ) the overlapping radiation from all the slits is composed of all the wavelengths of the original light but the radiation from each slit is retarded or advanced relative to the radiation from an adjacent slit by the distance $\sin\theta$ where d is the distance between each slit.

This means that only wavelengths for which $d \sin\theta$, is a whole multiple will be in phase with each other and will be constructively reinforced while all other wavelengths will suffer destructive interference to a greater or lesser extent. Hence the wavelength of radiation produced at an angle is

$$n\lambda = d \sin\theta$$

where n is a whole number.

At any specified angle the wavelengths that are whole multiples of the value $d \sin\theta$ will be enhanced and are known as first-, second- and third-order spectra. The unwanted wavelengths can be eliminated by the use of simple cut-off filters as previously described for interference filters. If a particular grating, for instance, has 1000 slits per millimetre and is viewed at an angle of 8.6°, the wavelengths of radiation enhanced will be 150 nm, 300 nm, 450 nm, 600 nm, etc.

Diffraction gratings are usually of the reflectance type rather than the transmission type mainly because of the loss of ultraviolet radiation that occurs during transmission through glass. Gratings are sometimes made by ruling a series of parallel grooves with a diamond in a layer of aluminium prepared on a glass blank, a process that is difficult because there may be as many as 2000 lines per millimetre.

A holographic grating, however, is much simpler to produce and is made by coating a glass blank with a thin layer of photoresist which is then exposed to the parallel interference fringes produced when two beams of light from lasers intersect. The photoresist is developed in an etching solution to yield a surface pattern of parallel grooves and is subsequently coated with a thin layer of aluminium to produce a reflectance grating. Because the dispersion is due to the geometry of the grating and is not a property of the material as it is with prisms, gratings may be produced with optimal characteristics for specific spectral ranges.

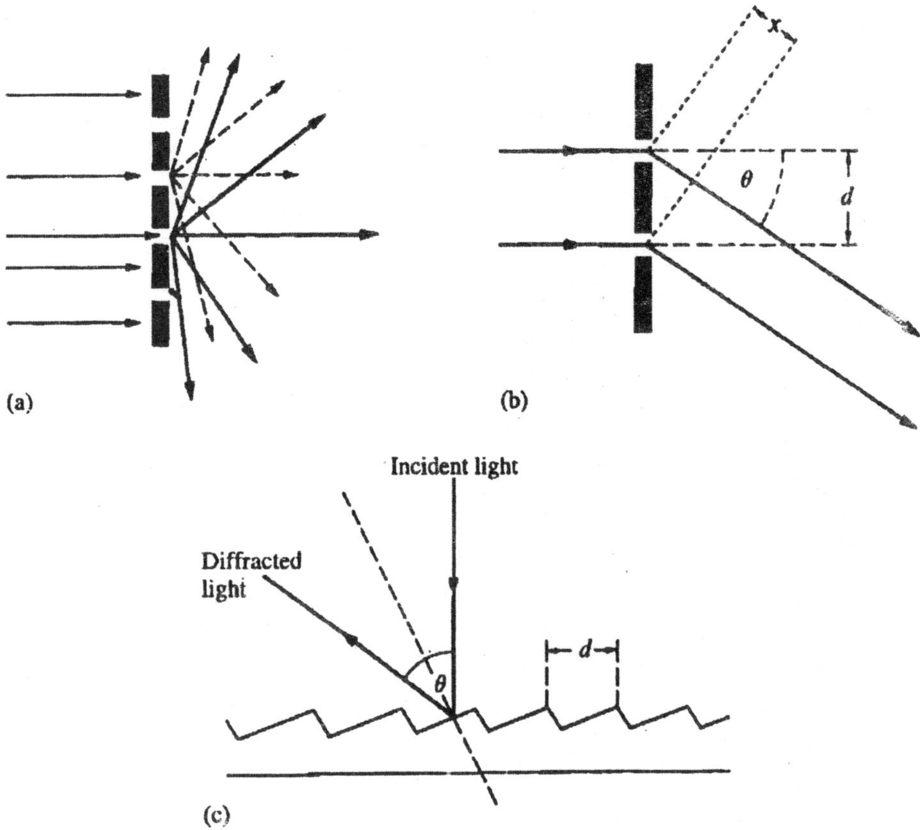

Fig. 8.23. Diffraction gratings. Each line in a grating acts as a separate source of radiation (*a*) but radiation transmitted at any angle θ is retarded relative to radiation from the preceding line (*b*) by the distance *x*. In the transmitted radiation some wavelengths will undergo constructive interference while the majority will suffer destructive effects. Reflectance gratings (*c*) are frequently used and the principles of monochromation are the same as for transmission gratings.

Such spectra show an equal distribution of wavelengths compared with those produced by prisms and this not only simplifies instrument design but also means that values for linear dispersion hold true for the whole wavelength range of a grating, often being as low as 1 nm per mm. The efficiency of monochromation can be considerably improved by the use of a double monochromator, in which the selected part of a spectrum from the first grating can be further resolved by a second grating, resulting in bandwidths as low as 0.1 nm without excessive reduction in the intensity of radiation.

Detectors

The materials and design of the various photoelectric detectors available are such that the absorption of radiation results in the displacement of electrons and hence in the development of a potential difference between two electrodes. The main types of photoelectric detectors may be classified as either photo-voltaic or photoconductive.

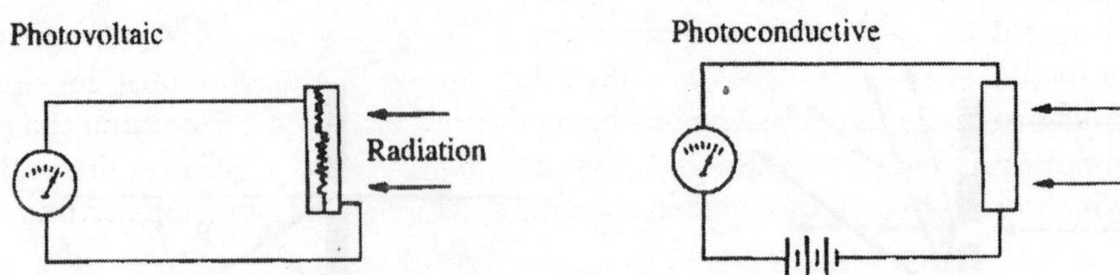

Fig. 8.24. Photoelectric detectors. Photovoltaic detectors measure the flow of electrons displaced by the absorption of radiation. Photoconductive detectors measure the changes in conductivity caused by the absorption of radiation.

Ultraviolet and Visible Region of the Spectrum

Photovoltaic devices are the most useful in this region of the spectrum and the barrier layer or solar cell (Fig. 8.25) may be used in simple photometers. These cells are only effective in the visible region of the spectrum and while giving a linear response to the intensity of the radiation over a limited range, they show a fatigue effect in which the signal decreases as the radiation continues to fall on the cell.

The current generated is very small and is difficult to amplify because of the low internal resistance of the cell. As a result such cells are not very sensitive. Developments in photoelectric cells have resulted in the production of photodiodes. These in principle are very similar to the barrier layer cell described above but the selenium is replaced by silicon products. Silicon is a semiconductor and with traces of boron or gallium added the trace element can accept electrons displaced from the silicon leaving it positively charged (p-silicon).

Conversely, trace amounts of arsenic will act as electron donors, giving the sili-con a negative charge (p-silicon): The displacement of electrons from silicon can be caused by the absorption of radiation and if the diode is designed in much the same manner as the barrier layer cell, with p-silicon as the face plate and *n*-silicon as the back plate, the absorption of radiation will result in a small current flowing between the plates of silicon.

Photodiodes can be produced as very small, solid-state devices which are effective over the wavelength range 200—1000 nm and the signals produced are amenable to computer handling. They are frequently produced in units in which a large number of diodes are arranged side by side, giving what is known as a diode array and as such can be used in spectrophotometers and monitoring systems for chromatography. Photoemissive tubes are necessary for work in the ultraviolet range and they show greater sensitivity and precision than photoelectric cells.

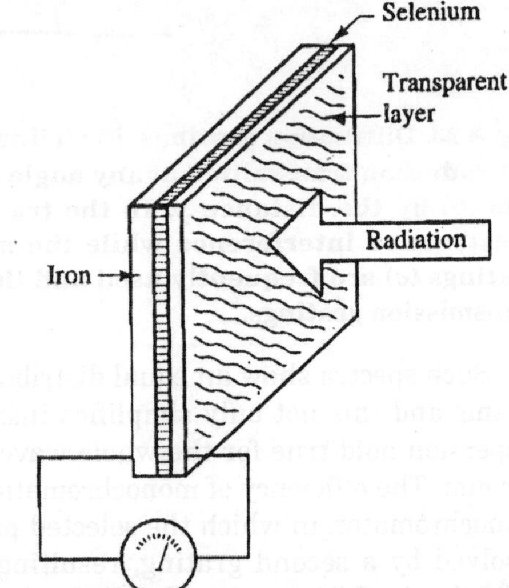

Fig. 8.25. Barrier layer cell. Radiation absorbed by the semiconductor, often selenium, causes electrons to be released and a small current flows which can be measured using a microammeter.

A simple photo-missive tube consists of two electrodes in a vacuum. A silver cathode coated with an alkali metal is maintained at a potential difference of about 100 V from the anode, which is a plain silver wire and serves to collect the electrons. A priotomultiplier consists of a series of electrodes (usually ten), the first one being a photo-missive electrode and the remainder being electron multiplier electrodes. Radiation falling on the first electrode results in the release of electrons which are collected by the next electrode, which is held at a slightly more positive charge.

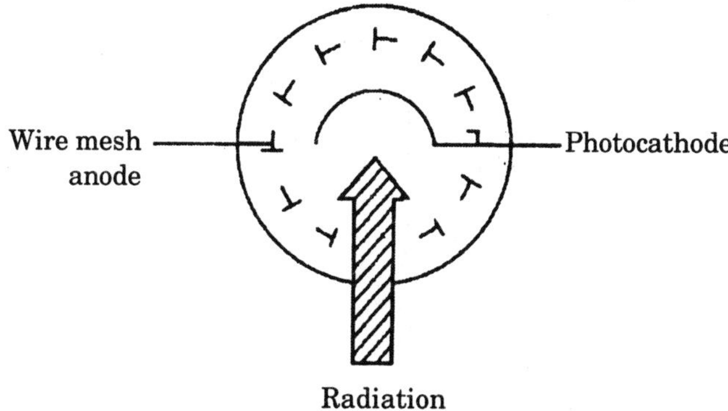

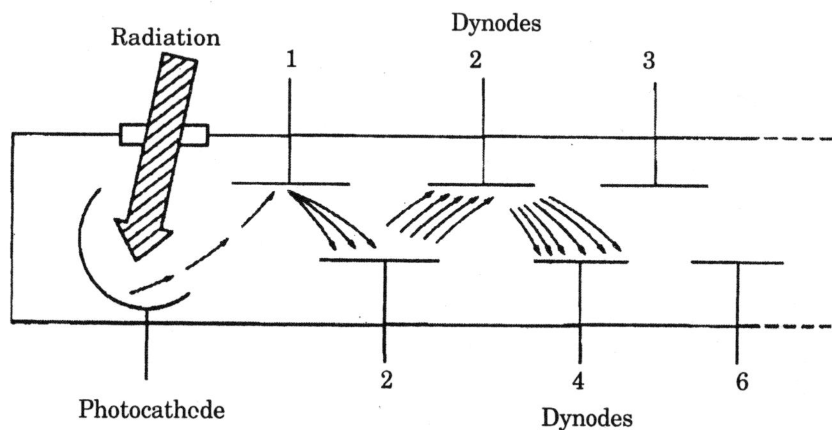

Fig. 8.26. Photoemissive tubes. Light enters a simple phototube (*a*) and causes the release of electrons from the photoemissive alloy of the cathode. Owing to the potential difference between the anode and the cathode, the electrons are captured by the anode and the resulting current can be amplified and measured. Photomultiplier tubes (*b*) are a development of simple phototubes and result in internal amplification of the current initially developed at the photocathode.

These electrons stimulate the release of more electrons from this electrode, increasing the electron flow by a factor of four or five. The successive electrodes are operated at voltages increasing in steps of about 100 V and the overall device requires a very stable high voltage power supply. The resulting amplification of the initial current is several million fold, making the photomultiplier plier tube a very sensitive photoelectric device. The spectral response of photo-missive tubes depends upon the composition of the cathode and the use of various mixtures of elements permits the production of a wide range of tubes of varying responsiveness.

Photomultiplier tubes can be used to detect low intensity radiation and even in the absence of any light will still generate a small current due to various emissions from the material of the tube, etc. This dark current' has to be compensated for in any measurements that are made.

Infrared Region of the Spectum

Photoconductive detectors, such as lead sulphide, are useful in the spectral region of 800-4000 run (*i.e.* the near infrared). They consist of a thin layer of a semi-conductor deposited on glass or another insulating material and protected against atmospheric damage by either a vacuum or a thin film of lacquer. The absorption of radiation increases the conductivity of the layer by displacing electrons. If a constant potential difference is applied across the layer, the current flowing will increase significantly when radiation is absorbed. Detection of the middle and far range of infrared radiation requires thermal detectors, the simplest of which is a thermocouple, in which the change in temperature at one junction of the thermocouple results in a small voltage being produce.

Although simple in design, thermocouples lack sensitivity. Bolometers are more sensitive and are based on the fact that as the temperature of a conductor varies so does its resistance. A suitable conductor (nickel plate or a thermistor) is incorporated in a Wheatstone bridge circuit through which a steady current is passed. When radiation strikes the sensing element, the electrical resistance changes and a corresponding change in the current can be measured.

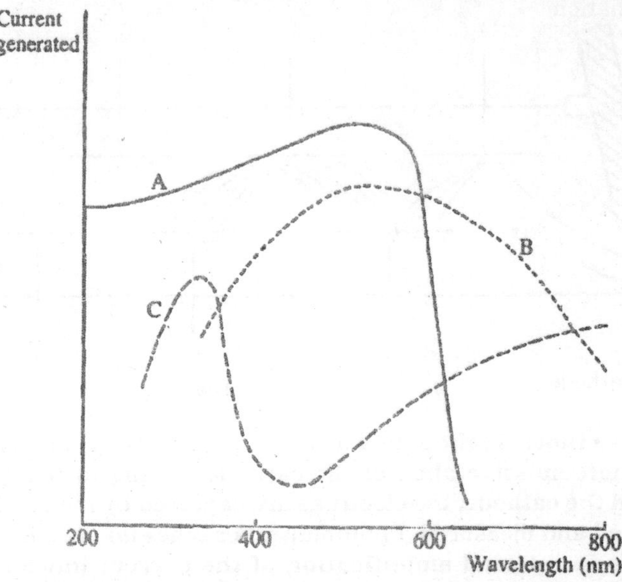

Fig. 8.27. Spectral response of cathode elements in photoemissive tubes. Various combinations of elements respond in different ways to specific wavelength ranges and are only suitable for the specified spectral regions: calcium-antimony (*A*); sodium-potassium-caesium-antimony (*B*); silver-caesium (*C*).

Optical Materials

Normal glass will only transmit radiation between about 350 nm and 3 μm as a result, its use is restricted to the visible and near infrared regions of the spectrum. Materials suitable for the ultraviolet region include quartz and fused silica. The choice of materials for use in the infrared region presents some problems and most are alkali metal halides or alkaline earth metal halides, which are soft and susceptible to attack by water, *e.g.*, rock salt and potassium bromide. Samples are often dissolved in suitable organic solvents, *e.g.*, carbon tetrachloride or carbon disulphide, but when this is not possible or convenient, a mixture of the solid sample with potassium bromide is prepared and pressed into a disc-shaped pellet which is placed in the light path.

Optical Systems

A basic photometer is a single-beam instrument in which there is only one light path and the instrument is standardized using a blank solution, which is replaced with the sample to obtain a reading. The major problem with a single-beam design lies in the fact that the absorbance value recorded for the sample may be a result of other factors as well as the absorbing properties of the sample.

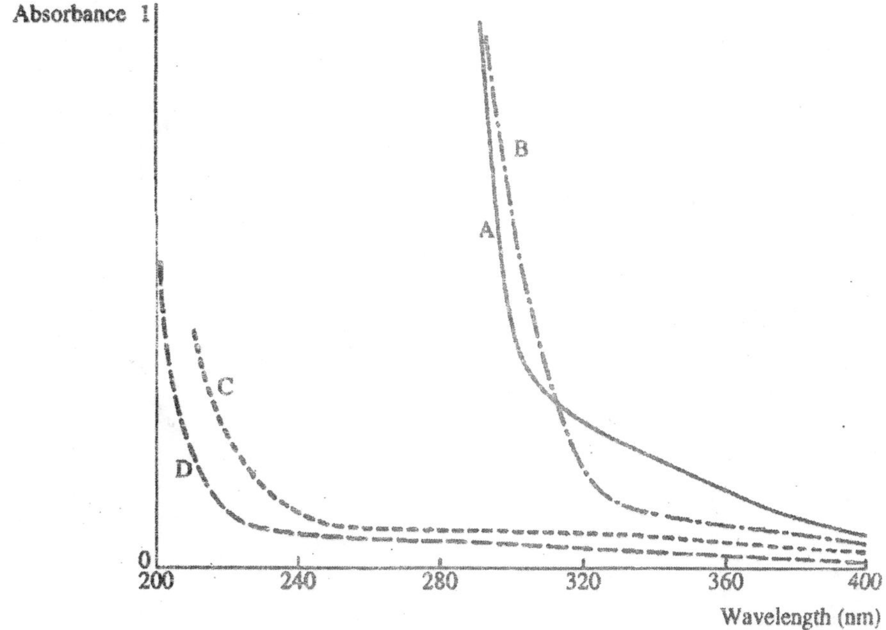

Fig. 8.28. Transmission characteristics of various optical materials: (*A*) glass, (*B*) plastic (*C*) silica glass, (*D*) quartz.

Fluctuations in the intensity of the light source or instability in the detection system will result in errors, while the fact that the absorption of the blank will vary at different wavelengths means that the setting-up procedure must be undertaken every tune the wavelength is changed. In order to minimize these potential errors and to permit automation or the measurement of absorbance at varying wavelengths it is necessary to introduce a constant reference signal by using a double-beam optical system. In a double-beam instrument the radiation is split into two beams, one passing to the detector through the sample and one through the blank. A series of half-mirrors may be used, permitting both light paths to function simultaneously or alternatively the beam may be 'chopped' mechanically, deflecting a beam of radiation alternately along each light path. The latter method is the most frequently used because the maximum intensity of the incident beam is maintained and it is usually accomplished by means of a rotating sector mirror. The detector receives alternate pulses of radiation which have passed through the blank and the sample and it produces alternating electrical signals which are directly proportional to the intensities of the radiation transmitted by both blank and sample.

In a ratio-recording photometer system this signal is resolved electronically into two voltages corresponding to the blank and sample signals and the former is used as the standardizing potential for a recording potentiometer. In an optical null balancing photometer, the alternating current is used to power a servo motor, which drives an attenuator into the blank light path until the intensity of the beam is exactly the same as the sample beam. The attenuator is designed so that the distance moved is proportional to the absorbance of the sample and is usually mechanically coupled to a pen recorder system.

In some situations measurement of the reflected, rather than the transmitted, radiation may be made to assess the amount of radiation that has been absorbed by the sample.

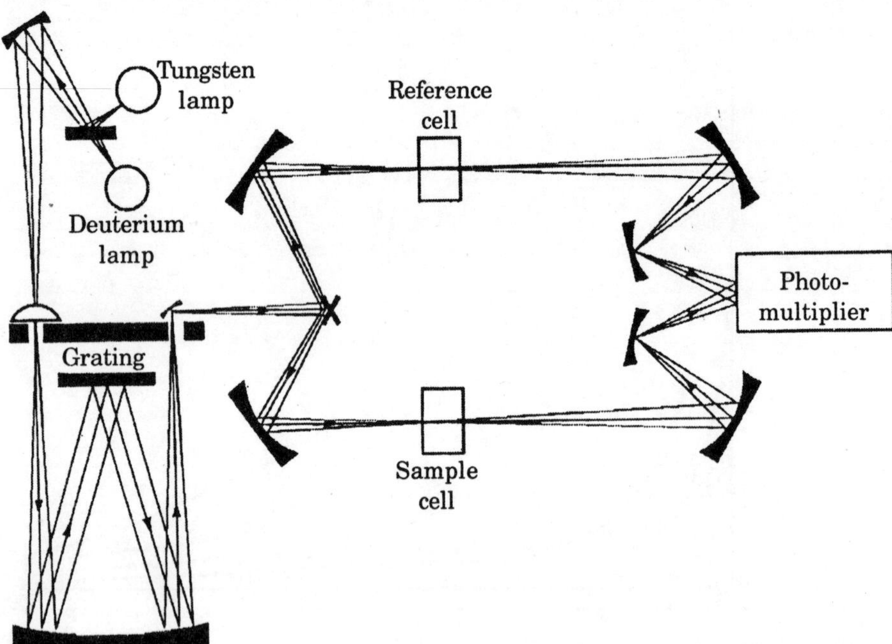

Fig. 8.29. Optical diagram of a double-beam spectrophotometer. Interchangeable lamps are available for work in the visible and ultraviolet regions and monochromation is achieved with a reflectance diffraction grating.

There are two main ways by which radiation might be reflected. Specular reflection is similar to the reflection by a mirror and, for quantitative work, the angles of the incident and the reflected radiation are important. Diffuse reflection is from within the layers of the material and the reflected light is disbursed over a range of 180°. This type of reflection is measured in the thin films used in dry chemistry systems. The term reflectance density is often used, which is defined in a manner comparable to absorbance: the logarithm of the ratio of incident to reflected light.

The components of the instrument are similar to absorptiometry but their relative positions are changed in order to catch the reflected radiation. In most of these instruments, rather than incorporating a monochromator, a light-emitting diode (LED) of a specific wavelength is used to produce the incident radiation. The principles of absorptiometry have been applied to the measurement of turbidity.

Suspensions of particles scatter incident radiation and, while there is no absorption of radiation by the analyte, the reduction in the transmitted radiation can be used as a measure of the degree of turbidity. Because absorption is not involved, there is no requirement for monochromation but the fact that the extent of light scattering increases as the wavelength of the incident radiation decreases explains the fact that some instruments do incorporate a simple monochromation system.

The sensitivity of the technique is considerably improve by measuring the amount of light scattered rather than the light transmitted.

In such instruments, known as nephelometers, the optical system is more similar to reflectance systems or fluorimeters in order to detect the scattered light rather than the transmitted light. The use of lasers appreciably enhances the performance of these techniques by allowing much more critical design of instrumental light paths and reducing the unwanted reflections from cuvettes, etc., which reduce the sensitivity of simpler instruments.

MOLECULAR FLUORESCENCE TECHNIQUES

Molecular fluorescence involves the emission of radiation as excited electrons return to the ground state. The wavelengths of the radiation emitted are different from those absorbed and are useful in the identification of a molecule. The intensity of the emitted radiation can be used in quantitative methods and the wavelength of maximum emission can be used qualitatively. A considerable number of compounds demonstrate fluorescence and it provides the basis of a very sensitive method of quantitation. Fluorescent compounds often contain multiple conjugated bond systems with the associated delocalized pi electrons, and the presence of electron-donating groups, such as amine and hydroxyl, increase the possibility of fluorescence. Most molecules that fluoresce have rigid, planar structures.

Instrument Design

Fluorescent radiation is emitted in all directions and use is made of this fact to avoid difficulties that may be caused by the transmission of incident radiation by the sample. By moving the detection system at right angles to the cell only fluorescent radiation is detected. Some instruments are designed to measure 'front face' fluorescence, *i.e.* the radiation that is emitted along a light path at an acute angle to the incident radiation. Such fluorescence measurements are comparable to reflectance measurements.Despite the measurement of the emitted radiation by these means it is still possible for scattered or reflected incident radiation to reach the detector.

To prevent this, fluorimeters require a second monochromating system between the sample and the detector. Many simple fluorimeters use filters as both primary and secondary monochromators but those instruments that use true optical monochromators for both components are known as spectroflurimeters. Other instruments incorporate a simple cut-off filter system for the emitted radiation while retaining the optical monochromator for the excitation radiation. Because the wavelengths of both excitation and emission are characteristic of the molecule, it is debatable which monochromator is the most important in the design of a fluorimeter.

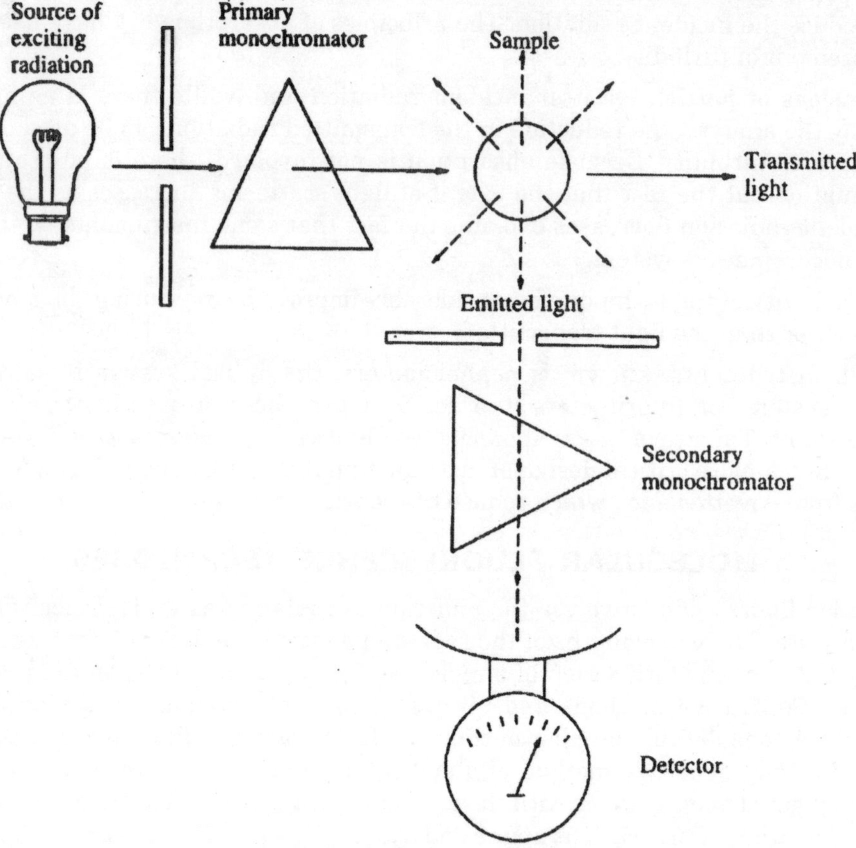

Fig. 8.30. Fluroimeter design.

Fluorimetric Methods

In order to select the operating conditions for any fluorimetric method, the excita-tion and emission spectra of the analyte must be determined. Fig. 8.9 illustrates the fact that an emission spectrum is an approximate mirror image of the excitation spectrum, the latter being similar to the absorption spectrum of the compound. The main advantage of fluorescence techniques is their sensitivity and measurements of nanogram (10^{-9} g) quantities are often possible.

The reason for the increased sensitivity of fluorimetry over that of molecular absorption spectrophotometry lies in the fact that fluorescence measurements use a non-fluorescent blank solution, which gives a zero or minimal signal from the detector. Absorbance measurements, on the other hand, demand a blank solution which transmits most of the incident radiation and results in a large response from the detector. The sensitivity of fluorimetric measurements can be increased by using a detector that will accurately measure very small amounts of radiation. Fluorescence measurements are quoted relative to a standard maximum intensity of fluorescence and it is usual to standardize the instrument on zero fluorescence (blank) and 100% fluorescence (highest standard).

The intensity of fluorescence of all intermediate standards and test samples is then measured and expressed as a percentage of the highest standard. Over a limited concentration range, the intensity of fluorescence is directly proportional to the concentration of the compounds but in practice it is advisable to produce a calibration curve for each assay. Fluorimetric techniques often demonstrate a greater degree of specificity than comparable absorption techniques for several reasons. Many compounds that absorb in the ultraviolet do not fluoresce and the presence of such substances would possibly interfere in the absorptiometric analysis of a sample but would not interfere in a fluorimetric assay. The converse is also true that the presence of a fluorescing substance in an absorptiometric assay would reduce the apparent absorbance by the amount of the fluorescence which reached the detector. This is a problem in spectrophotometers in which the monochromator is in front of the sample but in practice many instruments position the monochromator after the sample and before the detector. Probably the main reason for the improved specificity of fluorimetric techniques is the ability to identify and select the emission wavelength and this is particularly significant in situations where compounds absorb at the same wavelength but show different emission spectra. A major difficulty in fluorimetry is the fact that a wide range of substances and conditions can suppress or quench the emission of fluorescent radiation.

Many excited molecules tend to lose their energy by molecular collisions rather than by fluorescence and the proportion doing so increases as the temperature rises. Frozen samples often show considerably more intense emission than do liquid samples and the control of temperature during measurements is an important consideration. Similarly a decrease in the pH of the sample often reduces the intensity of fluorescence owing to the binding of protons by non-bonding electrons. Some compounds, particularly those containing electrophilic groups (*e.g.* carboxylic, azo and halides), can quench fluorescence, and their presence even in trace amounts can significantly reduce the intensity of emission. Because of this all equipment should be scrupulously clean and all reagents should be of the highest purity.

ATOMIC SPECTROSCOPY TECHNIQUES

The atoms of certain metals when heated emit radiation and methods of analysis have been developed which use the wavelength of emission for qualitative analysis and the intensity of emission in quantitative work (Table 8.4). Emission spectroscopy is most frequently encountered as flame emission photometry, which is almost entirely restricted to the visible region of the spectrum and to those elements that are easily excited at the temperature of a flame. Alternatively, excitation may be achieved by means of a high voltage arc struck between two electrodes.

More recently, temperatures as high as 10000 °C, produced by inductively coupled plasma (ICP) discharges, have been used to cause the excitation of atoms and offer improved levels of sen-sitivity. Atomic fluorescence is a variant of flame photometry in which atoms are excited by the absorption of radiation rather than thermal energy. The instruments used for such measurements require optical systems similar to those used in molecular fluorimetry. However, despite the great potential increase in sensitivity of such a method, it is limited to a few elements for which the necessary intense excitation sources of radiation are available.

Table 8.4. Emission Lines of Various Elements.

Element	*Wavelength most suitable for:*		*Other lines*	*Flames suitable for:*	
	Atomic absorption	*Flame emission*		*Atomic absorption*	*Flame emission*
Calcium	**422.7**	**422.7**	239	A-Ac	N-Ac
Copper	**324.7**	324.7	216, 222, 249, etc.	A-Ac	N-Ac
Iron	**248.3**	372.0	248, 252, 373	A-Ac	N-Ac
Lanthanum	550.1	**579.1**	418, 495, etc.	N-Ac	N-Ac
Lead	**283.3**	405.8	217, 261, 368	A-Ac	N-Ac
Lithium	670.8	**670.8**	323, 610	A-Ac	N-Ac
Magnesium	**285.2**	285.2	202	A-Ac	N-Ac
Mercury	**253.6**	253.6	—	A-Ac	N-Ac
Phosphorus	**213.6**	526.0	213, 214-	N-Ac	A-H
Potassium	766.5	**766.5**	769, 404	A-Ac	A-Ac
Sodium	589.0	**589.0**	589.6, 330	A-Ac	A-Ac
Zinc	**213.9**	213.9	307	A-Ac	N-Ac

The emission line in bold type indicates the method of choice.
Key to gases: A, air; Ac. acetylene; N. nitrous oxide; H. hydrogen.

Whereas flame emission photometry relies on the excitation of atoms and the subsequent emission of radiation, atomic absorption spectrophotometry relies on the absorption of radiation by non-excited atoms. Because the proportion of the latter is considerably greater than that of the excited atoms, the potential sensitivity of the technique is also much greater.

Flame Emission Photometry

A flame photometer is designed to cause atomic excitation of the analyte and subsequently to measure the intensity of the emitted radiation.

A monochromating system is essential to distinguish between the emission of the test element and other radiation from the flame. The *flame* combines both the source of radiation and the atomized sample and hence must be very stable if steady readings are to be obtained. The flame temperature must be high enough to excite the atoms under investigation; the hotter the flame, the greater the proportion that will be excited. If it is too hot, however, the atoms may be raised to higher energy levels and electrons may be removed altogether, resulting in ionization.

The sample is converted into an aerosol in an *atomizer.*

It then passes through an expansion chamber to allow a fall in the gas pressure and the larger droplets to settle out before passing to the burner, where the solvent evaporates instantly, the atoms remaining as a finely distributed gas. Atoms in the sample that are bound in molecules should be decomposed at the flame temperature so rapidly that the same effect is achieved. In practice only a small proportion of the sample (approximately 5%) is effectively atomized because the drop size of the remaining 95% is so large that the water is never effectively stripped away. In low temperature flames, for instance, only one sodium atom in about 60000 is excited but

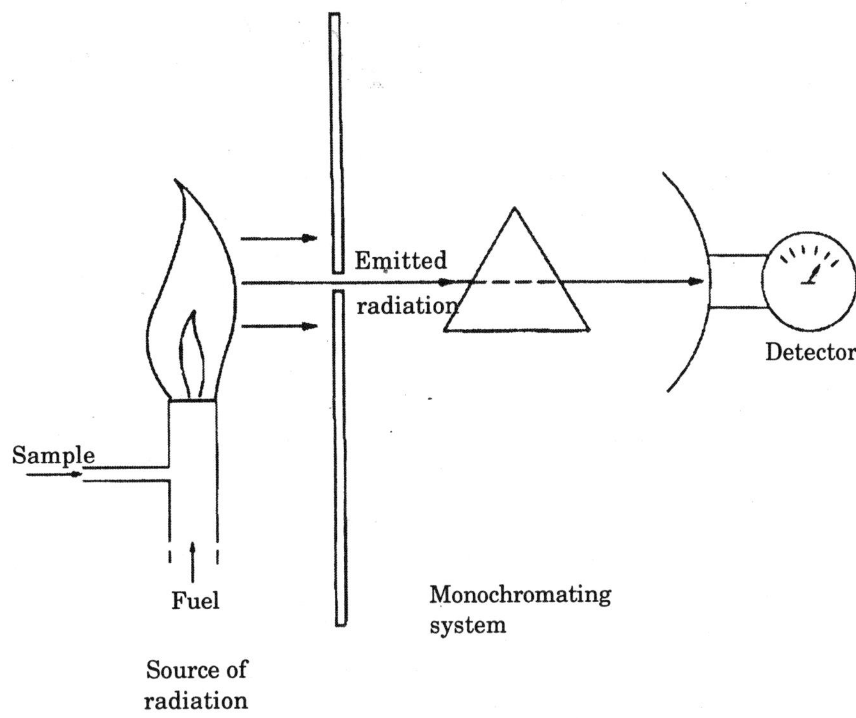

Fig. 8.31. Components of a flame photometer.

despite this apparently low efficiency the technique is very sensitive. Any of the *monochromating systems* described for absorptiometers may be used, although the cheaper models of flame photometer usually employ filter systems.

In these cases interference from other elements at wavelengths near to the test wavelength may be a real problem but owing to the intensity of emission it is possible to reduce the bandwidth by using very dense filters. In many instruments, the use of interference filters improves the specificity of the analysis.

Table 8.5. Fuels for Flame Spectroscopic Techniques.

Fuel	*Oxidant*	*Remarks*	*Temperature (°C)*
Propane	Ari	Low temperature flames suitable	
Butane	Air	For easily excited atoms, such As Na, Li, K and Ca	1900
Acetylene	Air	Medium temperature flame suitable for most emission analyses, *e.g.* Mg, Mn and Sr	2300
Acetylene	Nitrous Oxide	High temperature flame necessary for the more refractile elements, *e.g.* P	3000

The presence of high concentrations of other elements in the sample might not only stress the monochromating system but also cause suppression of the emission by the atoms under investigation, a phenomenon known as the matrix effect. In such situations it is often necessary to prepare standard solutions that contain equivalent amounts of the interfering elements.

Similar effects can also result from significant differences in the viscosity or density of the sample and standards, often caused by the presence of protein. Various anions, such as phosphate and aluminate, may complex the cations being measured and give falsely lowered readings. This can be countered by using organic chelating agents, *e.g.* EDTA or citric acid, which, although binding the cations themselves, are readily decomposed in the flame compared with the very stable phosphates and aluminates. The use of lanthanum chloride and, less effectively, strontium chloride, is perhaps the most convenient method of overcoming the effect of such anionic interference. These cations bind the interfering anions and are possibly precipitated when the aerosol is formed, leaving the test cations available for excitation.

In the absence of such protecting cations it is possible to decompose the phosphate and aluminate complexes thermally by using flames of higher temperature but care has to be exercised using this approach as unwanted ionization may occur. Such anionic interferences are often relevant to atomic absorption spectroscopy.

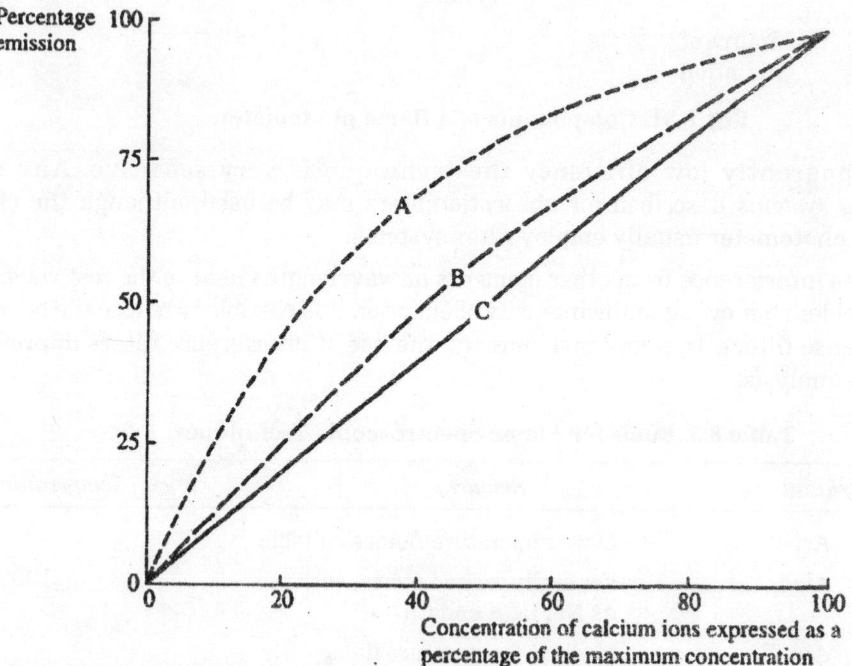

Fig. 8.32. Quantitative flame photometry. The linearity of the relationship between the concentration and the percentage emission becomes better as the concentration range becomes smaller. The maximum concentrations for the calibration curves are (*A*) 25 mmol l^{-1}, (*B*) 2.5 mmol l^{-1} and (*C*) 0.25 mmol l^{-1}.

Quantitative methods using flame emission photometry cannot be absolute because an unknown, although relatively constant, proportion of the sample will reach the flame of which only a further small proportion of atoms will actually be excited and subsequently emit radiation. Hence it is essential to construct calibration curves for any analysis. The radiation emitted by the flame when pure solvent is sprayed is used to zero the instrument and the maximum reading set when the standard with the highest concentration is sprayed.

A range of standard solutions is then sprayed and a calibration graph is drawn. As with fluorescence measurements, the intensity of emission is reported as a percentage of the emission of the highest standard solution, and over a limited concentration range there is a linear relationship between percentage emission and the concentration of the atoms (Fig. 8.32).

In addition to the emission due to the test element, radiation is also emitted by the flame itself. This background emission, together with turbulence in the flame, results in fluctuations of the signal and prevents the use of very sensitive detectors. The problem may be appreciably reduced by the introduction into the sample of a constant amount of a reference element and the use of a dual-channel flame photometer, which is capable of recording both the test and reference readings simultaneously.

The ratio of the intensity of emission of the test element to that of the reference element should be unaffected by flame fluctuations and a calibration line using this ratio for different concentrations of the test element is the basis of the quantitative method. Lithium salts are frequently used as the reference element in the analysis of biological samples.

Plasma Emission Spectroscopy

Atoms can be excited using the high energy levels associated with inductively coupled plasma (ICP) instead of a flame. Such a method of excitation is far more effective and permits the analysis of elements beyond the scope of simple flame emission techniques, such as the refractory elements of boron, phosphorus and tungsten.

The high temperature eliminates many of the interference effects and the instrumentation is designed with a series of photomultiplier tubes set for the different emission wavelengths of specific elements, permitting multi-element analysis of samples. The method is very sensitive and specific.

The ICP discharge is caused by the effect of a radio-frequency field on argon gas flowing through a quartz tube. The high power frequency causes a changing magnetic field in the gas and this in turn results in a heating effect. Temperatures of 9000-10000°C can be produced in this way.

Atomic absorption spectrophotometry

When atoms are dispersed in a flame, the vast majority remain in the ground state and only a very small proportion are thermally excited. If a beam of polychromatic radiation is passed through such a flame, atoms in the ground state will absorb the appropriate wavelength and, for each element, a series of absorption bands may be demonstrated in the transmitted radiation. These absorption bands will correspond to the excitation of the valence electrons and although the energy required for such excitation varies considerably from one element to another, the energy levels for a single element are very limited and result in very narrow absorption bands (*e.g.* 0.001 nm).

For atoms of metals or semi-metals, the energy can be supplied by radiation in the range of 200-900 nm but for all non-conducting elements (insulators) the energy required is very large and radiation of the far ultraviolet and X-ray regions would be required. The proportion of the incident radiation which is absorbed by the atoms in the flame (or vapour) is measured and related to the number of atoms in the flame in a manner directly comparable to molecular absorption spectrophotometry.

Instrument Design

The absorption bands due to atoms are very narrow and the use of white light as the incident radiation would swamp even the best monochromating system with unabsorbed radiation on either side of the absorption band. It is fundamental, therefore, to the technique of atomic absorption spectrophotometry that the incident radiation is of the correct wavelength, bandwidth and intensity. This radiation is produced by a lamp in which the cathode is coated with atoms of the element under investigation and which emits radiation of precisely the same wavelength as that which will be absorbed by non-excited atoms of the same element in the flame.

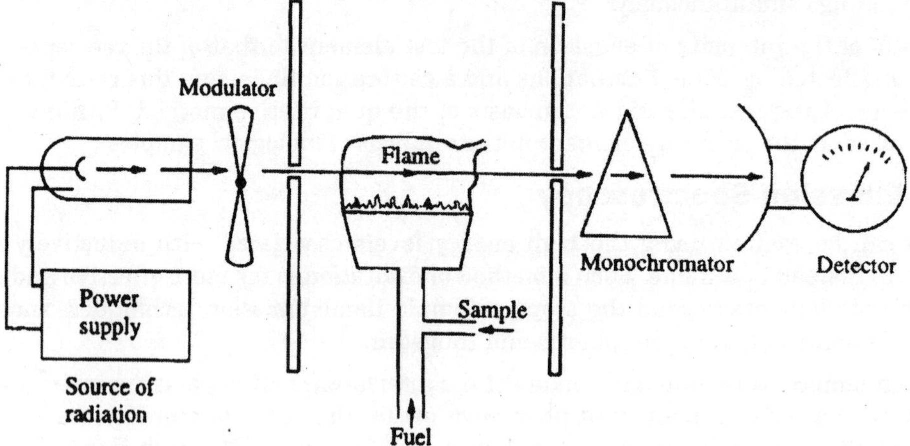

Fig. 8.33. Components of an atomic absorption spectrophotometer.

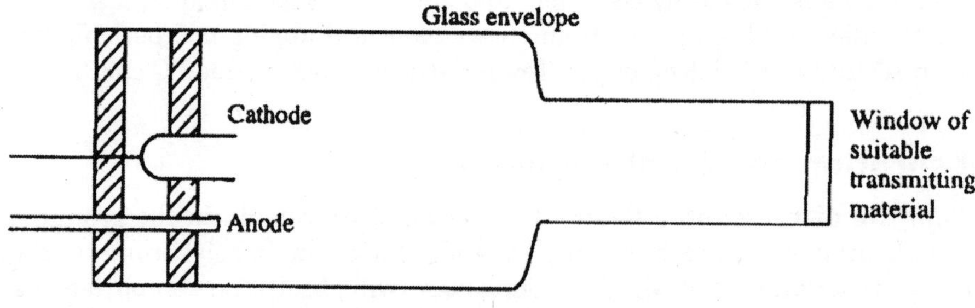

Fig. 8.34. A hollow cathode lamp.

A *hollow cathode lamp* consists of two electrodes sealed in a glass envelope filled with an inert gas, usually argon or neon. The end window of the lamp must be of an appropriate material in order to transmit the emitted radiation and is either quartz or silica. The cathode of the lamp, usually cup shaped, is either made of the element whose spectrum is required or coated with the element, and the application of a potential of between 300 and 400 V is usually required to cause excitation of the atoms and discharge of the appropriate radiation.

Lamps are available with cathodes which contain two or more elements with emission lines that are easily distinguishable. These are used for the analysis of several elements without changing lamps, although such multi-element lamps do tend to give less satisfactory performance than single-element lamps. For certain elements, electrodeless discharge lamps (EDLs) have been designed in which the excitation of atoms is achieved by radio frequencies that induce resonance effects and the energy liberated causes vaporization and excitation of the element. Double-beam atomic absorption spectrophotometers are designed to control variations which may occur in the radiation source but they are not as effective as double-beam molecular absorption instruments in reducing variation because there no blank sample in flame techniquest. The *flame,* as well as containing the unexcited atoms of the element, will also emit radiation due to the thermal excitation of a small proportion of atoms, and it is essential that the detector is capable of distinguishing between the identical radiation that is transmitted by the flame and that emitted from the flame.

This is achieved by introducing a characteristic signal or modulation into the incident radiation by means of a rotating segmented mirror (chopper) or an electrically induced pulse. This pulsed beam is detected as an alternating signal, which is superimposed on the relatively constant signal generated by the emission from the flame. The difference between the two signals is automatically measured by the instrument and only the light emitted from the flame is recorded. The design of the burner head and the method of atomization of the sample both influence the sensitivity which can be achieved. The burner head is designed to give a long narrow flame so that as many atoms as possible are presented in the light path. It needs to be kept spotlessly clean to minimize background emission and the position in the light path has to be adjusted to give maximum sensitivity, the precise position depending not only on the gases being burnt but also on their flow rate.

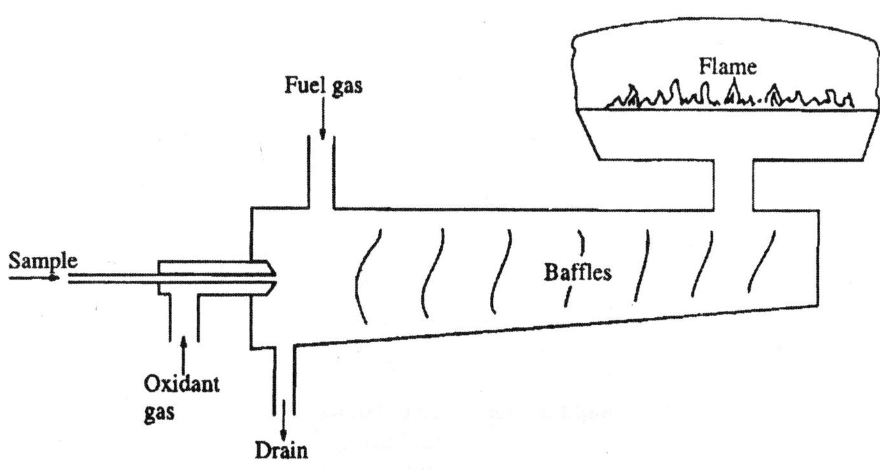

Fig. 8.35. A spary atomizer.

The proportion of fuel to oxidant alters the characteristics of the flame, a high proportion of fuel resulting in a flame with reducing properties while an excess of oxidant gives an oxidizing flame. For each element analysed the optimum proportions must be determined. The basic design of an atomizer is the same as that for flame emission spectroscopy. The method of producing an aerosol involves spraying the sample in air or oxidant gas. The larger drops precipitate on the baffles of the expansion chamber and flow to waste. The fuel gas is introduced and the components and aerosol ample are mixed before passing to the burner. quality *monochromating systems* are necessary to isolate the required emission line of the element from those emission lines due to the gases that are also present in the lamp.

Owing to the very narrow bandwidth of atomic emission lines, it is not adequate simply to select the required wavelength using the monochromator scale, and a procedure known as 'peaking up' has to be undertaken. The wavelength of the emission line required can be obtained from the literature and the monochromator is initially set to this value. With the lamp operating, the monochromator is adjusted to give maximum response from the detector. In some instances, owing to the presence of interfering elements, it may be necessary to use an emission line other than the mance line for the test element, despite the loss in sensitivity which this causes.

As distinct from molecular absorption spectrophotometry, the onus for specificity in atomic absorption spectrophotometry lies with the radiation source rather than with the monochromating system. Conventional flame techniques present problems when dealing with either small or solid samples and in order to overcome these problems the *electrothermal atomization* technique was developed. Electrothermal, or flameless, atomizers are electrically heated devices which produce an atomic vapour. One type of cuvette consists of a graphite tube which has a small injection port drilled in the top surface. The tube is held between electrodes, which supply the current for heating and are also water-cooled to return the tube rapidly to an ambient temperature after atomization.

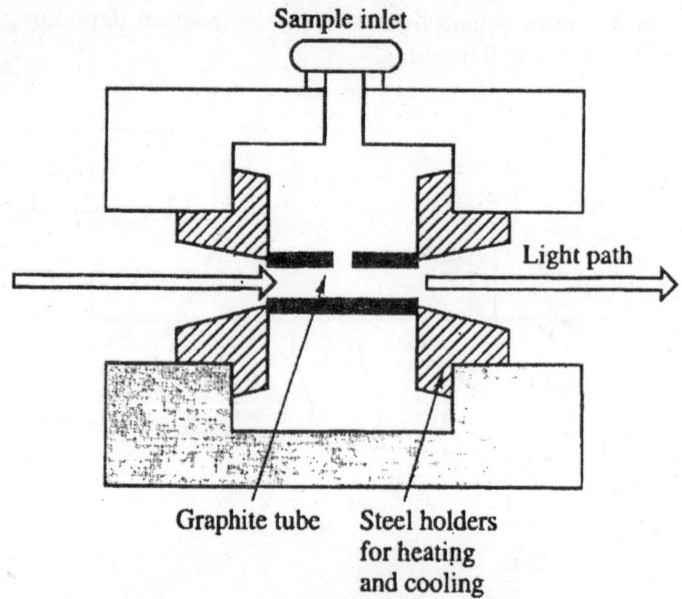

Fig. 2.36. An electrothermal atomizer.

The sample is introduced into the cuvette either using a micropipette or as a weighed portion. It is obvious that these manipulations are a potential source of error and considerable care must be taken at this stage. The first step in the analysis is the heating of the sample (ashing) to remove the solvents and to destroy the matrix. The presence of any organic material in the sample will produce smoke or carbon particles and interfere with any absorbance measurements being made at the same time.

Care must be taken in the choice of the ashing temperature to ensure that the sample matrix is destroyed and the resulting smoke eliminated without the loss of any of the test element before the analytical stage. This is particularly important with the more volatile elements such as mercury and lead. When ashing is complete, the temperature of the cuvette is rapidly raised to the atomizing temperature, which is usually about 1000°C higher than the ashing temperature. This causes vaporization of the sample and a resulting transient increase (2–5 seconds) in the absorbance.

Atomic absorption spectrophotometers must be able to detect and record this maximum absorbance value (Fig. 8.37). The cuvette is cleaned after the analysis by a further increase in temperature of several hundred degrees. When samples contain a large amount of protein or other organic material, repeated heatings or mechanical cleaning may be necessary. Although electrothermal atomizers have certain advantages, they are slower than flame techniques particularly when large numbers of samples have to be analysed, and the transient readings which result from such methods may show poorer precision than do the steady readings obtained by sample aspiration. A microsampling system known as the Delve's cup is a hybrid of flameless and flame techniques.

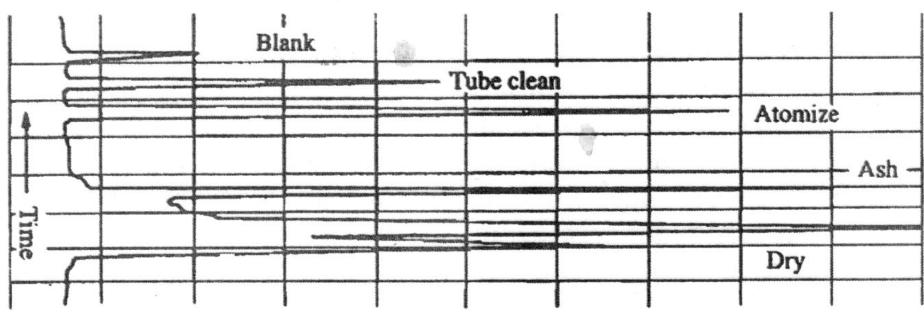

Fig. 8.37. Analysis of chromium by atomic absorption spectrophotometryusing electrothermal atomization.

Analysis sequence (monitoring at 357.9 nm)		
Dry	200°C	25 s
Ash	1400°C	30 s
Delay		10 s
Atomize	2950°C	5 s
Delay		10 s
Tube clean	3100°C	5 s
Delay		10 s
Blank	2950°C	5 s

The sample is placed in a small crucible, which is held in the flame by means of a wire loop. The sample is ashed in a cooler part of the flame and then moved to the hotter part in order to cause the rapid vaporization of the element. The cup is held beneath an opening in a nickel or aluminium tube which is in the light path of the instrument. The atomic vapour is trapped in the tube for a short period of time (up to 10 seconds) while the absorbance measurements are made.

Interference Effects

Atomic absorption spectroscopy is highly specific and there are very few cases of interference due to the similar emission lines from different elements. General interference effects, such as anionic and matrix effects, are very similar to those described under flame emission photometry and generally result in reduced absorbance values being recorded. Similarly, the use of high temperature flames may result in reduced absorbance values due to ionization effects.

However, ionization of a test element can often be minimized by incorporating an excess of an ionizable metal, *e.g.* potassium or caesium, in both the standards and samples. This will suppress the ionization of the test element and in effect increase the number of fest atoms in the flame. Any ions that need to be introduced into a sample either to prevent ionization or suppression effects (for example, a lanthanum and caesium mixture) or as an internal reference (for example, lithium) are usually incorporated in the diluting fluid in which the samples are prepared.

Quantitative Methods

The relationship between absorbance and the concentration of the sample as specified in the Beer-Lambert equation is only true over a limited concentration range and, in practice, a calibration graph may show some curvature. The main reasons for this lie in the interference effects and the stray light present with flame techniques due to the narrow bandwidth of emission lines and the broad bandwidth of the monochromator. Obviously other factors such as flame instability and sample distribution within the flame play a part. It is usual to record data in absorbance units and, although a straight line relationship is theoretically valid, the effective linear range does not usually exceed 1.0 absorbance unit and in many cases may be only up to 0.5 absorbance units.

In order to increase the versatility of an instrument, some manufacturers incorporate a concentration mode in which it is possible to alter the sensitivity of the instrument and so work over various concentration ranges. If a direct read-out of concentration is used, the problem of non-linearity becomes more serious and in order to try to overcome this, some instruments incorporate a curvature correction device.

MAGNETIC RESONANCE SPECTROSCOPY

The absorption of electromagnetic radiation resulting in molecular transitions is not restricted to the ultraviolet, visible and infrared regions of the spectrum; energy associated with the longer wavelengths, *i.e.*, microwaves and radiowaves, can be absorbed and result in molecular transitions. Nuclear magnetic resonance (NMR) spectroscopy and electron spin resonance (ESR) spectroscopy are examples of techniques that depend upon the magnetic properties exhibited by nuclei and electrons in certain molecular situations. They are useful tools in the study of molecular structures and biochemical reactions such as enzyme catalysis. Magnetic properties are only exhibited by molecules that have either atoms with an odd mass number or an uneven number of electrons.

These two features provide the basis for NMR and ESR spectroscopic techniques respectively. Nuclei with an equal number of protons and neutrons will have an even distribution of charge, as will an electron orbital occupied by the full complement of two electrons with opposite spins. An imbalance in either will give the atom or orbital a magnetic moment, i.e. it will have a positively charged section and a negatively charged section. Under normal conditions there will be no preferred orientation but if the molecule is placed in a magnetic field it will tend to occupy a position involving the least energy and will, like a compass needle, align itself with the magnetic field. In most instances the only other possible orientation is in a position directly opposite to the magnetic field, *i.e.*, at 180° to the original position.

The transition between these two orientations, which is due to the electric field effects, will require specific quanta of energy, which can be provided by the magnetic aspects of electromagnetic radiation in a comparable manner to absorption in the ultraviolet and visible region of the spectrum. The energy is provided by radiation in the microwave and radiowave regions of the spectrum with wavelengths in centimetres or metres, corresponding to frequencies of gigahertz (GHz) and megahertz (MHz). These are associated with electron (ESR) and nuclear (NMR) transitions respectively. A nucleus with its magnetic moments aligned with the magnetic field is said to show 'parallel' alignment, while the higher energy transition state is said to be 'antiparallel'. For electron magnetic moments the terms used are 'up-spin' and 'down-spin', the latter being the higher energy state.

Table 8.6. Elements Important in NMR Studies of Biological Compounds.

Group of compounds	*Relevant atoms*	*Natural occurrence (%)*
All	^{1}H	100
	^{13}C	1
Nucleotides		
Phospholipids	^{31}P	100
Phosphorylated compounds		
Amino acids	^{14}N	99.9
Peptides	^{15}N	0.4
Proteins	^{33}S	0.7
Substitute for hydrogen	^{2}H	0.2
	^{19}F	

There are an appreciable number of elements with odd atomic mass numbers although some are rare and have only a low natural occurrence. Those that are useful in biological applications of NMR are given in Table 8.6. These include the hydrogen atom (^{1}H) and the isotope of carbon (^{13}C), both of which are applicable to all biochemical compounds. The isotopes of phosphorus (^{31}P) and nitrogen (^{15}N) are useful in the study of nucleotides and amino acids or their derivatives. The hydrogen nucleus is particularly useful but does usually require the

NMR studies to be performed in a solvent that is free from protons, deuterated water, D_2O, being most frequently used. However, it is possible to analyse aqueous samples by saturating the signal due to water by applying a long pulse (1–2 s) of radiation at the water frequency.

This equilibrates the hydrogen nuclei between their various energy levels and after the saturating radiation is stopped, but before the hydrogen atoms revert to their ground state (relax), the sample is test scanned with a short pulse of radiation. The number of molecules with single electron orbitals, and therefore suitable for ESR, is limited due to the electron-sharing feature of the usual covalent bond. This tends to restrict its use to compounds containing transition metals and reactions involving free radicals. However, this does make ESR very useful for monitoring reactions involving metalloenzymes or free radicals.

Instrumentation

The basic technique of magnetic resonance spectroscopy involves placing the sample in a magnetic field generated between the poles of either a permanent magnet or an electromagnet. The energy for the transition is produced by a radio-frequency generator and the strength of the signal is measured by a radio receiver. The detection of radio signals does not demand an optical light path as in ultraviolet/visible spectroscopy and in some instruments the same probe (antennae/aerial) is used both as a transmitter and as a receiver, with very rapid changes from one function to the other.

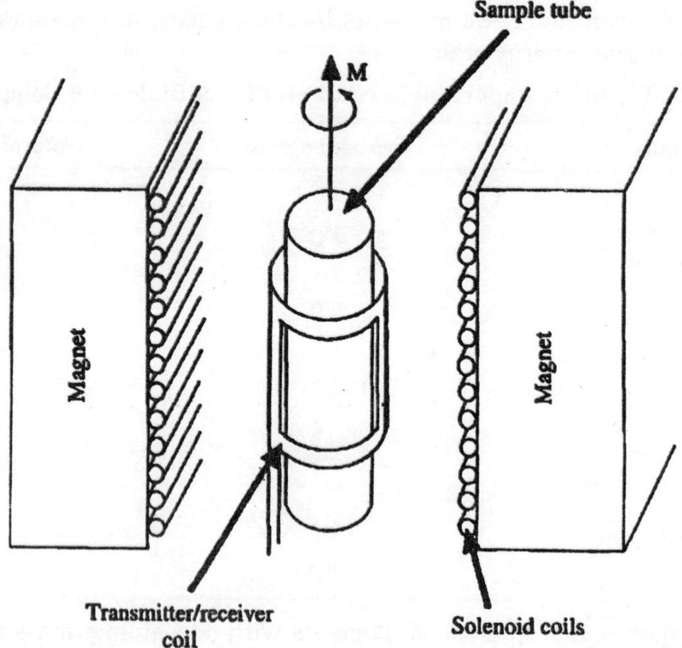

Fig. 8.38. Diagram of an NMR probe. The solenoid coils on the pole faces of the magnet produce the variable magnetic field. The direction of the magnetic field through the sample is indicated by M.

The energy required to cause a transition depends not only on the molecular/atomic features of the test substance but also on the strength of the magnetic field that has to be overcome during the transition. If the technique was directly comparable with ultraviolet/visible spectroscopy, the absorption characteristics would be monitored as the radio frequency was increased.

However, in practice, because of technical difficulties in the generation and detection of a signal of varying frequency, the frequency is held constant and as the magnetic field is changed the absorption is monitored. The magnets give a constant magnetic field and the variation required is introduced by coils mounted on the pole faces of the magnet; the current applied to the coils provides the basis for the x axis of the resulting spectrum, *i.e.*, comparable to wavelength in a visible absorption spectrum.

All magnets show some variation in the magnetic field across the face of the magnet and in order to average out the effects of these differences, the sample is rotated rapidly in the field. The NMR scan is basically a series of pulses and delays. A short pulse or radiation is applied to the sample as the magnetic field is scanned. If the signal due to water is to be suppressed there is an initial saturation signal at the water frequency as described earlier.

Before another test scan can be taken, the sample must be allowed to relax and during this period a further saturation of the water signal can be done and the process repeated. The sequence of events and collection of data are automated.

The Spectrum

It is technically possible, but very difficult, to measure the exact frequency of a radio signal, and in practice the frequency of the energy absorbed by a test compound (usually called the resonance frequency) is measured relative to that of a reference compound. This reference may be mixed with the sample (direct referencing), or if contamination of the sample is undesirable it may be placed in a separate container within the sample tube (external referencing).

In proton and ^{13}C NMR, the reference compound usually used is TMS (tetra-methyl silane) or its water-soluble derivative DSS (2,2-dimethylsilapentane 5-sulphonic acid). These compounds give a sharp proton peak at the right-hand side of a typical NMR spectrum. The difference between test peak and the reference peak is known as the 'chemical shift' of the test. This is a scale of frequency normalised to give the reference peak a value of zero.

$$\text{Chemical shift} = \frac{\text{sample frequency} - \text{reference frequency}}{\text{reference frequency}} \times 10^6$$

This value, being a ratio, has no units and the factor 10^6 makes it a manageable figure. As a result chemical shift values are reported as being 'parts per million' (ppm). The absorbing property of a nucleus is affected by the magnetic effects of adjacent nuclei. This results in what is known as 'spin-spin splitting' of the initial resonance peak.

In general, if there are n adjacent similar nuclei, the peak will be split into $n + 1$ components and this can give information about the manner in which the nuclei are arranged in the molecule. Fig. 8.39 shows a proton NMR spectrum of ethanol, a compound that has three different kinds of hydrogen nuclei, those in the CH_3, those in the CH_2 and the one in the OH group. These three kinds of nuclei absorb energy at different frequencies and as a result give three resonance peaks in the spectrum.

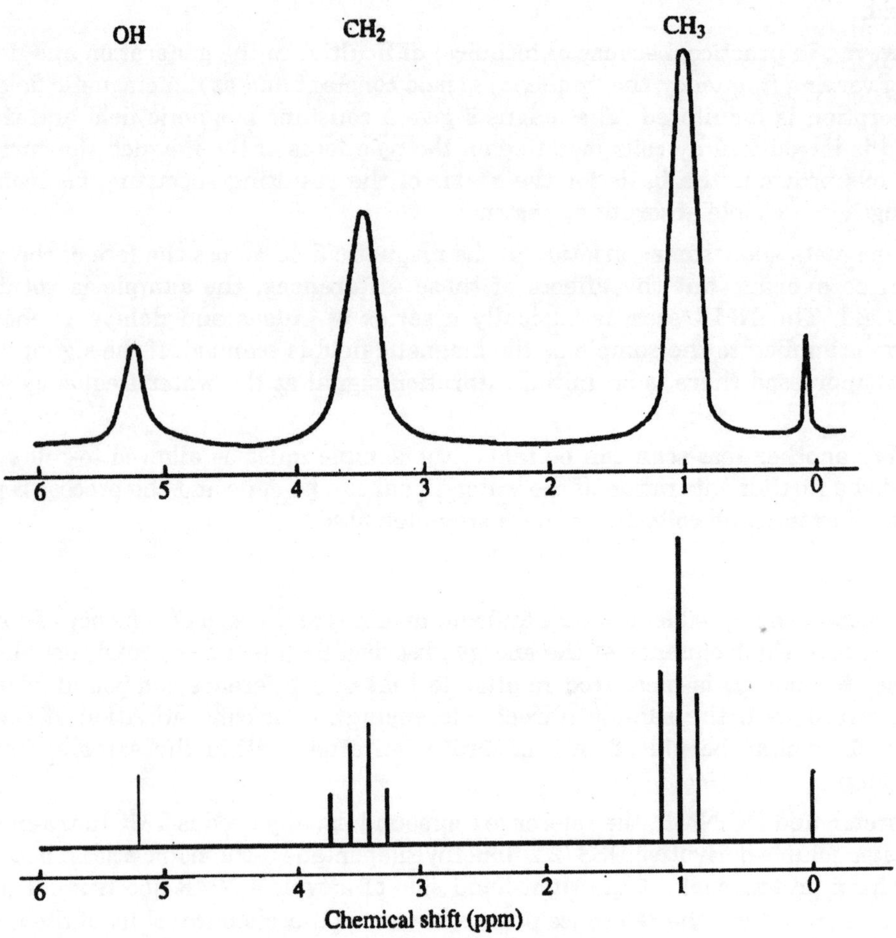

Fig. 8.39. NMR spectrum of ethanol (CH_3CH_2OH). The three peaks are due to the three groups in the compound, which can be split into multiple peaks using a higher resolution. The peak at zero chemical shift is due to the reference compond, DSS.

The integrated area under each peak is proportional to the number of protons in the group and the position of the resonance peak on the scale (the chemical shift) is characteristic of the particular molecular environment in which the protons are found, *i.e.* the specific group. The peak on the right-hand side of the spectrum with a chemical shift of zero is the reference peak of DSS. When a higher resolution spectrum is produced some peaks are split into multiple peaks.

The methyl protons, each giving the same chemical shift, are all affected by the two methylene protons, resulting in three peaks from the original one (*i.e.* $n + 1$). In a similar manner, the two protons of the methylene group are affected by the three protons in the methyl group, which results in the single methylene peak splitting into four.

Applications

The biological applications of NMR include the study of the structure of macro-molecules such as proteins and nucleic acids and the study of membranes, and enzymic reactions. Newer methods and instruments have overcome, to a large extent, the technical difficulties encountered with aqueous samples and the analysis of body fluids is possible, permitting the determination of both the content and concentration of many metabolites in urine and plasma.

NMR is not a very sensitive technique and it is often necessary to concentrate the sample either by freeze drying and dissolving in a smaller volume or by solid phase extraction methods. The principles of spectroscopy are very similar to NMR spectroscopy but the technique gives information about electron delocalizations rather than molecular structure and it enables the study of electron transfer reactions and the formation of paramagnetic intermediates in such reactions. In some situations, information regarding molecular structure can be obtained when suitable prosthetic groups are part of a molecule, *e.g.* FMN (flavin mononucleotide) in certain enzymes or the haem group in haemoglobin. Sometimes it is possible to attach suitable groups to molecules to enable their reactions to be monitored by ESR techniques. Such 'spin labels' as they are called, are usually nitroxide radicals of the type.

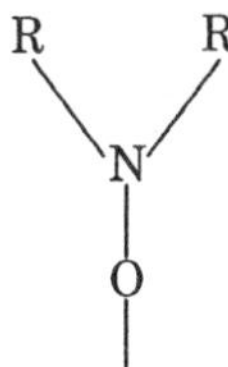

CHAPTER 9 Analytical Methods in Immunity

The roots of immunoanalytical methods are found in the 19th century. By 1890 von Behring and Kitasato demonstrated the neutralising characteristics of bacterial antitoxins. In an effort to identify the origin of the most common bacterial diseases at the turn to the 20th century, the Widal test to verify typhoid (1896) or the Wasserman test to confirm syphilis (1906) antibodies were used as analytical tools. Subsequently, the rapid progress in biosciences, together with a deeper understanding of the basic principles of the immune system, led to a vast increase in different methods taking advantage of the specific interaction between antibodies and antigens. It is beyond the scope of this chapter to describe all the laboratory methods, that rely on the antigen-antibody interaction. But a few methods that are currently used will be described in more detail.

Since the analysis of insulin by Berson and Yalow 623, several different immunoassays for determination of a wide variety of drugs, marker substances in blood, and biological proteins have been developed. To date, many assays are commercially available as kits being used routinely in automated clinical laboratories. When a kit cannot be purchased, part of this chapter should help to establish an immunoassay in your lab. Additionally, immunoanalytical methods become more and more important in our understanding of how cells function.

The development of flow cytometers makes multiparametric analysis of single cells possible. Furthermore, confocal laser scanning microscopic imaging of tissues helps us to visualize processes occuring within cells and tissues. Both techniques use immunoflourescence and provide for a powerful tool in life sciences in the future.

STRUCTURE AND CHARACTERISTICS OF ANTIBODIES

The antibodies are glycoproteins that belong to the group of immunoglobulins. Primarily, they are secreted by plasma cells in response to an immunogen. Each plasma cell secretes about 2000 antibody molecules per second, amounting to about 20% (w/w) of blood. Because blood can be sampled easily, antibodies provide for a powerful analytical tool as a result of their specific interaction with antigens. Structurally, antibodies are Y-like proteins composed of four polypeptide chains, which stick together by three disulfide-bridges and hydrophobic interactions. Each molecule contains two identical heavy chains, each composed of about 440 amino acids and two identical light chains, each containing 210 to 230 amino acids.

There are five types of different heavy chains termed as α-chains, γ-chains, δ-chains, ε-chains, or μ-chains. Accordingly, five classes of antibodies, namely IgA, IgG, IgD, IgE, and IgM, are distinguished (Tab. 9.1). These isotypes of antibodies can contain two identical types of light chains, either κ-chains or λ-chains. The heavy and light chains are made up of domains with sequence homologies comprising about 110 amino acids, which fold into parallel sheets linked by an intradomain disulfide bond. Thus, light chains contain two domains and heavy chains contain four domains.

Table 9.1. Isotypes of Immunglobulins. IgM and IgA Additionally Contain Joining Peptides to Connect the Y-units. IgA also Exhibits a Secretary Protein.

Isotype	*heavy chain*	*light chain*	*molecular formula*	*mass (kD)*	*valency*	*function*
IgG	γ	κ or λ	$\gamma_2\kappa_2$ or $\gamma_2\lambda_2$	150	2	Secondary response
IgM	μ	κ or λ	$(\mu_2\kappa_2)_5$ or $(\mu_2\lambda_2)_5$	950	10	Primary response
IgA	α	κ or λ	$(\alpha_2\kappa_2)_{1-3}$ or $(\alpha_2\lambda_2)_{1-3}$	180–500	2, 4, 6	Secretory Ig
IgD	δ	κ or λ	$\delta_2\kappa_2$ or $\delta_2\lambda_2$	175	2	Not known yet
IgE	ε	κ or λ	$\varepsilon_2\kappa_2$ or $\varepsilon_2\lambda_2$	200	2	Allergy-Ig

The amino-terminal domains of the heavy and light chains mediate antigen binding and are variable, whereas the remainder are constant regions. Each variable domain is composed of four framework regions and three hypervariable regions. The short loops of the hypervariable region of each heavy and light chain are referred to as complementary determining regions (CDR). Each CDR comprises 5–10 amino acids and protrudes, forming the site of interaction with the antigen. The area of interaction between the epitope, which represents the deteminant on the surface of the antigen, and the paratop which is the complementary structure formed by six CDRs of the antibody, was found to be about 1.6 × 0.6 nm.

The interaction between antigen and antibody is reversible and is in equilibrium with the free components. Thus, the precise fitting is a prerequisite for non-covalent interactions such as hydrogen bonds, van der Waals forces, hydrophobic bonds, and sometimes ionic interactions, which stabilize the immune complex. The high variability of antigen-binding site derives from about 45,000 different variable domains of the heavy chain, which are genetically encoded by 4 J(joining)-regions, 15 D(diversity) regions, and 250 V(variable) regions, which is extended by three kinds of combination.

The variable domain of the light chains is encoded by 250 V-segments and 4 J-segments, resulting in about 3000 different variable domains of the light chain. Linking the variability of heavy and light chains results in at least 1.35×10^8 antibodies with different specificity. Additionally, somatic mutations contribute to the diversity of antibody specificity. Exposition to proteolytic enzymes demonstrates the highly folded structure of antibodies. Upon treatment with papain, the heavy chains are cleaved N-terminally to the interconnecting disulfide bond at amino acid 224, yielding three fragments. Two single fragments with paratops at the N-terminus for antigen binding are known as Fab-fragments, corresponding to the arms of the Y. The third fragment that corresponds to the base of the Y was isolated by crystallisation and is known as Fc-fragment. This part of the immunglobulin is not involved in antigen-binding but plays a key role in immune regulation, *e.g.,* activation of the complement system and macrophages.

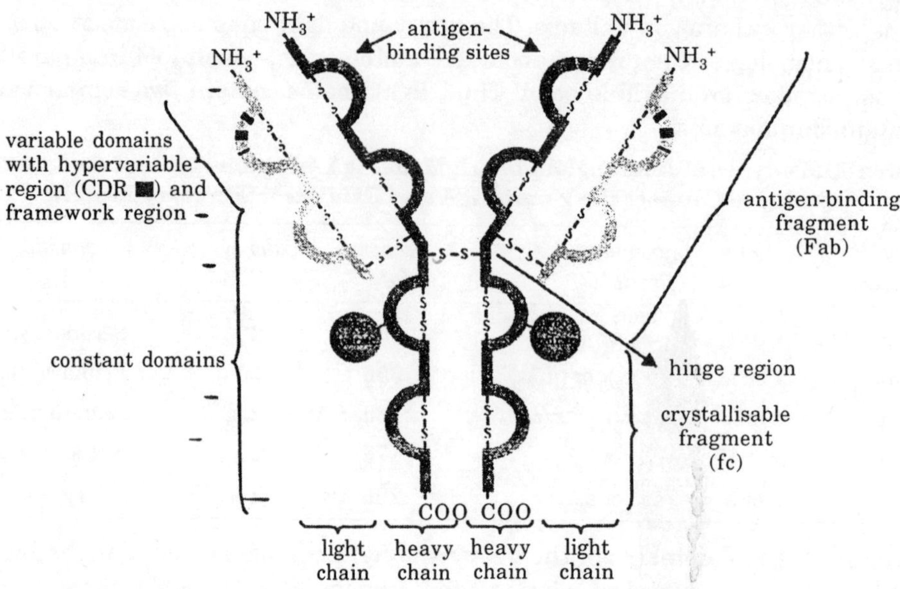

Fig. 9.1. Scheme of IgG-molecule

The region between the Fab and Fc-fragments is known as hinge-region, which provides for flexibility of the Fab-fragments. Lateral and rotational movement facilitates the binding of two determinants of different conformations of a certain antigen. Especially as a reagent in immunoanalytical techniques, the product of pepsin treatment becomes more and more important. Pepsin attacks the immunoglobulin at C-terminal positions below the heavy chain connecting disulfide bond at amino acid 234. Thus, a divalent fragment referred to as $F(ab)_2$-fragment is obtained, which possesses two antigen binding sites, but the Fc-portion is fully digested by pepsin. This $F(ab)_2$-fragment is a useful substitute for intact immunglobulins, as the Fc-fragment is known to interfere in many techniques, resulting in higher background staining or higher signal/noise ratio.

There are three terms closely related to each other that describe utility of antibodies in practice. Affinity describes the exactitude of stereochemical fit of the epitope of the antigen to the paratope of the antibody. Thermodynamically, it indicates the strength or energy of the interaction. Deriving from the law of mass action, it is described by an association constant K_a indicating the ratio of bound and unbound antigen (mol^{-1}). Thus, the affinity constant is the reciprocal value of that concentration of a certain antigen, where half of the antigen is bound to the paratopes. In practice, this is the amount of antigen-antibody complex at equilibrium. High affinity antibodies will bind larger amounts of antigen in a shorter period of time than low affinity antibodies. Consequently this term K_a is independent from the number of paratopes of an antibody molecule.

To characterize the overall strength of the antigen-antibody interaction, the term avidity is used. It includes affinity, valency of the antibody, and arrangement of the interacting compounds. Assuming that the affinity of an antigen to the paratope of IgG and IgM is equal, the avidity of IgM is five times higher than that of IgG because of the pentameric IgM. Thus, avidity is also expressed by an association constant K for a particular test system. When an antigen binds to an antibody, this interaction is termed to be specific. But some antigen determinants are shared between molecules, especially closely related ones. Thus, cross-reactivity describes the fact that common epitopes on different antigens are bound by the same antibody. On the other hand, cross-reactivity also can derive from binding of structurally different epitopes by the same antibody, especially in antisera.

FROM ANTIGENS TO ANTIBODIES

Basic Considerations

Antibodies used as reagents in immunoanalytical methods are host proteins that are produced by the immune system in response to foreign large molecules in the body. To provoke formation of antibodies by B-cells and cellular immune response mediated by T-cells, an immunogen is required. According to the clonal selection theory by Sir MacFarlane Burnet, the immunogen binds to well-suited antibodies located at the surface of virgin B-cells, which determine specificity of the antibodies finally obtained. The immunogen is degraded within the virgin B-cell, and its fragments are presented at the surface as a complex with class II proteins. This complex is recognised by T-helper cells, which induce differentiation of B-cells to antibody-secreting plasma cells.

For these reasons, epitope binding, degradability, and mediation of the B-cell – T-cell interaction require a minimum molecular weight of the immunogen of about 5 kDa. In this context, it has to be considered that animals usually used for immunization have developed some self-tolerance, eliminating well-suited B-or T-cells. Thus, the immunogen or carrier protein should be foreign to the immunized species.

ANTIGEN PREPARATION

Large Molecules

According to the requirements of epitope binding, degradability, and mediating interaction of B-and T-cells, high-weight proteins should elicit an immune response, which usually increases with a higher degree of conformation and structural rigidity. The antigenic sites at the surface of proteins molecules can be made up of different domains of amino-acid side chains, which are distant in sequence but close in space. In contrast to these conformation-dependent determinants, fibrous proteins possess sequiential determinants comprising up to six amino-acid residues.

Probably all proteins are immunogenic, but individual proteins may differ markedly in immunogenicity. Self-aggregation of proteins usually is associated with a slight change in its specificity, but with increased immunogenicity. On the other hand, large molecules can be made more immunogenic by denaturation. Heating or heating in the presence of sodium dodecyl sulphate will increase immunogenicity, but, concurrently, specificity will be altered because hidden epitopes will become accessible. Polysaccharides can act as immunogens in purified form, but only in humans and mice, not in rabbits or guinea pigs. Additionally, antibodies against nucleic acids are described, but they are elicited only in humans suffering from the autoimmune disease lupus erythematodes or in animals with similar diseases.

Small molecules

In contrast to large molecules, small molecules including drugs, steroids, or peptides do not elicit an immune response. As first shown by Landsteiner exemplified by 2,4-dinitrophenol, low-molecular-weight compounds termed as haptens have to be coupled to carrier proteins to provide for class II protein-T-cell binding, which results in formation of antibodies reacting specifically with the dinitrophenyl group.

To gain better access to virgin B-cell antibodies, antennary exposition of the hapten by covalent linkage via spacer molecules at the surface of the carrier molecule is favoured. The most common carrier proteins are serum albumins of various species, preferably the well-characterized bovine serum albumin or high immunogenic keyhole limpet hemocyanin from the mollusc megathura crenulata, but oval-bumin and thyroglobulin or fibrinogen also are described. For conjugation of hapten and carrier protein, different protocols are described in the literature, but the use of heterobifunctional cross-linkers and two-step mechansims is highly recommended to avoid formation of homopolymers and precipitation of the hapten-carrier complex.

Table 9.2. Reagents for conjugation of haptens to proteins

Hapten with Carboxylic Group–carrier Amino Group

Mixed anhydride method: Activation of the tributylammonium salt of the hapten in anhydrous milieu with isobuytlchloroformate to yield the rather unstable mixed anhydride. Coupling at alkaline pH, adjustment of pH necessary.

Notes : Racemisation omitted; no side products to be removed; isolation of activated hapten not necessary.

Carbodiimide method: Activation with DCC in anhydrous or EDAC in aqueous milieu at pH 4-5, coupling at pH 8.

Notes : Works well; probable formation of autopolymers; remove excess cross-linker; use water-miscible anhydrous solvents for activation such as DMF, THF, or dioxane.

Active ester method (carbodiimide/N-hydroxy-succinimde method): Activation and coupling is performed at neutral pH.

Notes : Works well; isolation of activated intermediate recommended; thiol-, hydroxyl-, and amino groups can react as well.

Hapten with Amino Group

Imidoester method : Homobifunctional cross-linker, activation and coupling at pH 9.

Notes : Isolation of activated intermediate required to omit polymerization, design of spacer-length possible (dimethyl-adipimidate 0.86 nm, dimethyl-pimelimidate 0.92 nm, dimethyl-suberimidate 1.1 nm).

Glutardialdehyde : Homobifunctional cross-linker, activation and coupling at alkaline pH.

Notes : Isolation of intermediate is recommended; reagent is a polymer; sometimes reduction by sodium-borohydrid is required for stability.

SPDP (N-succinimidyl-3-(2-propyldithio)-propionate: Heterobifunctional, hapten activated at pH 7–9, coupling to thiol-containing carrier after conversion of amino- to thiol-residues with 2-iminothiolan or cleavage of carrier disulfide bonds with dithiothreitol.

Notes : Mild conditions; degree of substitution measured by appearance of 2-thiopyridon.

Hapten with Hydroxyl-group Carrier with Amino Group

Hemisuccinoylation : Hapten converted to carboxylderivative by reaction with succinic anhydride in anhydrous milieu followed by use as carboxylated hapten.

Notes : Spacer is introduced.

Divinylsulfone : Derivatisation of hydroxyls at pH 11, coupling at pH 9.

Notes : Inactivation of excess reagent by addition of glycine.

Periodate oxidation : Mildly alkaline conditions required for cleavage and oxidation of cis-diol partial structures, coupling of aldehyde at pH 9.

Notes : Stability is increased by reducing the aldimin with sodium borohydride.

The amount of hapten coupled per molecule of carrier protein can be determined only after rigorous purification of the artificial antigen by dialysis, gel permeation chromatography, or precipitation with ammonium sulphate or any other useful method. Most simply, the degree of substitution can be determined by UV-difference spectroscopy of the carrier protein and the conjugate. Appearance of an additional peak at the maximum absorbance of the hapten in the spectrum of the conjugate in comparison to the carrier protein is a qualitative confirmation of successful coupling. When the solutions are carefully prepared by weighing accurate amounts of the lyophilised conjugate and the carrier protein and the solutions are adjusted to the same concentration, the hapten content of the conjugate can be estimated quantitatively.

An alternative is an indirect assay by determining the number of free amino groups of the carrier protein prior to and after coupling of the hapten. The difference points to the number of modified amino residues, which corresponds to the number of hapten molecules attached. A rather time-consuming method is the 2, 4-dinitro-2-fluorobenzene test, originally applied by Sanger for determination of insulin.

A simple and rapid alternative is the trinitrobenzene sulfonic acid test, which avoids total hydrolysis and extraction steps of dinitrophenylated amino acids. But it should be considered that non-covalent interactions also can contribute to hapten-fixation on the carrier protein, which amounts in the case of bovine serum albumin to about 2 to 3 hapten molecules. While bovine serum albumin has 59 lysine residues, only 35 are accessible at the surface for coupling at neutral pH. A conjugation rate of about 10 hapten molecules per carrier molecule is sufficient to obtain not only serum albumin-specific anti-bodies but also hapten-specific ones

Immunization

Before an immunization protocol is started, the legal responsibilities should be considered. In the countries of the European Union, as well as in the USA, a personal and a project license is required for all immunization and bleeding procedures. The choice of the animal species to be immunized greatly depends on the facilities. Most commonly, rabbits of 4 to 6 months of age are used, as about 25 ml of serum is yielded from a single bleed. Guinea pigs, which are often hard to bleed, rats, or hamsters yield 1 to 2 ml, while mice yield only 200 µl. For rabbits, two animals should be used as a minimum, whereas in the case of rodents three to six animals are recommended.

It is self-evident that the animals should be held under proper environmental conditions preventing any stress. Prior to immunization of an animal, the antigen is mixed with adjuvants.

Because adjuvants enhance the humoral immune response and activate unknown sectors of the immune system, immunization never should be done without adjuvants. Heat-killed bacteria containing lipopolysaccharides *(Borde-tella pertussis)* or muramyldipeptides *(Mycobacterium tuberculosis)* stimulate lymphokine production resulting in strong local inflammatory reaction, which mobilizes macrophages.

On the other hand, aluminium salts precipitate the antigen and mineral oils form water-in-oil emulsions. Both result in formation of a depot, which prevents rapid degradation of the antigen and guarantees sustained nonspecific stimulation of the immune response. Most commonly, the primary injection is given as an emulsion of the immunogen in complete Freund's adjuvant containing killed *Mycobacterium tuberculosis* and mineral oil, which provokes aggressive and persistent granulomas. But the boosts are administered as a mixture with incomplete Freund's adjuvant containing mineral oil and emulsifier only. As a rough basic rule, in rabbits about 0.5 to 1 mg antigen in adjuvant is administered, preferably subcutaneousely, on multiple sites at low volumes for the primary immunization, which corresponds to 50 to 100 μg antigen for mice. 'When only small amounts of poor immunogens are available, the antigen might be injected in rabbits in the poplietal lymph node at the lower leg to enhance the probability of an appropriate immune response. In this latter case, however any adjuvant must be omitted. The dose of the following boost injections about three to four weeks after the first injection might be one-third to one-half that of the first immunization. For deeper insight into immunization procedures, there is some useful literature available.

Sampling Serum and Storage

After the first injection of the immunogen, the primary response results in exponential increase of antibody secretion occuring within 30 days. In the majority of cases, the antibody concentration in the blood, termed as liter, reaches a maximum within 9 to 11 days so that the first test bleed can be collected. When using rabbits, the animal is placed in a small box and a small distal area of the rear ear is rinsed with a small amount of 70% ethanol.

When the marginal ear vein is not apparent, it might be dilated by gently rubbing or warming with a lamp, but avoid use of toluene or xylene. The area, which should be incised or opened by insertion of a thick injection needle, is covered with Vaseline. The blood is collected by droping in a clean test tube. When the blood flow stops by clotting, the incision area should be wiped with sterile warm water. After collecting usually 5 ml, but 20 ml at the maximum, the needle is removed and gentle pressure is applied to the cut for half a minute. Check to be sure that the blood flow has stopped. Before screening this first test bleed for presence of specific antibodies, the blood is allowed to coagulate for one hour at 37 °C.

The clot is removed with a Pasteur pipette or by simply rotating a skewer made from wood followed by centrifugation for 10 min. at 13,000 rpm to remove any remaining insoluble material. The clot might be allowed to retract overnight to yield an additional few milliliters of serum. Additionally, the serum can be heated to exactly 56°C for 10 to 20 min. to inactivate the complement proteins. These proteins can interfere with some immunoanalytical techniques by binding to the antigen-antibody complex. The serum can be stored for many years at –20 °C or below. When stored at 4°C for weeks, it is recommended to add 0.02% sodium azide or thiomersal at 1:10,000. But be aware of the fact that sodium azide is highly sonous and can interfere with some immunoanalytical techniques. Test bleeds can be taken from mice in a similar way or from rats by incision the tail veil.

The test bleed usually yields a volume of 200 to 400 µl and can be treated as described above. The first test bleed predominantly contains IgM and only few IgG, but antibody formation declines with time. About one month after the first immunization, another exposure to the immunogen provokes the secondary immune response, which yields a stronger response by activating memory cells. Moreover the antibodies of the secondary response belong predominantly to the IgG-type and often exhibit higher affinity, which might even be increased by the number of boosters.

The titer should be monitored by appropriate methods such as enzyme immunoassays or simple immunodiffusion assays. When a good titer has developed, regular boosts are preformed every 6 weeks, and up to 50 ml blood (rabbit) is collected 10 days after the booster injection. Through the course of time, affinity, specificity and cross-reactivity are altered.

The concentration of the antigen in the animal decreases with time so that lymphocytes producing low-affinity antibodies are not stimulated further, whereas those secreting high-affinity antibodies are still stimulated. By boost injections with low doses of antigen, the threshold value for stimulation remains high, and only those lymphocytes remain to be stimulated that produce high-affinity antibodies. The decrease in specificity with time probably results from early formation of antibodies with low affinity. Each determinant of the antigen provokes formation of specific antibodies with different lag times. Thus, shortly after administration of the immunogen the number of antibodies with different specificity is low, but it increases with time. Additionally, cross-reactivity increases by time. In response to different determinants located at the surface of an antigen, a higher number of different antibodies is formed. Thus, it is likely that certain antibodies also recognize antigens with similar conformation.

Polyclonal Antibodies

According to the clonal selection theory, a clone of B-cells produces antibodies of only one specificity. When an animal is immunized with an antigen, a random number of clones is activated and, consequently, a polyclonal antiserum is obtained. This strategy of the immune system to produce polyclonal antibodies yields important advantages in terms of affinity and specificity. A major advantage of polyclonal antibodies is the formation of large insoluble immune complexes with polyvalent antigens. At a molar ratio of 4 Mol antigen mol antibody, the multivalent antigens and the bivalent antibodies forms a three-dimensional network, which is opaque and can be determined quantitatively. Additionally, polyclonal antibodies are usually well-suited for cell staining, immunoblotting, and immunoassays with labeled antigens, but it is difficult to perform immunoassays with labeled antibodies. In the latter case, use of monoclonal antibodies is recommended. When usmg antibodies as a reagent tor immunoanalytical techniques, it should be considered that there is a batch-to-batch variation of polyclonal antibodies, even those deriving from the same animal.

The antibodies are heterogenous with regard to specificity, isotype composition, and optimum binding conditions so that each assay has to be adapted. Additionally – with exception of specific pathogen-free animals – the antiserum contains antibodies against all antigens to which the animal was ever exposed. Thus, the antiserum contains at most 20% to 30% immunglobulins, of which about 10% are specific antibodies against the desired immunogen.

Monoclonal Antibodies

In 1984 G. Kohler and C. Milstein were awarded with the Nobel Prize for development of the hybridoma technique, which yields monoclonal and, consequently, monospecific antibodies. The rationale for this technique is to fuse antibody-secreting lymphocytes from the spleen, which produce polyclonal antibodies, with immortal myeloma cells. Thus, the capability of antibody production deriving from the B-cells and the immortality deriving from the B-lymphocyte tumor cells are joined. In order to render the polyclonal antibodies in the supernatant of the fused cells monoclonal, certain antibody-secreting cells are selected by cloning.

The growing hybridomas are diluted and grown in microplate wells as far as the antibodies secreted might stem from the same fused, antibody-secreting cell. Production of monoclonal antibodies requires mice, usually the inbred strain BALB/c, for immunization and plasmacytoma cells such as the non-secreting, HAT-sensitive P3-X63-Ag8.653 cells. Additionally, good tissue culture facilities including a 5% CO_2/95% air incubator, a sterile workbench, and an inverted microscope for visual control of the cultures are needed. In the following some basic items for production of monoclonal antibodies are given.

There are many different protocols for production of monoclonal anti-bodies; for detailed information; the reader is referred to the literature. Per immunogen, a minimum of three RALB/c mice is immunised by subcutaneous injection of about 25 μg immunogen in complete Freund's adjuvant at multiple sites at the back. Four weeks later, half of the amount of immunogen mixed with incomplete Freund's adjuvant is administered intraperitoneally. About one week after the booster injection, the animals are sacrified and the spleen is removed under aseptic conditions.

Using a stainless sieve, the spleen is homogenised, and clotted cells of the connective tissue are removed by centrifugation. Erythrocytes are lysed by osmotic shock, and the spleen cells are recovered in the pellet after centrifugation. The myeloma cells P3-X63-Ag8.653 represent a variant of BALB/c plasma-cytoma cells that do not secrete antibodies and that are sensitive to HAT medium. They should be in the logarithmic phase of cell growth. While fusion of the cells in early work was achieved by inactivated Sendai virus, PEG 4000 is commonly used today. Fusion of the lymphocytes with myeloma cells by PEG requires strict compliance with the selected procedure with regard to time and amounts of reagents added.

After fusion, the cell suspension contains hybridoma cells, non-fused lymphocytes, and myeloma cells. Whereas lymphocytes are mortal and lose their viability during further propagation, the myeloma cells are immortal. In order to eliminate the myeloma cells, the cell suspension is cultured in a selective culture medium (HAT medium) containing hypoxanthin, aminopterin, and thymidin. The myelomn cells suffer from a deficiency of the enzyme hypoxanthin-guanine-phosphoribosyl transferase (HPGRT). Thus, they are not able to synthesize DNA in presence of the HAT medium and lose viability. In contrast, in hybridoma cells the enzyme deficiency is compensated py the parent lymphocytes and they still grow. About two weeks later, the cells are cultured in HAT medium and cells from dense cultures are propagated further on feeder layers containing macro-phages from the peritoneum of mice.

They are obtained by rinsing the abdominal cavity of adult mice with sterile saccharose-solution. Upon propagation of the cell suspension containing hybridomas on feeder-layers, non-growing hybridomas (99%) are degraded by the macrophages, whereas the growing hybridomas (1%) take advantage of the growth factors provided by the macro-phages. Different methods are available to select the hybridomas secreting relevant antibodies.

Most commonly, enzyme-linked immunosorbent assays are used. After coating the wells of a microplate with antigen and washing, an aliquot of the supernatant of the hybridomas is added and washed after incubation. The antigen-bound antibodies are detected with goat-anti-mouse IgG/IgM labeled with horseradish peroxidase. After removal of excess second antibody by washing and addition of substrate solution the absorption of the cleaved, stained substrate is determined photometrically after stopping the enzyme reaction, by addition of acid. According to the results of screening the hybridoma-containing wells for antibody secretion, relevant hybridomas are cloned by limited dilution rather than by the agar technique.

Table 9.3. Characteristics of Polyclonal Antisera and Monoclonal Antibodies

Characteristics	*Polyclonal antisera*	*Monclonal antibodies*
Purity of immunogen	Important	Not important
Time and expense	Low	Initially high
Specific antibodies	0.1–1.0 mg/ml	5-25 mg/mL in culture supernatants mg range in bioreactors
Nonspecific antibodies	About 10 mg/mL (> 80%)	None in serumfree supernatants
Specificity	Recognizing all determinants of immunogens the animal	Recognizing a single determinant of the imunogen
Affinity	Mixture of high and low Affinity antibodies	Hingh or low-according to the screening assay
Avidity	High	Low
Cross-reactivity	Yes	Usually not, but cannot be excluded
Antigen-precipitation	Yes	No
Isotypes and subtypes of immunoglubulins	Typical spectrum	One isotype and subtype
Physical stability	Usually high	Differs markedly
Utility for:		
Cell staining	Good	antibody-dependent, but excellent with pooled antibodies
Immunopreciptation	Good	Antibody-dependent, but excellent with pooled antibodies
Immunoblots	Good	Antibody-dependent, but exellent with pooled antibodies
Immunoaffinity Purification	Poor	Antibody-dependent, poor with pooled antibodies
Immunoassays with Labeled antibody	Difficult	Good, excellent with pooled antibodies
Labeled antigen	Good	Antibody-dependent, but excellent with pooled antibodies

Usually limiting dilution is done in three subsequent series, ensuring that the antibodies in the supernatant derive from fusion of a single B-cell and a myeloma cell. A problem sometimes encountered with production of monoclonal antibodies is loosening of chromosomes. Thus, a significant number of initially positive clones can be lost. This can be prevented by omitting overgrowth of the hybridomas in the wells, but this phenomenon occurs as the hybridomas become monoclonal. The hybridoma cells can be propagated continuosly *in vitro,* and the supernatant contains homogenous antibodies, which recognize only one or a few closely related antigens.

In contrast to polyclonal antibodies, monoclonal anti-bodies are homogenous molecules with regard to specificity, affinity, and isotype. Additionally, in contrast to polyclonal antisera, the predominant protein in the supernatant of hybridomas is the specific antibody that might be used as a homogenous reagent for nearly all immunoanalytical techniques especially for antibody-labeled techniques. Because of their monospecificity, monoclonal antibodies exhibit only poor antigen-precipitating properties as single reagents. Thus, indirect measurement of antigen-antibody-binding by use of a label is possible, which in turn represents the more sensitive assays.

IMMUNOASSAYS

Immunoassays take advantage of the specific interaction between antigen and antibody. They obey the law of mass action:

$$\text{antibody} + \text{antigen} \leftrightarrow \text{antigen} - \text{antibody complex.}$$

At equilibrium, the antigen-antibody complex is formed as described by the association rate constant K1 and concurrently dissociated to yield again its free components according to the dissociation rate constant K2. But when at a constant amount of one free component the amount of the second free component is increased, the equilibrium of the interaction is driven toward complex formation. Thus, immunoassays enable detection and quantification of antigens and antibodies. Formerly, non-labelled techniques such as immunoprecipitation, agglutination or light scattering were used, but they still suffered from low sensitivity in the micromolar range.

Today labelled immunoassays are used that exhibit at least 1000 fold higher sensitivity. Antibodies or antigens are decorated with labels that impart measurable signals such as radioactivity, fluorescence emission, or enzymic activity. The extent of complex formation is then determined by measuring the amount of labelled antigen or antibody. Depending on the assay design, the signal intensity quoted is directly or inversely proportional to the analyte of interest. In order to describe the quality of an immunoassay, the terms sensitivity and accuracy are used. In a narrower sense, sensitivity corresponds to the change in response per amount of reactant, whereas sometimes the term is used to characterize the detection limit of the assay. Accuracy relates the amount of substance determined by the assay to the true value.

Design of an Immunoassay

The criteria to classify immunoassays can be as follows: (1) determination of the antigen or antibody, (2) labelling of antigen or antibody, (3) homogenous or heterogenous assay, and (4) competitive or non-competitive assays.

Homogeneous Assays

Homogenous methods are based on changes in enzyme activity, which is mediated by formation of an immune complex. Consequently, modulation of enzyme activity reflects the degree of the immunochemical reaction resulting in enhancement or inhibition. Thus, measurement of the degree of complex formation is carried out without any physical separation of bound and free compounds.

Enzymes used as labels in homogenous assays are lysozyme, malate dehydrogenase, and β-galactosidase. An assay design using enzyme-labelled antigen or analyte is referred to as enzyme multiplied immunoassay technique (EMIT) and was developed for determination of antiepileptic drugs, cardioactive drugs, anti-asthmatic drugs, and drugs of abuse. The EMIT assay system requires an active enzyme-hapten-conjugate. Upon binding of hapten-specific antibodies, the activity of the conjugate is inhibited by inducing or preventing changes in conformation of the enzyme as necessary for enzyme activity.

Thus, free hapten as analyte and hapten-enzyme conjugate compete for antibody binding. In the presence of excess conjugate, the enzyme activity is directly proportional to the concentration of the hapten. Generally, homogeneous methods for detection of small or large analytes are less sensitive than heterogeneous assays. A major drawback of homogeneous assays is reproducible and time-consuming preparation of well-characterised conjugates.

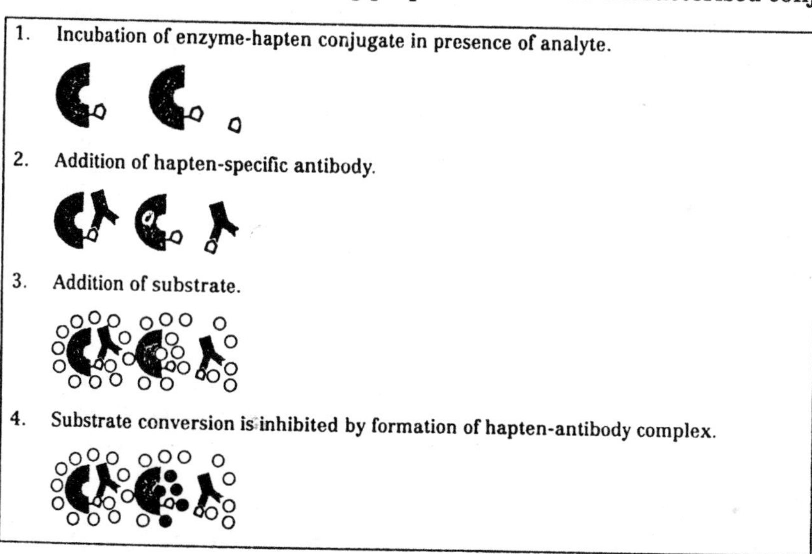

Fig. 9.2. Scheme of EMIT using enzyme-hapten conjugate, free hapten as analyte, and specific antibodies for competitive binding.

In order to modify the enzyme activity by complex formation, the hapten needs to be exposed to guarantee unrestricted interaction with the antibody. On the other hand, a separation step as a source of imprecision is omitted. When a homogeneous method is established, the assay still consists of mixing and measuring facilitating automatisation.

Heterogeneous Assays

In contrast to homogeneous methods, in heterogeneous systems the activity of the labelled ligand is not altered upon formation of the antigen-antibody complex. Thus, most radioimmuno-

and enzymeimmuno-techniques require separation of bound and free-labelled antibody or antigen in order to be able to discriminate between bound and free-labelled ligand, which in turn allows quantification of the analyte of interest. Usually the activity of radiolabelled reactants is not affected by complex formation, whereas the activity of enzyme-labelled immunoreactants can be altered upon binding to the complementary antigen or antibody depending on the site of coupling the enzyme.

In heterogenous enzyme immunoassays, usually horse radish peroxidase, alkaline phosphatase, (β-galactosidase, or glucose-oxidase are used as labels. Methods for separation include use of solid-phase-bound antigen or antibody as well as second antibody precipitation. When enzyme-labelled antigen or antibody is used for detection of the reciprocal solid-phase-bound reactant, this is referred to as enzyme-linked immunosorbent assay (ELISA). The amount of the bound or free fraction is determined by enzyme-driven conversion of a colourless or non-fluorescent substrate to a intense coloured or fluorescent substrate, which is quantitated by photometry or fluorimetry.

Competitive Assays

In competitive assays, the competition in binding to a fixed antibody between a constant amount of labelled antigen and unknown amounts of sample antigen is measured. Of course, this assay can be reversed for measuring the antibody. As the labelled analyte competes with unlabelled analyte for solid-phase binding, the reaction product measured is inversely proportional to the test antigen or antibody. Competitive assays are easy to perform but require high accuracy in dispensing. Additionally, high purity of the labelled ligand is required, but the result is less influenced by contaminants such as cross-reactive and non-specific binding compounds.

However, the measurement of extremely low concentrations of analytes can be affected by non-specific adsorption. Furthermore, optimum conditions must be established for each step of the assay. This includes optimisation of the concentration of the solid-phase ligand and the labelled ligand, the type as well as amount of substrate, and incubation time and temperature. The main advantage of competitive assays is that the detection limit is found in the picogram-range and they are easy to perform once established. Therefore, in most laboratories this type of assay is the first design that is applied upon establishing an assay for a new compound to be quantified.

Non-competitive Assays

These are often called immunoradiometric assay (IRMA) in the case of radioactive tracers or immunoenzymometric assays. In this assay excess of immobilized antibody is incubated with standard antigen or sample antigen. Formation of the antigen-antibody complex is measured in a second step. When multivalent antigens are measured, the antigen can be detected by a second, labelled antibody. Because at least three layers of immunoreactants corresponding to antibody-antigen-antibody are built up, this technique is called sandwich immunoassay. Since the measured signal derives from the amount of antigen-bound, labelled second antibody,

1. Absorb specific antibody to the solid phase.
2. Remove excess of unbound antibody by washing.
3. Saturate remaining binding sites by incubation with blocking buffer.

1. Adsorb specific antibody to the solid phase.
2. Remove excess of unbound antibody by washing.

3. Saturate remaining binding sites by incubation with blocking buffer.
4. Remove excessive blocking agents by washing.

5. Add test antigen in the presence of constant amounts of enzyme-labelled antigen.

6. Remove excess reactants by washing.
7. Add substrate solution and incubate for a certain period of time at appropriate temperature.

8. Stop enzyme activity and quantify converted substrate, which is inversely proportional to the test antigen.

Fig. 9.3. Competitive ELISA with solid-phase antibody using enzyme-labelled antigen for quantification of test antigen.

4. Remove excessive blocking agents by washing.
5. Add test antigen in the presence of constant amounts of enzyme-labelled antigen
6. Remove excess reactants by washing.
7. Add substrate solution and incubate for a certain period of time at appropriate temperature.
8. Stop enzyme activity and quantify coverted substrate, which is inversely proportional to the test antigen.

there is a direct proportionality between analyte and signal quoted. Conversely, this design is also applicable to immobilized antigen, and the antigen-antibody complex can be detected by a second, heterologous-labelled antibody, which recognizes the Fc-part of the first antibody. The results obtained by sandwich ELISA are comparable to those of radio-immunoassays in terms of sensitivity. Additionally, optimisation steps are required as described above.

1. Adsorb specific first to the solid phase.
2. Remove excess of unbound antibody by washing.
3. Saturate remaining binding sites by incubation with blocking buffer.
4. Remove excessive blocking agents by washing.

1. Adsorb specific first antibody to the solid phase.
2. Remove excess of unbound antibody by washing.

3. Saturate remaining binding sites by incubation with blocking buffer.
4. Remove excessive blocking agents by washing.

5. Incubate with multivalent, unlabelled antigen.
6. Remove unbound antigen by washing.

7. Add excess of second, enzyme-labelled antibody.

8. Remove unbound reporter antibody by washing.
9. Add substrate solution and incubate under appropriate conditions.

10. Stop enzyme activity and quantify converted substrate, which is directly proportional to the concentration of the antigen.

Fig. 9.4. Sandwich ELISA for determination of polyvalent antigens with second enzymelabelled, antigen-specific antibody.

5. Incubate with multivalent, unlabelled antigen.
6. Remove unbound antigen by washing.
7. Add excess of second, enzyme-labelled antibody.
8. Remove unbound reporter antibody by washing.
9. Add substrate solution and incubate under appropriate conditions.
10. Stop enzyme activity and quantify converted substrate, which is directly proportional to the concentration of the antigen.

PRACTICAL CONSIDERATIONS

Antigen and Antibody

It is highly desirable to have pure antigen at one's disposal. When two monoclonal antibodies or affinity-purified antibodies are available, which recognize two independent epitopes of the antigen, the sandwich assay is appropriate to detect even impure antigen preparations. To perform competitive assays for detection of antigen, pure or partially purified labelled antigen is required. On the one hand, the solid support is coated with antiserum or culture supernatant, and subsaturating amounts of a mixture containing labelled and non-labelled antigen is allowed to compete for binding to the antibody.

On the other hand, the antigen is immobilised and a mixture of antigen and labelled antibody is added. Consequently, high amounts of antigen inhibit binding of the labelled antibody to the immobilized antigen. If pure antigen is not available, the wells are coated with crude antigen preparation followed by detection with saturating amounts of labelled antibody. For this reason and because of possible interference with compounds of the antiserum or the biological matrix containing the analyte, it is sometimes necessary to purify and concentrate the antibodies, especially polyclonal anti-sera. This can be done most simply by ammonium sulphate precipitation. IgG usually precipitates at 33% to 40% ammonium sulphate saturation. This can be performed simply by mixing equal volumes of antiserum and PBS-buffer, followed by addition of two volumes of saturated ammonium sulphate solution.

After stirring for 30 min. the pellet is collected by centrifugation and washed with half-saturated ammonium sulphate solution. Finally, the pellet is dissolved in PBS (one volume or less) and dialysed against PBS. Thus, the antibody preparation is purified and concentrated. An alternative method is to adsorb serumproteins on a cation-exchange matrix. The IgG is neutral or slightly positively charged at pH 6.5, whereas most of the serum proteins are negatively charged. Consequently, the serum proteins are bound to the DEAE-matrix, whereas IgG passes the column or is detected in the supernatant of the batch. Large amounts of ion exchanger are needed and the preparation is diluted.

Furthermore, the pH should be adjusted to 8 as soon as possible to prevent denaturation of the immunoglobulin. Much more comfortable is isolation of IgG by purification with protein A immobilised on beads that are commercially available. Because the Fc-proportion of IgG is bound to the surface protein of *Staphylococcus aureus* at pH 8, all contaminants are removed by washing the column. The IgG is dissociated from the beaded carrier by elution at pH 2. Thus, pure and concentrated IgG is yielded by elution in counter-current flow, but the pH of the eluate should be adjusted to 8.5 immediately. Only minimum amounts of the commercially available matrix are necessary, as the IgG-binding capacity of protein A is high (about 20 mg IgG/mg protein A). For preparation of monospecific antibodies from polyclonal antisera, affinity chromatography on immobilized antigen is performed.

This rather time-consuming method requires synthesis of an affinity matrix containing covalently bound antigens via 6C-12C spacers. The antiserum is bound to the matrix at physiological pH and, after removal of impurities by washing the column, the monospecific antibodies are recovered by lowering the pH, enhancing the ionic strength of the buffer, or addition of chaotropic agents. Again, physiological conditions in the eluate should be restored as soon as possible. In general, the sensitivity of immunoassays increases with purity of antibody and antigen preparation.

Diluents

The diluent should provide optimum conditions for formation of the antigen-antibody complex. Furthermore, extremely low concentrations of analytes are applied, as they are susceptible to surface adsorption on the reaction vessels and denaturation. Whereas monoclonal antibodies and enzyme labels are more sensitive to ionic conditions and temperature, for polyclonal antibodies a pH near neutrality and an ionic strength of 0.15 is recommended, usually PBS pH 7.4. Addition of bovine serum albumin up to 1% minimizes adsorption to the surface of the vessel and stabilizes antigens and antibodies, especially in highly diluted solutions.

However, in assays for haptens addition of albumin should be avoided, as it binds many drugs and hormones. Washing buffers can contain up to 0.5% Tween 20 to enhance solubilisation of non-specifically adsorbed's compounds. Inclusion of bacteriostatic agents, *e.g.*, 0.02% sodium azide, enhances storage stability of antibody solutions but sometimes hardly interferes with enzymic detection. To improve storage stability of concentrated antibody solutions upon freezing, addition of glycerol up to 50% is recommended. Prior to the assay, the solution should be highly diluted with working buffer.

Solid Supports

A solid matrix with immobilised antigen or antibody greatly facilitates manipulation during immunoassays. Formerly, beads were used as carriers, but today 96-well microplates are used in labs routinely for many purposes. Sometimes the reactants are immobilised covalently, but non-covalent interactions are used more frequently. In the latter case, the solid support can be made from polyvinylchloride or polystyrene, each exhibiting a protein-binding capacity of up to 300 ng/cm^2. Because polyvinylchloride microplates are not translucent, they are not suited for enzyme immunoassays measuring conversion to coloured substrate.

Translucent polystyrene microplates with flat bottoms are recommended for enzyme immunoassays, as they are commercially available with high and low protein-binding capacity. Infrequently, protein-A-Coated beads serve as solid support. As protein A interacts with the Fc-fragment of the IgG-molecule, the site-directed immobilisation of the antibody renders the antigen-combining sites freely accessible. Furthermore, so-called "magneto-beads" are commercially available and contain a core made from iron, which facilitates separation by magnetic fields omitting centrifugation.

Detection

There are three ways of detection in immunoassays: (1) direct detection using labelled antigen or antibody; (2) indirect detection using a secondary labelled analyte, most frequently a second antibody recognizing the Fc-fragment of the first antibody; and (3) use of the biotin-avidin system comprising detection of biotinylated antigen or antibodies with enzyme-tagged or radiolabelled avidin. Direct detection requires fewer steps in the assay procedure and background problems are minimal. However, pure analyte is required and needs to be conjugated covalently with the label. In the case of radioimmunoassays, labelling procedures are well established, but for enzyme immunoassays this can be crucial to do in the lab.

Additionally, direct detection is less sensitive than the indirect way. The main advantage of indirect detection is commercial availability of labelled antibodies, which might be used to detect a wide range of antibodies with different specificity. The interaction between the glycoprotein streptavidin and the vitamin biotin is characterised by a dissociation constant of 10^{-15} mol/L, which corresponds to a half life of 160 days and approaches covalent interactions. Streptavidin consists of four identical subunits, each binding one molecule of biotin. Thus, an enhancement effect can result from the multivalency of streptavidin, which might be amplified further by a multilayer design using avidin-biotin complexes.

Reagents for biotinylation of proteins containing spacer molecules are commercially available, and avidin is easily coupled to enzymes. The type of the label needs to be carefully selected prior to the experiment. Radiolabelling is easy to perform by the chloramin-T iodination or by the lactoperoxidase method, but safety regulations concerning use, storage, and disposal need to be fulfilled. In radioimmunoassays low amounts of radioactivity are necessary, about 10 to 100 µCi/µg in the case of exhibiting a half life of 60 days, and about 50,000 cpm/well are required for reliable results. Consequently, danger to health is minimal if handled correctly, but special equipment is a prerequisite. In comparison, labelling with enzymes is safe, but it is emphasized that some cross-linking agents and substrates are hazardous.

The equipment necessary for enzyme immunoassays is inexpensive and the reagents exhibit a long shelf-life. Additionally, enzyme immunoassays allow a wide variety of assay designs, and detectability is comparable to that of radio-immunoassays, which is supported by the amplification effect of enzymes and by use of fluorogenic substrates. Usually 5 to 50 ng enzyme-labelled antibody well yields a colour development by substrate conversion corresponding to an absorption of 0.2 to 2.0. In practical terms, direct visualisation of the result on the microplate is possible.

For further enhancement of detectability, fluorogenic substrates might be used. A high quantum yield provides for the best detection limit, a large Stoke's stift should minimise the interaction between excitation and emission, and naturally occuring fluorophores in the sample should not interact. Quenching effects need to be characterised for each individual assay design. A major disadvantage of enzyme immunoassays is the possibility of interference of enzyme activity with compounds of the biological matrix of the sample, which might be reduced by selection of an appropriate assay design.

PROTOCOLS

Quantification of Cortisol by ELISA

Protocol 1 : Antigen preparation

1. A solution containing 5.5 mmol cortisol and 22 mmol succinic anhydride in anhydrous pyridine is refluxed under protection from light and humidity for 4 h.
2. The solvent is completely removed by evaporation under vacuum, and the oily residue is dissolved in chloroform followed by extraction with distilled water. The extraction is repeated three times and the organic phase is dried with anhydrous sodium sulphate.

   Formation of cortisol-21-hemisuccinate is confirmed by thin-layer chromatography on KGF254 in chloroform/methanol/water (65+30+6, v/v/v) indicated by hRf 49 as compared to cortisol with hRf 65. IR-spectroscopy and mass spectroscopy confirm these results.

3. The antigen is prepared by the mixed anhydride method.

   Under a hood, a solution containing 1.6 mmol cortisol-21-hemisuccinate in 32-ml anhydrous dioxane is cooled to 13 °C and 1.6 mmol tri-N-butylamine is added under stirring. After addition of 1.6 mmol isobutylchloroformate, the solution is incubated for 30 min. followed by addition of a solution containing 880 mg BSA in a mixture of 26 ml distilled water pH 10 and 17.2 ml dioxane. Immediately, the pH is adjusted to 7.5-8.5 by addition of 1 M NaOH and maintained for the next 2 h.

4. The BSA-cortisol conjugate is dialysed extensively against distilled water followed by lyophilisation. For immunization of mice, the conjugate is further purified by chromatography on Sephadex G-25 using TRIS-buffer pH 8.6 for elution to remove even traces of the free immunosuppressive drug.

The degree of substitution is determined by UV-difference spectroscopy. At the absorption maximum of cortisol at 246 nm, the difference in absorption of accurately equimolar solutions of conjugate and carrier protein is determined and calculated from the calibration curve of cortisol to be 19.4 mol cortisol/mol carrier protein.

Protocol 2 : Immunization and preparation of monoclonal antibodies

1. Three Balb/c mice were immunised with 100 μg antigen emulsified in 1 ml complete Freund's adjuvant for the first immunization and 50 μg antigen in 1 ml incomplete Freund's adjuvant for the subsequent two boosters.
2. Fusion of isolated spleen cells and 2×10^8 myeloma cells was performed with PEG as described above, following a protocol reported in the literature. The primary hybridoma cultures (one drop of fused cells in 2 ml medium) were grown in 24-well culture dishes as described above, yielding 75 growing clones.
3. After reaching confluency, the supernatant was screened for cortisol-specific antibodies yielding six positive clones, which were propagated further.
4. Cloning was done by limited dilution. The primary dilution series was done in 96-well microplates at a density of 10^4 cells/well in 200 μl HAT medium. The secondary dilution series was done accordingly from positive clones.
5. Screening of the supernatants resulted in eight cortisol-antibody secreting clones. These clones were further propagated in 24-well microplates.

Protocol 3 : ELISA protocol

1. The wells of a 96-well microplate were coated with 100 μl antigen solution δ (0.003% in PBS containing calcium and magnesium) overnight at 4 °C.
2. After washing three times with 100 μl PBS each, the surface of the wells was saturated by incubation with 100 μl 1% BSA-solution in PBS for 2 h at room temperature.
3. After washing three times with 100 μl PBS each, a mixture of 50 μl hybridoma-supernatant and 50 μl cortisol in PBS (serial dilution, 5–200 ng) was added and incubated for 2 h at 37 °C.
4. The wells were washed three times with PBS. After addition of 100 μl goat-anti-mouse-IgG-conjugated peroxidase, the microplate was incubated for 2 h at 37 °C.

5. The supernatant was removed and the wells were washed again three times with PBS. 100 μl of freshly prepared substrate solution consisting of 10 mg o-phenylendiamin hydrochloride, 9 ml 0.11 M aqeous disodium-hydrogen-phosphate, 1 ml 0.5 M aqueous citric acid, and 10 μl 30% H_2O_2 were added.
6. After incubation for 30 min. under protection from light substrate, conversion was stopped by addition of 100 μl 3 M sulfuric acid. Immediately, the absorption was read at 450 nm using a microplate reader.

The samples were assayed in triplicate at the minimum, including blanks (well-filled with PBS), negative controls (as above, but without addition of cortisol-antibody and cortisol), and positive controls (as above, but without addition of cortisol).

Cross-reactivity of the antibody with structurally related steroids can be determined by the same assay, but add the steroid of interest instead of cortisol.

IMMUNOFLUORESCENT ANALYSIS

Because of the fascinating facilities concerning the characterisation of cells, immunofluorescence techniques in combination with flow cytometric analysis have gained more and more interest over the past years. Application of fluorescent-labelled monoclonal antibodies directed against specific proteins or polysaccharides located at the cell surface or with some limitations-selected cytoplasmic, as well as nuclear, ligands offers the possibility to obtain direct information on the state of activation, proliferation, and differentiation of the cells.

In comparison to microscopical methods, the advantage of flow cytometry is the rapid quantification of the amount of a particular stain per cell, which offers a good statistical evaluation because of the large number of cells usually assessed. The objective of this chapter is to give an overview of the basic principles of flow cytometric measurements and to provide readers with basic considerations concerning immunofluorescence techniques. Upon understanding the important parameters, the specificity and sensitivity of the staining methodology can be optimized for a particular application.

Principles of flow Cytometry

The basic components of a flow cytometer consist of a fluidic system to transport the cells across the microscopic field; an optical system for illumination and detection; and the electronics for light collection, data management and control. The most important part of the *flow system* is the flow chamber, where the microscopic observation takes place. An important parameter for successful flow cytometric analysis is the hydrodynamic focusing of the cells in this flow chamber providing for individualisation and correct positioning of the cells under investigation. Furthermore, the sample suspension is injected into a particle-free "sheath fluid", and the diameter of the sample flow is reduced dramatically at the injection point of the flow chamber to guarantee serial alignment of the individual cells during optical analysis. The optical system consists of three parts, the first is the *illumination optics,* which represents the light source for the flow cytometry and is made up of either a conventional lamp or a laser, generally an argon ion laser. Via prisms and mirrors, and if necessary monochromatic filters, the illuminating light is directed to the liquid stream and focused by several lenses to the point of microscopic observation. The other two components belong to the detection system of the flow cytometer.

The *forward scatter collection optics* is a microscope with low numerical aperture observing the liquid stream from opposite the illuminating light. The "forward scatter" (FS) collects the light scattered by the cells in the flow stream in the range of 2° to 20° off the axis of the illuminating light. The scattering signal measured by the FS corresponds to the diameter and, there-fore, the size of the cell under investigation. The second detection component includes the fluorescence and side scatter collection optics, comprising a second microscope with a long working distance and high numerical aperture, which analyses the light scattered perpendicular to the illumination beam as well as the liquid stream.

The "side scatter" (SS) signal is dependent on the surface characteristics and the granularity of the measured cells, whereas the fluorescence detectors collect the light emitted by fluorescent compounds associated with the analyse cells. The electronic system is necessary for the correct collection and processing of the light signals gained. Conversion of optical into electrical signals is done by semiconductor photodiodes or/and photomultiplier tubes.

To analyse two or more fluorescent compounds with overlapping emission spectrum (*e.g.*, fluorescence isothiocyanate and phycoerythrin) simultaneously, the flow cytometer should be equipped with an electronic fluorescence compensation network for correction of fluorescence overspill and proper calculation of the different fluorescence signals. For data calculation, either a linear or a logarithmic amplifier can be used, depending on the dynamic range of the parameter under investigation. In addition, a trigger circuit provides for correct separation between signals and electronic noise or debris by setting a threshold level for one parameter (*e.g.*, forward scatter). Each analog pulse acquired by the photomultiplier tubes that exceeds this trigger threshold is converted into a digital signal and transmitted to a computer for further storage and data evaluation.

Data Evaluation

Today, state-of-the-art flow cytometers offer the possibility to analyse five parameters per cell simultaneously: forward and side scatter as well as up of three fluorescence light parameters. Normally, the intensity of the scattered light is analysed on the basis of linear amplification, but the intensity of immunofluorescence is analysed upon logarithmic amplification over four decades. Thus, a single gain setting can be used for data acquisition allowing direct comparison between relative fluorescence intensities of various cells. For evaluation a multitude of data plots is available. Usually a two-dimensional dot plot is used to characterize the scattering properties of a cell population by plotting the FS signal *versus* the SS signal.

One-dimensional histograms are best suited for illustration of the intensity of fluorescence staining emphasising the quantitative aspects of flow cytometry. In order to exclude non-specifically stained particles (*e.g.*, dead cells, debris) or conglomerates of cells from analysis, cells for data evaluation can be preselected by setting lower and upper thresholds for certain parameters ("gating").

Statistical calculation of data is done by the cytometer software, making the. evaluation of immunofluorescence assays rather simple. In this statistical evaluation, the cell populations are usually described by the relative mean and the coefficient of variation (CV) of a particular fluorescence parameter as well as the number of cells in the respective population.

The mean is defined as the arithmetic mean in the case of linear scale and geometric mean for logarithmic scale. The CV is calculated as half the width of the distribution at 0.6 times maximum height. Usually the distribution of fluorescence intensities on a logarithmic scale varies symmetrically around the mean because of both biological variation (*e.g.*, size of cells, density of target antigen and staining antibody) and variation of measurement (*e.g.*, focusing, orientation). Absolute quantification of the amount of antigen molecules per cell is difficult because of the necessity to include a reference method for calibration of the relative immunofluorescence.

Basic

The area of immunofluorescence covers the staining of cellular structures with antibodies and other specific ligands directly or indirectly labelled with fluorescent dyes. During an immunofluorescence analysis, all the steps prior to the flow cytometric measurement, including preparation of cells, staining, and fixation, are most critical.

Preparation of Cells

In order to obtain appropriate data from flow cytometric analysis, attached or connected cells have to be individualized prior to staining and measurement. Preparation of a single-cell suspension is done easily by repeated re-suspension in the case of natural single cells derived from blood, bone marrow, or non-adherent cultured cells. Adherent growing cells or cells from tissues can be separated mechanically or enzymatically either prior to or after the staining procedure.

Upon use of enzymes for disintegration of cells, care has to be taken in order to ensure that the antigen of interest or the attachment of the immune label is not affected by the dissociation procedure. Recent protocols often include pre-enrichment steps, especially in the case of blood samples, *e.g.*, the depletion of erythrocytes for analysis of leukocytes. Among these procedures, density gradient centrifugation and lysis are used frequently. Pre-enrichment of rare cells is advantageous because of the reduction of staining volume and elimination of cells that could be stained unspecifically, such as dead cells, resulting in improved discrimination of the cells of interest.

Staining of Cells

The staining procedure can be performed either directly or indirectly. In the case of direct staining, the fluorescence-label is attached covalently to the antibody against the specific cellular antigen under investigation and used for analysis. Indirect staining is done by incubating the cells with an unconjugated primary antibody followed by detection of the antibody with a secondary, fluorescent-labelled, usually polyclonal anti-antibody. Direct staining is preferred because of easier handling and better control of the staining parameters.

Especially in routine use and for multiparametric analysis using a combination of various markers in one staining step, control and handling are important to consider. Prior to use the labelled primary antibodies of interest have to be purified. Indirect staining methods facilitate application of the primary antibody without any previous labelling and purification. A further advantage of indirect staining protocols is the possibility to improve the sensitivity of detection by increasing the number of secondary antibodies bound to one primary antibody, thus increasing the number of fluorescence markers per specific antigenic determinant.

But particularly in multiparameter analysis, control of proper indirect staining is not trivial and requires highly purified isotype-specific secondary antibodies or complete blocking of free binding sites of the secondary antibody prior to staining of the next parameter to avoid any cross-reactions. To combine high specificity and high sensitivity, an indirect staining protocol using primary antibodies conjugated to haptens such as digoxigenin or biotin can be followed. In these systems the detection is pursued very specifically by fluorescence-labelled secondary antibodies directed against the hapten or streptavidin, while the sensitivity is controlled by the number of marker molecules per mol secondary antibody/streptavidin. In order to avoid any loss of or damage to cells, the number of steps in the staining protocol should be kept to a minimum.

Additionally, reproducible results can be expected only if the concentrations of cells and antibodies used for detection are standardized. In so doing it is important to apply concentrations that are as low as necessary for predominance of specific high-affinity antigen-antibody interactions over low-affinity cross-reactions but are high enough to ensure that staining is as bright as possible to obtain good quality of data. The optimum concentration of cells depends mainly on the nature of cells and the antibodies used for detection and should be determined in preliminary studies.

The lower limit amounts to approximately 10^5 cells in 10 µl providing for flow cytometric analysis of 10^4 cells. Thus, 10 µl is the minimum volume necessary for proper staining in order to avoid significant changes in the concentration of the staining antibody during the washing procedure. Usually staining volumes of 20 to 100 µl containing 0.7 to 5×10^6 cells are used. All these concentrations-refer to viable positive cells; additional non-stainable cells can be neglected.

However, for flow cytometric measurement of rare cells the number of negative cells has to be considered upon optimization of the staining protocol. To stain larger numbers of cells, it is better to increase the volume of the cell suspension than the concentration of cells. To obtain results that guarantee adequate statistical evaluation, the optimum concentration of staining reagents should be determined by titration. For that matter, it should be kept in mind that too-low concentrations resulting in poor discrimination between positive and negative cells are as problematic as too-high concentrations resulting in non-specific staining of negative cells or subpopulations due to low affinity cross-reactions.

Titration can be performed by flow cytometric analysis of a mixture of positive and negative cells stained with varying concentrations of the fluorescent-labelled antibody using a standard protocol. By plotting the mean cell-associated fluorescence intensity against the concentration of the staining reagent, optimum conditions in an antibody concentration exist where the difference in cell binding between positive and negative cells is highest. If there are any problems with cross-reactions, it might be useful to apply lower concentrations. Higher concentrations are recommended in the case of low antigen density. The incubation with the antibody should be kept as short as possible because high-affinity antigen-antibody reactions occur very fast.

Extending the incubation time may lead to enhanced low-affinity cross-reactions and non-specific adsorption. Since with most staining reagents about 90% of the maximum staining is achieved within 5 min., doubling the time for safety reasons results in a standard staining time of 10 min. For immunofluorescent staining of intracellular components, a prolonged staining period after fixation and permeabilisation of the cells is necessary. To guarantee appropriate diffusion of the staining antibodies into the cytoplasm to reach the antigens of interest, staining times of about 1 h are used. To avoid any problems deriving from cell physiology, staining of live cells is done at an incubation temperature of 4 °C. At this temperature, the metabolism of the cells is minimized and the fluidity of the cell membrane is reduced, resulting in a preferred binding of the staining reagents to the cell surface.

In the case of staining cytoplasmatic parameters of fixed and permeabilized cells, the temperature during staining should be raised to 37 °C in order to enhance diffusion of the staining antibodies. Staining of live cells at higher temperatures may lead to altered results that are due to modulated expression of the antigenic receptors (*e.g.*, lymphocytes can "cap" their receptors with the staining antibodies, throw off the cap into the medium, and then appear "negative"). Blocking cell physiology by addition of 0.003% sodium azide in the staining medium may reduce these problems.

Washing and Fixation of Cells

Prior to flow cytometric analysis, unbound staining antibodies have to be removed by washing. Therefore, the cell suspension is spun down at 1000 rpm followed by removal of the supernatant and resuspending the cells in fresh medium or isotone buffer. Because under even optimum conditions 10% of the cells are lost per washing step, washing should be reduced to a minimum. Applying an adequate low concentration of reagents during staining can help to restrict washing of the cells to one or two steps, particularly in direct staining protocols. In the case of indirect staining, proper washing of the cells is very important to improve staining results and to avoid any reaction between primary and secondary antibody in solution.

In these protocols, the specificity of the antigen-antibody interaction often is improved by addition of 1% BSA to the washing solution for blocking of residual non-specific binding site. For routine analysis cells usually are fixed prior to flow cytometry using para-formaldehyde (2% in PBS) to avoid time-dependent alterations of the cell staining as well as the risk of infections, thus standardizing analysis parameters. However, like all other manipulations, fixation of cells may lead to unpleasant side effects. Especially for analysis of rare cells, discrimination between rare positive cells and non-specifically stained dead cells is crucial.

After fixation it is very difficult to distinguish whether a cell was dead before staining or not. In such cases it is more convenient to analyse live cells without fixation immediately after the staining procedure in the presence of propidiumiodide, gating out dead cells according to the scatter and the propidiumiodide fluorescence. In contrast, staining of cytoplasmatic or nucleic substructures requires fixation and permeabilisation of cells because of the inability of antibodies to penetrate the cell membrane of live cells. For this reason, cells are fixed either by addition of ice-cold methanol (10 min. at –20°C) and subsequent rehydratisation in PBS or with para-formaldehyde (2% in PBS, 10 min). In the case of para-formaldehyde, fixation has to be followed by an extra permeabilisation step using Triton X-100 (0.1% in PBS, 10 min.).

Measurement and Data Evaluation

Prior to flow cytometric analysis, the sample usually amounting to 50 to 200 µl has to be resuspended in 1 ml Cell Pack. Amplification of the fluorescence signals during flow cytometry should be adjusted in such a way that the auto-fluorescence signal of unlabelled cells is put in the first decade of the 4-decade log range. For each measurement between 5000 and 10,000 cells of interest are accumulated. Generally data evaluation is done by using a forward *versus* side scatter plot for gating and the relative mean of the particular fluorescence parameter for further calculation, as described above in detail.

Staining and Flow Cytometric Analysis

Protocol 4 : Direct staining of surface antigens

1. Prepare 50 µl of a single-cell suspension containing 1×10^6 cells.
2. Add 50 µl of the specified fluorescence-labelled antibody solution and incubate for 10 min. at 4°C. The adequate concentration of antibody depends on antigen density on target cell, fluorescein/protein ratio of the labelled antibody, concentration and nature of cells, etc., and has to be determined by titration in a preliminary study.
3. Add 100 µl PBS and spin down the cell suspension at 4°C and 1000 rpm. Discard 150 µl of the supernatant and resuspend the cells using 150 µl of PBS. This washing step can be repeated once. Finally the volume is adjusted to 100 µl.
4. If fixation is required, add 100 µl of a para-formaldehyde solution (4% in PBS) and incubate for 10 min. at r.t.
5. Dilute the sample with 1 ml Cell Pack and analyse by flow cytometry, accumulating 5000-10,000 cells per measurement.

Protocol 5 : Indirect staining of surface antigens

1. Prepare 50 µl of a single cell suspension containing 1×10^6 cells.
2. Add 50 µl of the specified primary antibody solution followed by an incubation for 10 min. at 4°C. The adequate concentration of antibody: depends on antigen density on target cell, fluorescein/protein ratio of the labelled antibody, concentration and nature of cells, etc., and has to be determined by titration in a preliminary study.
3. Add 100 µl 1% BSA in PBS and spin down the cell suspension at 4°C and 1000 rpm. Discard 150 µl of the supernatant and resuspend the cells in 150 µl of 1% BSA/PBS. This washing step can be repeated once. The volume should be adjusted to 50 µl.
4. Add 50 µl of the corresponding fluorescence-labelled secondary antibody solution and incubate again at 4°C for 10 min. For considerations on the concentration of antibody, see above.
5. Repeate the washing procedure as described above.
6. If fixation is required, add 100 µl of a para-formaldehyde solution (4% in PBS) and incubate for 10 min. at r.t.
7. Dilute the sample with 1 ml Cell Pack and analyse by flow cytometry accumulating 5000-10000 cells per measurement.

Protocol 6 : Staining of cytoplasmatic or nucleic tructures

1. Prior to staining, the single-cell suspension has to be fixed and permeabilized:
2. By addition of ice-cold methanol (70%) and incubation for 10 min. at –20°C followed by rehydratisation at 4°C.

A. By addition of para-formaldehyde in PBS (final concentration 2%) and incubation for 10 min. at r.t.

B. Use of para-formaldehyde for fixation requires additional permeabilisation of the cells using 0.1% Triton X-100 in PBS and incubation for 10 min. at r.t.

3. Staining is performed as described above, but incubation time is extended up to 1 h.
4. During washing the cells should be incubated after resuspension for about 15 min. to guarantee proper removement of antibody and blocking of residual binding sites.

All in all, immunofluorescence techniques in combination with flow cytometric analysation offers exciting possibilities to determine specific antigens at the surface of cells or even cytoplasmatic and nucleic structures. The advantages of flow cytometric analysis are based mainly on the ability to determine cell-associated fluorescence intensities at the single-cell level and the rapid assessment of a large number of cells leading to excellent statistical data. Thus, flow cytometry is best qualified to improve characterisation of cells in all areas, ranging from basic immunology and molecular biology to clinical medicine and development of new strategies for pharmaceutics, as well as discrimination between even very close subsets of mammalian cells as applied to detect and monitor certain diseases.

Confocal laser scanning microscopy has been established within the last 15 years as a valuable tool for obtaining high-resolution images and three-dimensional reconstructions for a variety of biological specimens. Convential light microscopy creates images with a depth of field of 2 to 3 µm, and the real information in the focal plane is overlayed with "out-of-focus" information from optical planes above and below. This is especially a problem in fluorescence miroscopy where "out-of-focus" fluorescence creates a diffuse haze around the objects of interest. The aim of confocal light microscopy is to use only the information of the plane that is in focus and to divide three-dimensional objects into multiple optical slices. These images can then be digitized and reconstructed into three-dimensional objects using modern computer technology.

Components of a Confocal Laser Scanning Microscope

Conventional epifluorescence microscope: Objective (oil or water immersion).

stage, focus motor (for optical slicing).

Confocal unit: laser, excitation filter, illumination pinhole, scanning unit (oscillating mirrors), dichroic mirror, detection pinhole, reflection filter, photo-multiplier.

Optical Principle

The principle of a confocal laser scanning microscope is shown in Fig. 9.5. A laser light beam from a strong laser source is expanded to make optimar use of the optics in the objective. This improves spatial resolution by the objective compared to that of conventional light sources. To generate a two-dimensional image, the laser light beam is scanned across the specimen. This is achieved by an *x-y* deflection mechanism (reflection from galvanometer-driven oscillating mirrors). This enables the scanning laser beam to focus by an objective lens on a very small area of the fluorescent specimen termed the excitation light. Using a high numerical aperture objective lens, the crossover of the conical beam of laser light is approximately 0.2 µm in diameter. The fluorescent specimen emits light of a higher wavelength that has to pass the objective lens, the oscillating mirrors, the reflecting dichroic mirror, an excitation light filter, and a confocal pinhole and ultimately is focused onto a photodetector. The emitted light from the fluorescent specimen is "de-scanned" by the same oscillating mirrors scanning unit. The reflecting dichroic mirror or beamsplitter allows only the emitted light up to a defined wavelength to pass through and reflects light with higher wavelengths (which includes all the excitation light coming back from the specimen). The confocal pinhole is constructed so that it eliminates all light from "out-of-focus" planes. This is especially important when evaluating thick specimens. A "confocal spot" is the term used for the area that is detected by the laser beam.

To generate a two-dimensional image of a small area of the specimen or "optical section", a raster sweep of the specimen is done at one particular focal plane. As the laser scans across the specimen, the analog light signal is detected by the photomultiplier and converted into a digital signal. The combination of scans creates a pixel-based image (256 × 256 pixels, 512 × 512 pixels, or 1056 × 1056 pixels) that is displayed on a computer monitor. The relative intensity of the fluorescent light emitted from the laser hit point on the fluorescent specimen corresponds to the intensity of the resulting-pixel in the image (8-bit grayscale). Colour can be added to the 8-bit grayscale images to define reacting proteins in a double or triple labeling and prior to three-dimensional reconstruction to highlight regions of interest.

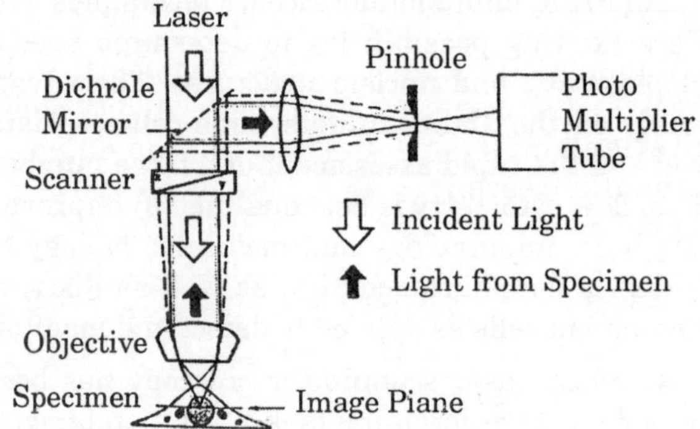

Fig. 9.5. Confocal laser scanning microscope.

Most frequently, assigning colours by thresholding is used to make images look clearer. This is done by assigning various intensity ranges to different colours. The fluorescence intensity of each assigned colour is ramped from dark to light to reflect fluorescence intensity, termed pseudocolor. The colours are usually shown in an 8-bit colour look-up table (LUT), which is included in the image legend.

Pinhole and Confocality

The confocal microscope produces an optical slice through a specimen of defined thickness as a consequence of the pinhole. The numerical aperture of the lens influences the thickness of the optical slice: Oil and water-immersion lenses are most suitable and can generate slices as thin as 1 μm. The diameter of the pinhole inversely correlates with confocality. For example, a large pinhole increases the signal but decreases the axial resolution of a plannar object and can result in a toss of confocality resulting in conventional microscopy.

A small pinhole limits brightness. In practice, microscope users often open up the size of the prinhole to increase the signal from a weekly fluorescent or scattering object. For each confocal microscopy specimen, it is necessary to determine an optimal pinhole diameter for slice thickness and brightness.

Fluorophores

Specimens are stained using antibodies labeled with fluophores for imaging with CLSM. This technique is called immunofluorescence histochemistry. However, other techniques using direct staining with specifically reacting substances like phalloidins also can be used. Fluorescent probes are available for a broad range of biomolecules, permitting the use of the CLSM to image macromolecular structures (proteins, lipids, carbohydrates, and nucleic acids) and physiological ions (*e.g.*, calcium, protons) for static fluorescent specimens, as well as for dynamic quantitative imaging and measurements. The choice of fluorophores used for immunofluorescence is critical because auto-fluorescence intrinsic from the tissue can mimic the appearance of the fluorophores. To exclude auto-fluorescence interfering with the fluorophore emission, the unstained specimens are examined with the same set of filters for excitation that are used with the specific fluorophore.

Auto-fluorescence is usually the highest, with excitation between 400 and 450 nm, and decreases with increasing wavelengths Additionally, the choice of fluorophores depends on their excitation/emission wavelength, the laser source, and the filter sets available. Immunofluorescence can be performed by either direct or indirect methods. Direct immunofluorescence involves the conjugation of the primary anti-body with a fluorophore such as fluorescein or rhodamine. For indirect immunofluorescence, the primary antibody is detected by a secondary fluorescence-labeled antibody that is directed against the first antibody.

This method is more commonly used because it is easier to label secondary antibodies than primaries and it is more sensitive, *i.e.*, more than one molecule of the secondary antibody can bind to the primary antibody. The specificity of the staining antibodies for an indirect staining protocol is especially important for multi-colour immunofluorescence because of cross-reactivity. Each individual secondary antibody must recognize only one primary antibody and must not cross-react with intrinsic proteins in the specimen or with other secondary antibodies. Unspecific staining of the tissue by a secondary antibody can be determined by staining the tissue in the absence of the primary antibody.

To ensure that a secondary antibody recognizes only the primary antibody, it is necessary to determine whether the secondary antibody binds to the primary antibody. To avoid cross-reactivity, it is ideal to use secondary antibodies that have been raised in species different from the species in which the primary antibodies have been derived. Some of the important considerations for choosing a laser for a CLSM system are cost, wavelength of emission, output power at each wavelength, efficiency, and stability. Laser lines below 400 nm are expensive to achieve, therefore, only a few CLSM systems with such lasers are available commercially.

Table 9.4. Commonly used Laser Sources.

Source	*Excltation lines*		
He Ne	543 nm	633 nm	
Argon lon 1	488 nm	514 nm	
Argon lon 2	458 nm	488 nm	529 nm
Argon Krypton	488 nm	568 nm	647 nm

Use of filters in a CLSM system: Gray filters are used for the attenuation of the laser power, and single lines are selected by excitation filters. Ideally, these filters should have minimal loss of laser intensity and should permit the detection of the fluorophores around their maximal emission wavelength. For multi-labelling experiments, fluorophores should have only minimal overlap in their excitation/emission wavelengths. To eliminate channel "crosstalk", it is necessary to use sequential excitation and imaging of each fluorophore to spectrally isolate their fluorescence emission. This method requires special filter combinations based on each of the fluorophore's excitation and emission wavelengths and is time-consuming because the specimen must be scanned several times with different excitation and emission wavelengths.

Anti-fade Reagents

Because of the high intensity and focus of the laser beam, the labeled specimens undergo significant bleaching/fading. The bleaching effect is more severe than in conventional epifluorescent microscopy, where the entire specimen is observed under lower-power excitatory light. This results in accelerated degradation of fluorophores and makes longer observation periods impossible. Several factors influence fading, among them the fluorescence intensity, pH, and base solution of the embedding medium. Antifade reagents can slow down the degradation process of the fluorophore, resulting in longer periods of observation, fluorimetry, and pattern recognition.

P-phenylenediamine (PPD) : This is one of the most effective antifade reagents, but it is photo/thermo sensitive and cannot be used for *in vivo* studies because of its toxic effects. The optimal PPD antifade mixture consists of 90% glycerol, 10 PBS, PPD concentrations between 2 mM and 7 mM, pH 8.5–9.0 [51].

N-propylgallate (NPG) : NPG is non toxic and photo- and thermostable. Though less effective as an antifade, it can be used for *in vivo* studies. Optimal working concentrations: glycerol base, 3 mM–9 mM NPG.

Standard Protocols for Labeling

Direct labeling : Staining F-Actin with fluorescent phallotoxins (protocol from Molecular Probes, modified).

Preparation of Stock Solution

Fluorescent phallotoxins are available from Molecular Probes, The Netherlands or Eugene, Oregon. The content (5 μg) is dissolved in 1.5 ml methanol (200 units/ml). Molecular Probe defines one unit as the amount of material used to stain one microscopic slide of fixed cells. This stock solution can be stored at 4°C. Prior to staining, the necessary volume is taken from the stock solution and transferred to a test tube, evaporated, and then dissolved again in the appropriate buffer immediately before use.

Protocol 7 : Procedures for staining slides

This procedure has yielded consistent results in most instances. Modifications may be necessary for particular specimens. For cells grown on glass coverslips (for each coverslip):

1. Fix coverslips or slides in 4% paraformaldehyde in PBS pH 7.3 for 3 min. and then 10 min. at room temperature.
2. Wash several times with PBS.
3. Place the coverslip in a glass petri dish and permeabilize the cells with ice-cold acetone.
4. Air dry sample.

5. Evaporate 5 μl of the stock solution in a small test tube and dissolve it again in 150 μl of PBS. Place on the coverslip for 20 min. at room temperature.
6. Wash twice rapidly with PBS.
7. Mount the coverslip on a slide with the cell side down in FluoroSave.
8. Seal the edges of the coverslip with rubber cement.
9. Specimens prepared in this manner retain actin staining for at least 5 to 6 months when stored in the dark at 4°C.

Indirect Labeling

Protocol 8 : Method for immunofluorescence on coverslips or slides

1. Fix coverslips or slides in 4% paraformaldehyde in PBS pH 7.3 for 3 min. and then for 10 min. at room temperature.
2. Wash several times with PBS.
3. To stain intracellular proteins, the membrane must be permeabilized prior to the labeling by rapid dehydration/rehydration using different alcohol concentrations (70%, 80%, 90%, 100%, 90%, 80%, 70%)
4. Wash once with PBS.
5. Block nonspecific staining by incubating in PBS + 0.1% bovine serum albumin (BSA) and normal serum from the host species of the secondary antibody, for example 5% natural goat serum (NGS) for 1 h at room temperature. To detect intracellular antigens, it is necessary to add 0.05% Saponin to this buffer and keep the BSA and Saponin through the procedure.
6. Incubate the primary antibody at room temperature between 1 and 3 h (depending on the characteristics of the primary antibody). Use approximately 100 ml per coverslip, more for slides. Wash several times with PBS, BSA, and Saponin if permeabilized.
7. Incubate with fluorescent-labeled second antibody for 1 h at room temperature in the dark to protect fluorescence (add 5% NGS to the secondary antibody solution).
8. Wash several times with PBS and once with distilled water.
9. Mount in Fluorosave™(embedding and antifade reagent) or other embedding media containing antifade agents.

Preparation of solutions

- **4% paraformaldehyde in PBS :** Heat 8 g paraformaldehyde in 100 ml PBS until dissolved (must be prepared in a fume hood). Remove particles using a paper filter in a glass funnel. Adjust to the original volume (100 ml), then mix 1:1 with 2x PBS to obtain a 4% paraformaldehyde solution in PBS. (This fixative works best when used fresh.)
- **Phosphate buffered Saline (PBS), 10x :** Dissolve 2.0 g KCl, 2.0 g KH_2PO_4,1.0 g $MgCl_2.6H_2O$, 80,0 g NaCl, 14.34 g $Na_2HPO_4 \cdot 2H_2O$, and 1.0 g $CaCl_2$ (add last) in 1000 ml distilled water, adjust pH to 3. Before diluting PBS to use as 2x or Ix solution, adjust pH to 7.2 with 1M NaOH.

Multi-labeling

For multi-labeling of fixed cells with antibodies, the basic protocol must be performed twice. During the labeling, the coverslips must stay in PBS + 0.1% BSA + 5% NGS. It is important to use the right combination of primary and secondary antibodies. If primary antibodies of similar origin (*e.g.,* mouse) are used, cross-reactions can occur during the detection with secondary antibodies. This can be avoided with appropriate blocking steps. If possible, use direct staining with directly labeled probes, such as Texas red-X-Phalloidin or DAPI. List of primary and secondary antibodies used in the multi-labeling experiment to stain the activation of cytoskeletal proteins:

1. Labeling: Anti-Phosphotyrosine, polyclonal, rabbit IgG (#06-427, Upstate Biotechnology); secondary antibody: FITC (Fluorescein isothiocyanate) anti rabbit IgG (Ex = 496/Em = 516).
2. Labeling: Anti-paxillin, monoclonal, mouse IgG (#P17920, Transduction Laboratories); secondary antibody: TRITC (Tetramethylrhodamine isothiocyanate) anti-mouse IgG (Ex = 543/Em = 571).
3. Labeling: direct, for cytoskeletal F-actin: Texas red-X-Phalloidin, (Ex = 594/ Em = 608).

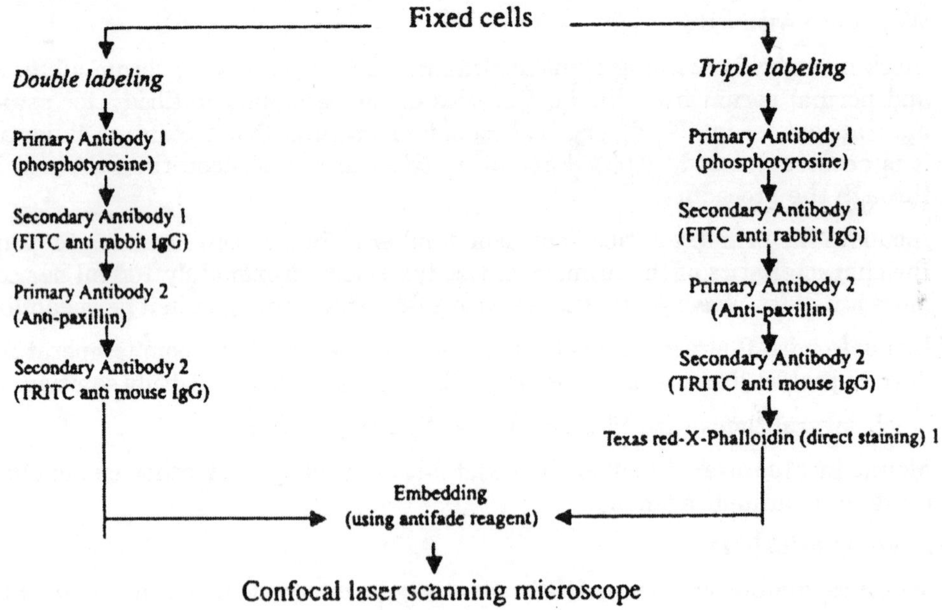

CONCLUSIONS

CLSM for multicolour immunofluorescence is a complex task involving detailed knowledge of immunohistochemistry, confocal microscopy operation and instrumentation, and image presentation and analysis. The development of CLSM systems with multi-wavelength excitation and powerful microcomputer systems allows confocal imaging to be used with triple-labeled specimens.

CHAPTER

10 Immunoassay

Antibodies are proteins produced by an animal, in a process known as the immune response, as a result of the introduction of a foreign substance into its tissues. Such a substance is known as an antigen or immunogen and the anti-bodies that are produced are capable of binding with that antigen when allowed to react in an appropriate manner. Some substances, known as haptens, are not capable of initiating an immune response by themselves but when conjugated with a protein may act as an epitope, resulting in the formation of antigen-antibody reactions have provided the basis of very useful methods of qualitative and semi-quantitative analysis for many years, particularly in microbiology.

Antibodies have been used in the identification of bacteria and often antibody preparations were given names that described a demonstrable feature of their reaction with the antigen. Thus lysins were antibodies that caused the disruption of cell membranes, opsonins rendered the antigen susceptible to phagocytosis, while agglutinins and precipitins caused the flocculation of cellular and soluble antigens respectively. It was originally thought that the antibodies that caused these different effects were quite different from each other and it was not appreciated at the time that the same antibody might have different effects depending on the environmental conditions and the presence of other substances. Major progress in the analytical use of antibodies occurred with the development of several fundamental techniques: the ability to detect cell-bound antibody-immunoprecipitation in gels radioimmunoassay and monoclonal antibodies. These, together with a considerably improved understanding of the immune response and antibodies, paved the way for development of many very sensitive quantitative methods of analysis.

GENERAL PROCESSES OF THE IMMUNE RESPONSE

The formation of antibodies is only one mechanism by which an animal may protect itself from substances or microorganisms that are potentially harmful. A *mechanical protection* against infection is provided by the presence of an intact skin surface and membranes together with the secretion of mucus from many internal membrane surfaces. The acids secreted by the stomach and skin have a bactericidal effect as does the presence in many body fluids of certain enzymes, particularly lysozyme. Invasion of the tissues by an infective agent initiates an *inflammatory response* in the animal. This is non-specific and is mediated primarily by substances released from tissues that are damaged as a result of gither trauma or the toxic effects of the infective agent.

The major mediator is the vasoactive amine histamine, which causes an increased local blood flow and capillary permeability, resulting in local oedema. A major aspect of the inflammatory response is the involvement of large numbers of phagocytic cells, particularly the polymorphonuclear leucocytes. These are chemotactically attracted to the inflamed tissues and are mainly responsible for the elimination of particulate material. This often results in the destruction of many of these cells and the formation of pus. One very important group of cells known as the lymphocytes, which are widely distributed throughout the tissues, appear in increased numbers during an inflammatory response and are primarily responsible for the *immune response*. This is a specific response to the invading substance or agent by the animal and involves the production of cells and antibodies with the ability to recognize and bind the invading substance.

Cells Involved in the Immune Response

The *lymphocytesy are* a very heterogeneous group of cells, almost identical when studied using light microscopy methods and only showing slight differences by electron microscopy techniques and yet the group contains cells with many different roles. Although large numbers of lymphocytes can be detected in the circulating blood and body fluids, the majority of lymphocytes are to be found in the group of tissues known collectively as the reticulo-endothelial system. This includes such tissues as the liver, spleen, bone marrow, thymus and lymph nodes, all of which are important in the immune response.

Experiments involving the removal of various tissues from experimental animals have indicated that there are two different features to the immune response. The removal of the thymus, a small gland located behind the sternum, impairs the ability of a young animal to reject skin transplants but does not affect to the same extent its ability to produce antibodies. This aspect of the immune response is known as *cell-mediated immunity* and is due to a sub-population of lymphocytes called *T lymphocytes* (thymus derived).

Experimental work with fowls has demonstrated that the removal of a lymphoid tissue nodule in the gut known as the bursa of Fabricius results in a reduced ability to produce antibodies but does not significantly alter the response to skin grafts. It was subsequently demonstrated that the production of antibodies, a feature known as *humoral immunity,* is associated with another sub-population of lymphocytes known as *B lymphocytes* (bursa derived). The bone marrow acts as the source of B lymphocytes in man. The introduction of an antigen into the tissues of a susceptible animal results initially in increased proliferation of lymphocytes in the tissues of the reticulo-endothelial system particularly the lymph nodes and the spleen. An animal will show different responses if the antigen is completely new to it (a primary response) or if the antigen has been encountered on a previous occasion (a secondary response).

The secondary response shows a reduced lag period and a considerably increased rate of anti-body synthesis compared with the primary response and the antibody persists for a longer period. The kinetics of the response vary depending upon the antigen and the animal but the relationship between the primary and secondary responses is characteristic. There are three major types of lymphocyte which, when stimulated, are directly responsible for the effects of an immune reaction.

B lymphocytes during proliferation develop into *Plasma cells,* which are the antibody-producing cells, the process referred to earlier as humoral immunity. An antibody protects in

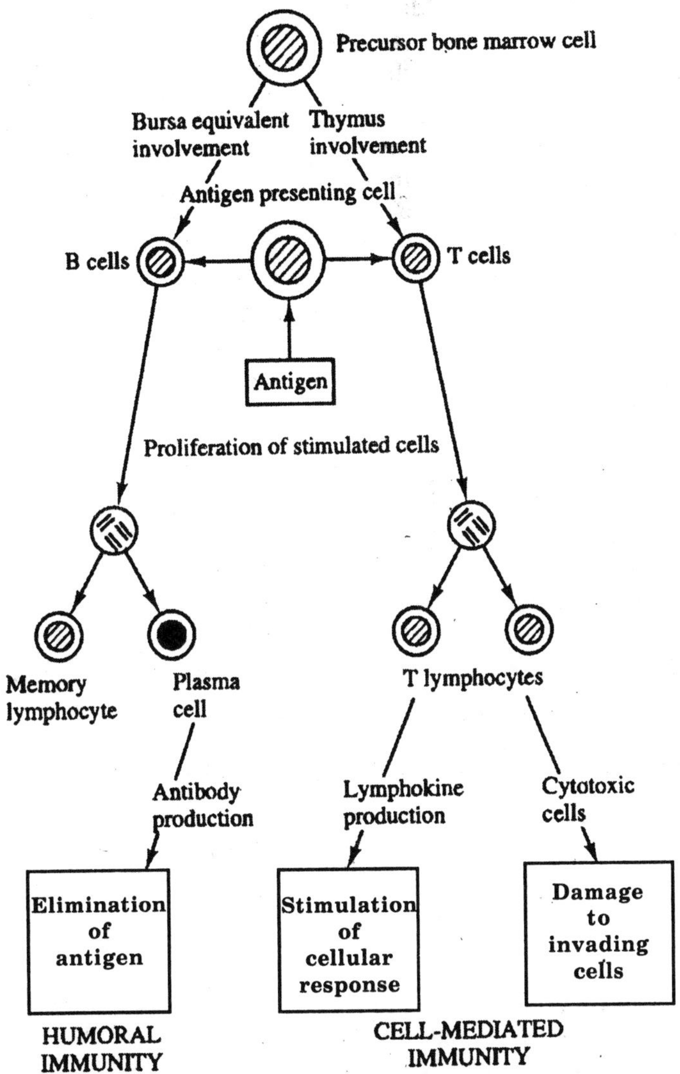

Fig. 10.1. Cellular processes of the immune response. A simplified summary of the sequence of events in both the humoral and cell-mediated response to an antigen.

a variety of ways. It can render the free antigen more susceptible to elimination by normal cellular processes such as phagocytosis or it may block the effects of a toxic substance. It can also function in conjunction with a series of plasma proteins known collectively as *complement* to cause lytic damage to the membrane of a target antigenic cell. There are two types of *T lymphocyte* involved: the cytotoxic cells (T_c cells) when stimulated are capable of binding to the antigen cell and causing irreversible lytic damage to the cell membrane; other T lymphocytes (T_d cells) release soluble substances known as lymphokines, which damage the invading antigen cells and stimulate other aspects of the host immune system. The effects of T_c and T_d lymphocyte activity are features of cell-mediated immunity. Immunologically competent cells, whether they are *T* or *B* lymphocytes, have membrane receptors that are specific for an antigen.

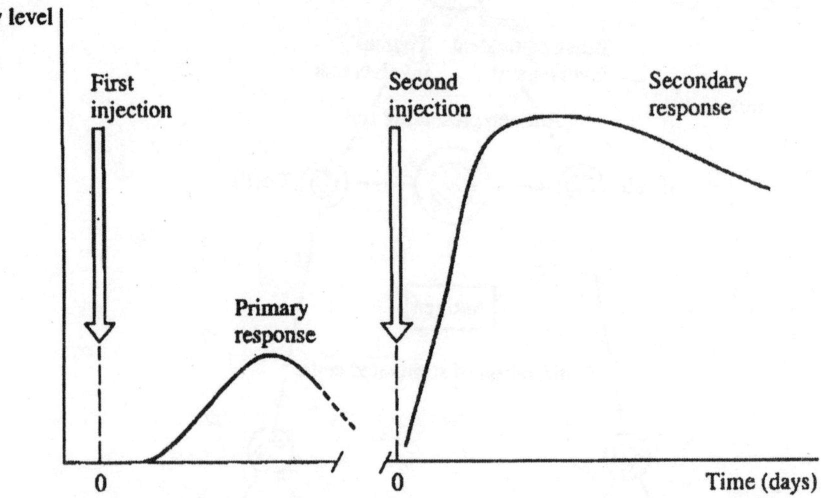

Fig. 10.2. The kinetics of the immune response. The injection into an animal of a second dose of an antigen several weeks after the first injection will result in a response that is more rapid and more intense than the first.

It is basically the binding of the antigen to the specific receptor on the appropriate lymphocyte which initiates the whole process, stimulating the cell to proliferate and producing a clone of identical cells, a process known as 'clonal selection'. The nature of the secondary response is due in the main to this large number of cells now available. The processes of antigen recognition and immune stimulation are extremely complex and involve a sequence of steps. Upon entering the tissues, the antigen is taken up by macrophages or other 'antigen-presenting cells' (APCs) and, after modification, is exhibited on the outer membrane of the cells. This membrane-bound antigen can be recognised by lymphocytes via the antigen-specific receptors carried on their membranes.

A key group of cells in the process are the T helper cells (T_h cells). These cells recognize the antigen presented by the macrophage and when stimulated by this recognition process they are capable of stimulating the specific B, T_c or T_d cells. In all cellular interactions in the immune process, as well as antigen recognition being a major feature, there are complex stimuli resulting from the secretion of a range of cytokines by the different cells involved. In the immune process, because the interactions are between leucocytes, these cytokines are known as interleukins. They are specific for the type of cell on which they can act and are necessary, often in conjunction with antigen recognition, to stimulate the target lymphocyte to act in the required manner. This action may involve protein synthesis, replication or differentiation and is an essential step in mounting and controlling a full immune response.

An individual lymphocyte is capable of being stimulated by only a limited number of antigens (probably only one but at the most only two or three) and in order to provide a large number of different antibodies it is essential that there are an equivalent number of different lymphocytes in the tissues.

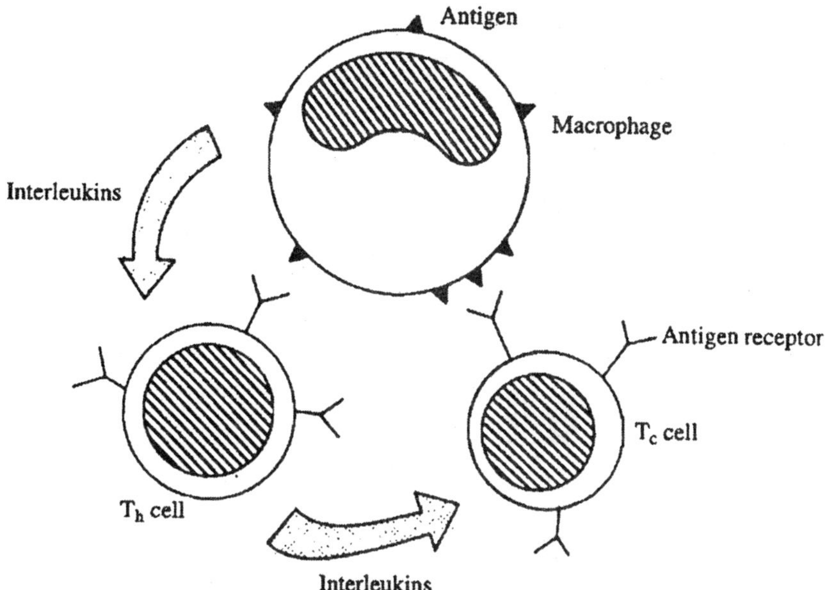

Fig. 10.3. Process of antigen recognition and lymphocyte stimulation.

Antibody Structure

Antibodies are members of a group of proteins known collectively as immunoglobulins. The name is derived from the observation that during electrophoresis of blood plasma the proteins associated with antibody activity migrate with the gamma globulin fraction. Immunoglobulins are glycoprotein molecules composed of four polypeptide chains linked covalently by disulphide bonds (Fig. 10.4). The four polypeptides consist of two identical chains of relative molecular mass 23000 and 50000-75000 and are designated light and heavy chains respectively. In addition to the interchain disulphide bonds, each polypeptide contains a number of intrachain disulphide bonds which divide up the polypeptide chain into a series of domains. In addition, immunoglobulins possess an area known as the hinge region which allows for flexibility of the chains in relation, to one another.

Comparison of the amino acid sequence of the heavy chains of a particular class of immunoglobulin reveals that approximately three-quarters of each chain from the *C*-terminal end show very-similar sequences (the constant section). The remaining quarter of the peptide chain (the variable section) shows considerable variation in the amino acid sequence and corresponds to that part of the chain associated with the antigen-binding site. Similar constant and variable regions are also demonstrable in the light chains although in this case each involves approximately half of the peptide. The variation is particularly noticeable at three distinct sections (hypervariable sections) in the heavy chains and it is suggested that these sections when associated with three similar sections in the light chains are responsible for the antibody activity and specificity of an immunoglobulin. Five major sub-classes of immunoglobulin (isotypes) have been recognized and the differences between the classes lie in the heavy chains, which, although of approximately the same size for all classes, vary considerably in the amino acid sequence.

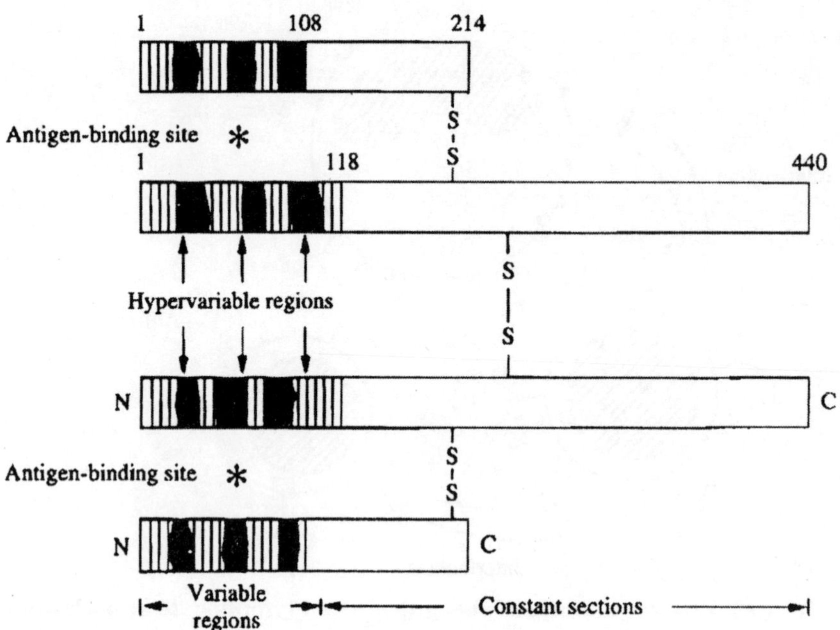

Fig. 10.4. Basic structure of an IgG molecule. Two heavy chains (440 residues) and two light chains (214 residues) are joined by disulphide bonds and each shows a relatively constant amino acid sequence in one section (C-terminal end) and a variable sequence section (N-terminal end). The variable regions of both heavy and light chains are involved in the formation of the antigen-binding site.

The heavy chains of immunoglobulins are designated as λ, μ, α, δ, or $\in$ and when two identical heavy chains are combined in an immunoglobulin, the molecule is designated as being either IgG, IgM, IgA, IgD or IgE respectively. The heavy chain isotypes can be further subdivided into several subtypes. There are four sub-types of IgG, ($\gamma_1, \gamma_2, \gamma_3, \gamma_4$), two of IgA ($\alpha_2$, α_2) and two of IgM. (μ_1, μ_2). The light chains do not show such variation and only two main types are demonstrable, known as the kappa (*k*) and the lambda (λ) chains.

Table 10.1. Classes of Immunoglobulins.

Immunoglobulin	*RMM*	*Number of basic four-chain units*	*Heavy chain*	*Heavy chain sub-classes*	*Antigenic valency*	*Percentage in normal serum*
IgA	1.5 ´ 10^5	1 or 2	α	$\alpha_1\,\alpha_2$	2-4	13
IgD	1.8 ´ 10^6	1	δ		2	1
IgE	2.0 ´ 10^6	1	$\in$		2	0.002
IgG	1.6 ´ 10^5	1	γ	$\gamma_1\,\gamma_2\,\gamma_3\,\gamma_4$	2	80
IgM	1.0 ´ 10^6	5	μ	$\mu_1\,\mu_2$	10	6

The role of Antibodies

IgG comprises some 80% of the total immunoglobulin in plasma and because it is relatively small it is capable of crossing membranes and diffusing into the extravascular body spaces. It can cross the placental membrane and provides the major immune defence during the first few weeks of life until the infant's own immune mechanism becomes effective.

IgM is a large molecule composed of five units, each one similar in structure to an IgG molecule. The tetramer contains an additional polypeptide, the J chain (relative molecular mass 15 000), which appears to be important in the secretion of the molecule from the cell. It is an effective agglutinating and precipitating agent and, although potentially capable of binding ten antigen molecules, it is usually only pentavalent. It does not cross membranes easily and is largely restricted to the bloodstream.

IgA is associated mainly with seromucous secretions such as saliva, tears, nasal fluids, etc., and is secreted as a dimer with both a J chain and a secretor piece (relative molecular mass 70000), the latter apparently to prevent damage to the molecule by proteolytic enzymes. Its major role appears to be the protection of mucous membranes and its presence in blood, mainly as the monomer, may be as a result of absorption of the degraded dimer.

IgE is known as a *cytophilic immunoglobulin* because of its ability to bind to cells, which may account for its low concentration in body fluids. When IgE reacts with an antigen it causes degranulation of the mast cell to which it is bound, with the release of vasoactive amines such as histamine. This process may well be helpful in initiating the inflammatory response but in allergic individuals the reaction is excessive and leads to a hypersensitive or over-reactive state.

ANTIGEN-ANTIBODY REACTIONS

An antibody combines specifically with the corresponding antigen or hapten in a manner that is very similar to the binding of an enzyme to its substrate and involves hydrophobic and electrostatic interactions. The bonding between an antibody and antigen, however, involves no subsequent chemical reaction and its stability depends upon the complementary shape of the antigen and the binding site of the antibody.

Because of the relative weakness of the forces that hold antibody and antigen together, these combinations are reversible and the complex will dissociate, dependent upon the strength of binding:

$$\text{Ag} + \text{Ab} \Leftrightarrow \text{AgAb}$$

where the binding is strong the equilibrium will lie to the right; where weak, to the left. The strength of the binding of an antibody to an antigen is referred to as its affinity and is defined by the equilibrium constant K, where

$$K = \frac{[\text{AgAb}]}{[\text{Ag}][\text{Ab}]}$$

where more than one antibody in an antiserum can combine simultaneously with an antigen, the sum of the binding strength is defined as the avidity of the antiserum. Most antigens are large and may have many antigenic characteristics or epitopes (determinants) and as a result serum taken from an animal which has been immunized against that antigen will contain several different antibodies against the different antigenic determinants.

It is possible that another antigen may share some similar antigenic determinants with the original antigen with the result that some of the antibodies in an antiserum will bind to both antigens. Such an antiserum will show cross-reactivity between the two antigens and is said to lack specificity. Antiserums used for analytical purposes should be specific and it is essential that every antiserum should be thoroughly tested prior to its use.

Production of Antibodies

The raising of a specific antiserum by the immunization of an animal usually involves intramuscular injection of the pure antigen, although intravenous injection may be appropriate for particular antigens. Because of the primary and secondary responses to antigens, a series of injections involving small amounts of antigen is likely to be more effective than a single large injection. The precise sequence and timing can significantly affect the quality of the antiserum pro-duced. An initial injection is normally followed by several booster doses given at two-to four-week intervals. Too frequent injections, although possibly giving a quicke response, may result in an antiserum of reduced avidity. The species of animal used should be as different as possible from the animal source of the antigen.

It should be relatively easy to handle and yet provide enough serum to make the process worthwhile. The animals most frequently used for the production of antibodies are the guinea pig and rabbit but for larger supplies of serum, goats, sheep and horses are used.

Antigens vary considerably in their ability to initiate an immune response and it is usual to incorporate an adjuvant into the sample before injection. An adjuvant is a mixture of substances that stimulates an inflammatory response and prevents the rapid removal of the antigen from the tissues by the normal drainage mechanism. Freund's adjuvant consists of an emulsion of dead mycobacteria in mineral oil but simpler alternatives of aluminium phosphate or hydroxide have a similar effect. Although proteins are generally immunogenic, they do need to have a relative molecular mass of at least 4000 and some structural rigidity to be effective.

In order to raise antibodies against a non-immunogenic molecule (a hapten) it is necessary to link it to a carrier protein which is capable of initiating a response. Bovine or human serum albumin is frequently used for this purpose as well as synthetic polypeptides such as poly-L-lysine. The hapten should be linked covalently with the carrier protein, a process readily achieved using a variety of chemistries and often involving the incorporation of a spacer molecule between the carrier protein and hapten. Antiserums produced in this manner are known as *polyclonal* because they contain many antibodies produced by different lymphocytes, each one responding either to different antigenic determinants of the original antigen or to other immunogenic substances in the injected material.

Such a range of antibodies reduces the specificity of the method. The development of techniques for the production of *monoclonal* antibodies by Kohler and Milstein has enormously expanded the potential of antibodies as analytical and therapeutic agents. A monoclonal antibody is one that is produced by a clone of cells all derived from a single lymphocyte. Any lymphocyte can probably produce only a single immunoglobulin and hence the antibody produced by a clone of identical cells is very restricted in the antigens to which it will bind, making it a very specific reagent. The production of monoclonal antibodies starts with the immunization of an animal (usually a mouse) in the traditional manner.

However, instead of allowing the immune system of the mouse to generate antibodies, lymphocytes are separated from the spleen of the mouse and fused *in vitro* with myeloma cancer cells growing in cell culture. A fusogen, usually polyethylene glycol, is used and the resulting fused cell is known as a hybridoma. The cell suspension is diluted and distributed among a large number of sub-cultures in order to achieve single-cell distribution. The original cancer cell has the ability to synthesize immunoglobulin non-specifically but when fused with a lymphocyte, which was stimulated by the injected antigen, produces the immunoglobulin for which the lymphocyte has the genetic information. The hybridoma cells are initially grown in a medium that will not main-tain the growth of the cancer cells; these therefore die, as do non-fused lymphocytes, leaving only the fused cells.

As the hybridoma cells grow, the supernatant fluid is tested for the presence of antibodies. Those cultures producing the desired antibody are further cloned and either grown in bulk or as tumours in animals and the monoclonal antibodigs harvested. In addition to their improved specificity, monoclonal antibodies offer other significant advantages over polyclonal antiserums: there is an indefinite supply of antibodies with constant characteristics together with relative ease in purification.

The Effects of Antigen-antibody Complex Formation

The combination of an antigen and an antibody usually results in the formation of a lattice-type structure. If this structure is large enough it will sediment and be measurable in some way. If the antigen was originally soluble, this process is known as precipitation but if the antigen was cellular or particular, the process is known as agglutination.

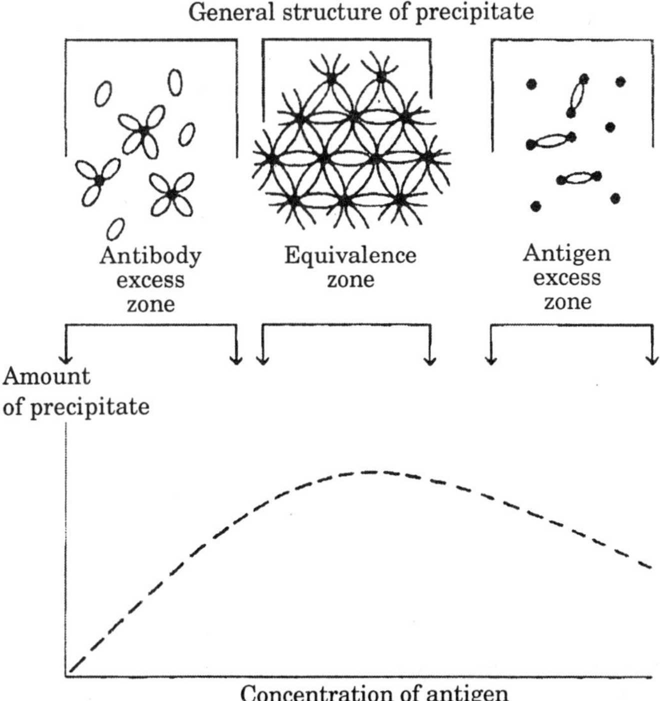

Fig. 10.5. Antigen-antibody reaction. Maximum precipitation occurs when the antigen and antibody are present in equivalent amounts and is due to the formation of large lattice structures. On either side of the equivalence zone the amount of precipitation is reduced because the aggregates are smaller and more soluble.

It is extremely difficult to measure the amount of antibody present in absolute terms but its activity or concentration can usually be related to a demonstrable aspect of the antigen-antibody reaction.

A study of the effect of the concentration of the two components indicates that the proportion of antibody to antigen is often critical in order to produce the detectable effect. Fig 10.5 shows the effect of varying amounts of antigen in the presence of a fixed amount of antiserum on the extent of precipitation. As the amount of antigen increases from zero, so the amount of precipitate increases in an approximately proportional manner up to a maximum value. However, if the amount of antigen is further increased, the amount of precipitate often diminishes.

This means that instead of working with excess reagents, as is often the *case* in chemical methods of analysis, it is often essential to work with optimal or equivalent pro-portions of antigen and antibody in order to produce the maximum effect. In some cases, particularly if the antigen is cellular, the combination of antigen and antibody will result in a biological effect rather than a physical effect. Cells may be damaged, resulting in either lysis or the inhibition of their natural processes, such as motility. Viruses may be neutralized making them non-infective. Many, tests involving such mechanisms are qualitative and do not readily lend themselves to quantitative analysis. Table 10.7 summarizes the major types of analytical methods involving antibody reactions.

Table 10.2. Immunologlcal Methods of Analysis. Demonstrable Reactions Between Antibodies and Antigens.

Natural reaction	*Nature of antigen*	*Analytical value*	*Example*
Agglutination	Particulate	Mainly qualitative or cellular	Blood grouping Detection or Semi-antibodies
Cytotoxicity	Cellular	Qualitative or semi-quantitative	Cel typing Complement fixation tests
Precipitation	Soluble	Qualitative or Quantitative	Double diffusion Single radial diffusion
Complex formation	Labelled Molecules	Quantitative or qualitative	Radioimmunoassay Enzyme immunoassay Immunohistochemistry Immunoblotting

ANALYTICAL TECHNIQUES—PRECIPITATION REACTIONS

Many qualitative and quantitative methods for soluble antigens are based on their precipitation by antibodies. It is important that the conditions for each assay are carefully optimized in order to achieve the desired end.

Immunoprecipitation in Solution

The number and size of the immune complexes formed by the reaction of antigen with antibody is dependent on the relative concentrations of the reactants. The formation of antigen-antibody complexes may be observed by measuring the apparent absorption of light when incident light is scattered by the complexes or by direct measurement of the scattered light.

The reaction is carried out in an excess of antibody and a calibration curve is prepared by measuring the turbidity of a series of standard solutions of the antigen. The measurement of the turbidity of a solution containing antigen-antibody complex, in which the amount of incident light lost by scattering is measured, has the advantage of simplicity but generally shows poor levels of sensitivity and precision.

The method may be considerably improved by the use of a nephelometer rather than a simple photometer. A nephelometer is similar in design to a fluorimeter in that the detector is placed at right angles to the incident light beam and thus measures the light scattered by the complex. The extent of light scattering by a turbid solution increases as the wavelength of incident radiation decreases and this, together with a requirement to minimize the internal reflection of incident light, has led to the increasing use of lasers as a light source. Immunoprecipitation is an extremely valuable technique because it permits the quantification of a specific protein in the presence of many other similar proteins.

It is capable of a good level of sensitivity, the lower limit being mainly determined by the clarity of the sample blanks. In order to achieve maximum sensitivity the antiserum must show a high avidity for the antigen and be capable of being used at low concentration (high titre), the latter being important not only from a cost perspective but also to minimize turbidity due to the reagents. In addition, the development of particle-enhanced light-scattering assays in which the antibodies are immobilized onto latex particles has overcome many of the limitations of basic immunoprecipitation in terms of improved sensitivity and assay range. A major advantage of immunoprecipitation is that it may be readily automated, a feature that has been particularly exploited using centrifugal analysis instrumentation.

Immunoprecipitation In Gels

Gels are used in immunoprecipitation techniques to stabilize the precipitate, enabling both the position and the area of the precipitate to be measured. The point has already been made that maximum precipitation occurs when the equivalent proportions of both antigen and antibody are available. Hence, if a high concentration of antigen is permitted to diffuse into a gel that contains a uniform concentration of antibody, at some point in the concentration gradient of antigen that is formed there will be optimum concentrations of both reactants and a precipitate will form. The dimensions of the gradient will depend on the original concentration of the antigen and hence the distance between the precipitate and the original starting point of the antigen will be proportional to its initial concentration. This principle forms the basis of single radial immunodiffusion (SRID), a technique first developed in 1965 by Mancini. SRID involves pipetting a measured volume of antigen into holes cut into a buffered agar gel containing the antibody.

The loaded gel is placed in a moist chamber at room temperature for at least 18 hours to permit diffusion. Rings of precipitate form around each well, the precipitate being maximal at the periphery and less intense towards the well. The time taken for samples containing a high antigen concentration to arrive within the zone of equivalence will be longer than for lower concentrations of antigen and so the diameter of the formed rings will larger.

The radial nature of the diffusion is such that the diameter of the precipitation ring will be related to the concentration of the antigen. In practice, the diameter of the ring is measured and a plot of the square of the diameter against concentration generally gives a straight line relationship. Two measurements of the diameter are taken at right angles in order to allow for any slight irregularities in the shape of the rings. SRID techniques provide useful and specific methods for the quantitation of individual proteins. The limit of sensitivity is about 5 mg l^{-1} and it is dependent on the ability to detect and measure very small precipitation rings. The concentration of antibody in the gel has to be carefully selected; reducing the amount of anti-body in the gel will increase the average ring size but will result in less precipitate and a compromise has to be made between ring diameter and precipitate intensity. The visualization of the precipitate may be improved by staining the protein with a suitable dye. One of the drawbacks to SRID is the time for the reaction to take place. Laurell, in 1966, sought to the overcome this difficulty by using electrophoresis rather than diffusion to produce the concentration gradient. In electro-immunoassay the antibody is incorporated into the gel in a similar way to SRID electrophoresis is usually performed at pH 8.6, conditions under which the antibodies do not migrate significantly but the test proteins do. Because the samples are driven towards the equivalence zone by electrophoresis rather than by diffusion the precipitation lines characteristically appear as peaks or 'rockets' in the gel. A calibration curve for the quantitation of unknown samples is constructed by plotting peak height against the concentration of antigen. Although the buffer systems are designed to restrict the migration of antibodies during electrophoresis, immunoglobulins may still be measured by this technique if they are first carbamylated or formylated. These procedures are designed to increase the ionic character of the immunoglobulins without significantly altering their immunological properties.

Carbamylation involves a reaction with potassium cyanate, and formylation a reaction with formaldehyde. These techniques are notable for the fact that they rely on the diffusion of only one component of the antigen-antibody reaction; in the technique of double diffusion both antigen and antibody are placed in wells and allowed to diffuse. Precipitation also occurs at the equivalence point but in this case the precipitation is in the form of a line or arc in the gel. The position of the line is characteristic of the antigen and allows its identification in a complex mixture of similar proteins. A variety of similar qualitative techniques (immunoelectrophoresis, crossed immunoelectrophoresis) have been developed.

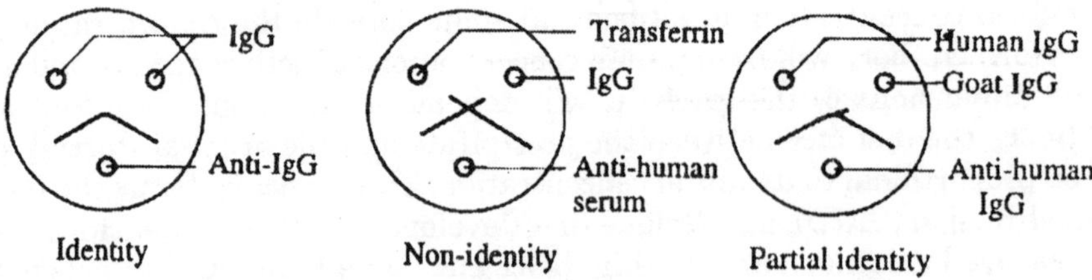

Fig. 10.6. Double diffusion in gels showing reactions of identity, non-identity and partial identity.

Immunohistochemistry

The specificity of antibodies can be exploited in order to probe the *in situ* organization of cells and tissues. Cellular antigens can be identified both in viable cells and in frozen or fixed tissue sections. Antibodies are used to identify the appropriate antigen in the section and then the position of this primary antibody may itself be detected either directly if it was initially labelled or indirectly using another secondary antibody or molecule to attach to the antibody. Samples need to be carefully washed after addition of the primary or labelled antibody in order to prevent any non-specific reactions. Labels that have been successfully linked to antibodies include the following :

1. *Fluorescent dyes.* Immunofluorescence using fluorescein or rhodamine has been very successfully employed for immunohistochemistry. Fluorescein, when illuminated with UV light, emits a characteristic green fluorescence while rhodamine gives an orange colour.
2. *Enzymes.* Enzymes too are useful labels and several have been employed, with peroxidase and alkaline phosphatase being the most popular to date. One important feature of this technique is for the enzyme to be able to convert a soluble substrate into an insoluble product in order to localize the antigen properly. Several suitable substrates are available with 3′ 3′-diaminobenzidine (DAB) and 3-amino-9-ethylcarbazole (AEC) being used with peroxidase and 5- bromo-4-chloro-3-indoylphosphate/nitroblue tetrazolium (BCIP/NBT) with alkaline phosphatase.
3. *Colloidal gold.* This is a useful label for both direct and indirect staining methods since it requires no further reagent additions. Gold probes may be readily seen by light microscopy when coupled with silver enhancement and in addition, being electron dense, offer excellent sensitivity for the electron microscope.

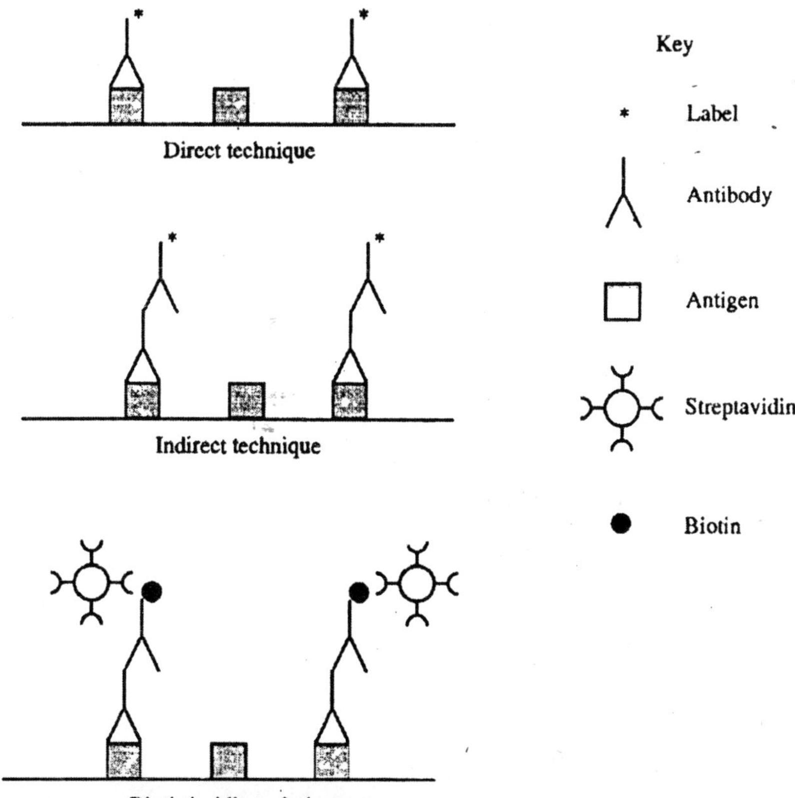

Fig. 10.7. Some examples of antibody detection techniques. (*a*) Direct labelling of the primary antibody, (*b*) indirect using labelled secondary antibody, (*c*) indirect using biotinylated secondary antibody and labelled streptavidin.

Immunoblotting

Antibodies may also be used to determine the presence or identity of soluble antigens by a process known generally as immunoblotting. In a technique known as 'dot blotting', soluble antigens are applied to a nitrocellulose membrane in the form of 'dots', antibody is then applied to the membrane, the membrane is washed to remove unbound antibody, and the presence or absence of the antigen is determined by similar means to that employed during immunohistochemistry. Alternatively, as in Western blotting, soluble antigens may be separated by electrophoresis on polyacrylamide gel either with or without prior treatmenir with SDS.

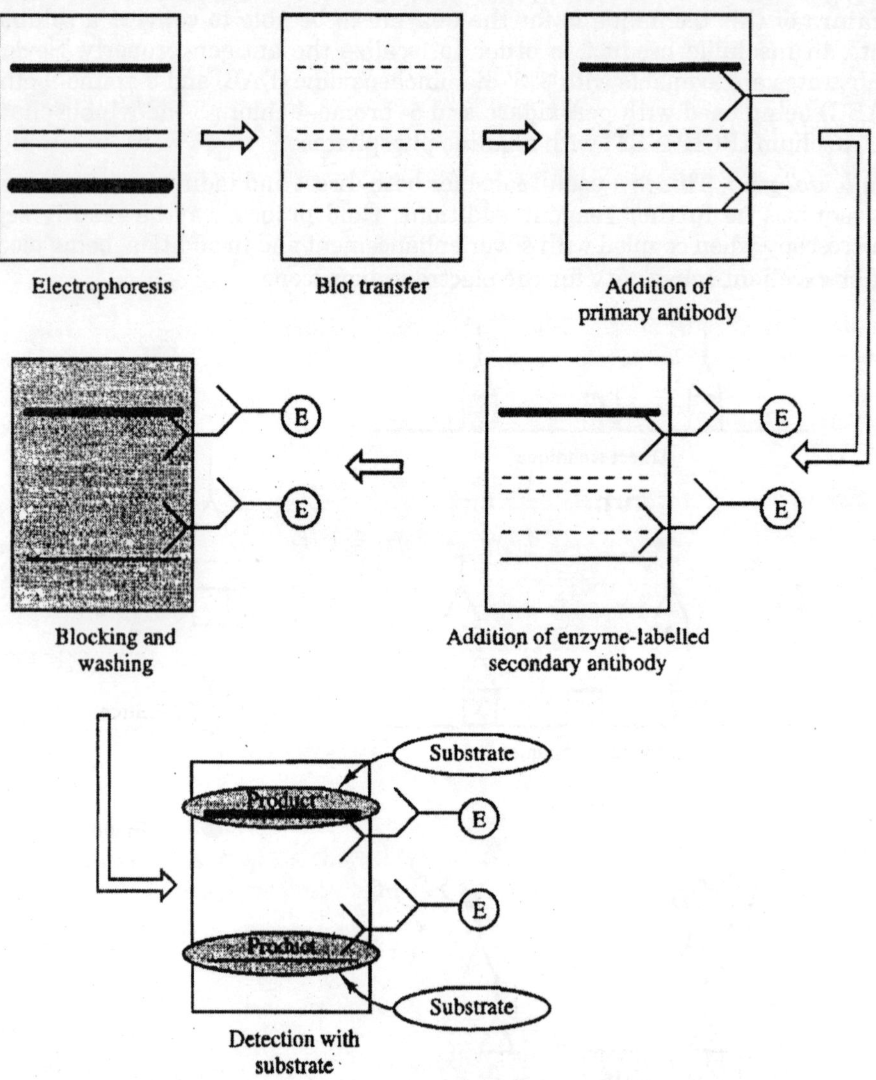

Fig. 10.8. Immunoblotting. A representation of the stages in the technique.

After electrophoresis the antigens are transferred from the gel to a nitrocellulose membrane. Once on the membrane antibodies may be used to probe for the presence of particular antigens either directly or indirectly as in immunohistochemistry. Non-specific binding sites may be 'blocked' using other non-specific proteins such as bovine serum albumin or casein before washing to remove unbound antibody. The labelled molecule may be visualized by the techniques used for immunohistochemistry or by autoradiography.

ANALYTICAL TECHNIQUES-IMMUNOASSAY

Immunoassay as an analytical technique was introduced by Rosalind Yalow and Solomon Berson in 1960 with their use of anti-insulin antibodies to measure the concentration of the hormone in plasma. This advance, for which Rosalind Yalow was awarded the Nobel prize, was probably the most important single advance in biological measurement of the following two decades. Examples of the use of immunoassay may now be found in almost all areas of analytical biochemistry. Despite many novel developments in immunoassay design the principles are confined to two broad approaches: those that rely on the competition between antigens labelled with a molecule which may be readily observed (for example, a radioisotope) and unlabelled antigens for a limited number of antibody binding sites; and those in which the antibody is available in excess and for which there is no competition for binding sites. These principles have been exploited by commercial manufacturers to provide 'kits', which are packages of reagents, designed to analyse samples for a wide variety of analytes. The commercial exploitation of immunoassay technology and the concurrent development of suitable automated instrumentation has had a profound effect on the analysis of biological samples. Kits can provide a wide range of tests offering complex technology and consistent reagents allowing even small laboratories to perform a full range of analyses.

Competitive Binding Immunoassay

The basis for this technique lies in the competition between the test antigen and a labelled antigen for the available binding sites on a fixed amount of antibody. While the binding sites are traditionally associated with an antibody, any source of specific reversible binding sites may be used to create an assay in this format. Examples of such are specific transport proteins such as thyroxine-binding globulin and certain cellular receptors such as opiate or benzodiazepine receptors. Under these circumstances the equilibrium mixture may be represented thus:

$$\begin{matrix} Ag \\ \\ [Ag^*] \end{matrix} \quad + [Ab] \quad \underset{K_2}{\overset{K_1}{\rightleftharpoons}} \quad \begin{matrix} Ag\,[Ab] \\ \\ Ag^*\,[Ab] \end{matrix}$$

Given that the quantities of labelled antigen [Ag*] and antibody [Ab] are fixed when no unlabelled test antigen (Ag) is present then the labelled antigen has free access to the binding sites and the subsequent equilibrium state will represent the maximum amount of labelled antigen that may be bound.

When unlabelled test antigens are present there will be competition between the labelled and the unlabelled antigens for the available binding sites with the result that at equilibrium the amount of labelled antigens bound will be reduced by an amount proportional to the amount of unlabelled antigens in the system. The concentration of antigens in unknown samples may be determined by comparison of the proportion of bound labelled antigens with the proportion bound when a series of standards of known antigen concentration is used.

Non-competitive or Immunometric Immunoassay

This approach to immunoassay is characterized by the fact that the antibody is present in excess and is generally also labelled. Because the labelled antibody is in excess there is no requirement for the setting up of an equilibrium since all of the test antigen may be sequestered by the excess of antibody.

$$Ag + Ab^* \longleftrightarrow AgAb^* + Ab^*$$

In immunometric assays, unlike competitive systems, the amount of labelled antibody bound is directly proportional to the amount of unlabelled antigen present rather than inversely proportional.

For each of these types of immunoassay, in order to observe the ratio of bound to free in the final reaction mixture, a method has to be employed which will separate these fractions and various ingenious ways have been devised to do this. Techniques which require the separation of free from bound fractions are referred to as heterogeneous assay systems.

COMPONENTS OF IMMUNOASSAY SYSTEMS

The Antibody

The properties of the antibody used in an immunoassay will in large measure define its usefulness as an analytical technique. Both polyclonal and monoclonal antibodies have been in immunoassays. It is essential to determine the required optimum antibody concentration when using a competitive assay. By convention this is the dilution of antiserum which will bind 50% of the labelled antigen in the absence of unlabelled antigen although this convention has no sound theoretical basis and many workers question its validity.

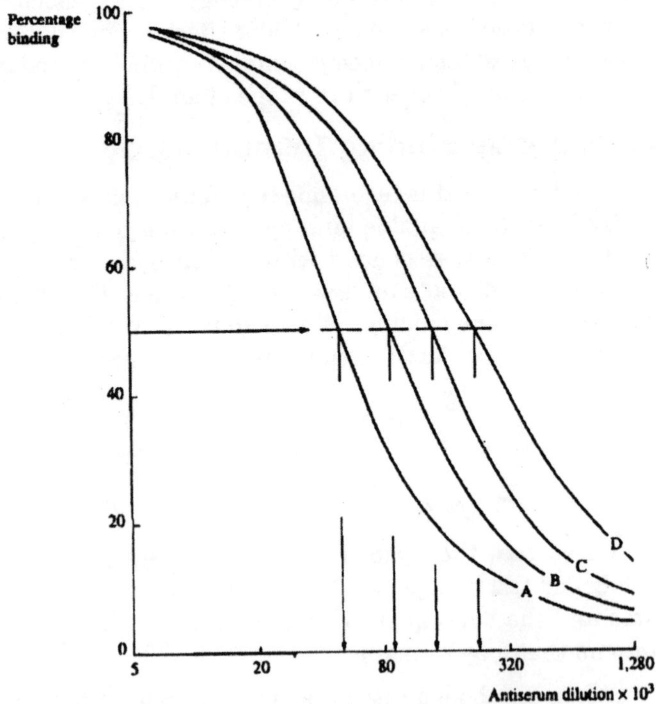

Fig. 10.9. Radioimmunoassay of insulin-titration of antiserum.

This dilution of antiserum is known as its titre but this is a proxy unit because in a polyclonal antiserum is not generally possible to determine the mass of immunoglobin involved in antigen. Fig. 10.10 also illustrates the effect that decreasing the mass of labelled antigen has on the assessment of the titre. Once the titre of an antiserum is known, its specificity may be assessed.

Curve	Amount of labelled reference antigen (pg)
A	200
B	100
C	50
D	12.5

Where related molecules are recognized by an antibody, they are said to cross-react and such antibodies are likely to be unsuitable for some assays. For example, where measurements of a particular drug are made, metabolites of the drug having similar structures may also react with the antibody. The extent of cross-reactivity may be assessed by comparing the concentration at which a 50% displacement of the related compound is obtained with that required for the antigen.

Table 10.3. Detector Molecules Commonly used in Immunoassay

Detector system	Example
Radioisotopes	Iodine-125, carbon-14, tritium
Enzymes	Horseradish peroxidase, Alkaline phosphatase, β–galactosidase
Chemiluminescence	Luminol, acridinium esters, adamantyl dioxetane.
Bioluminescence	Luciferase/luciferin
Fluorescence Fluorescence in, rhodamine	Europium chelates

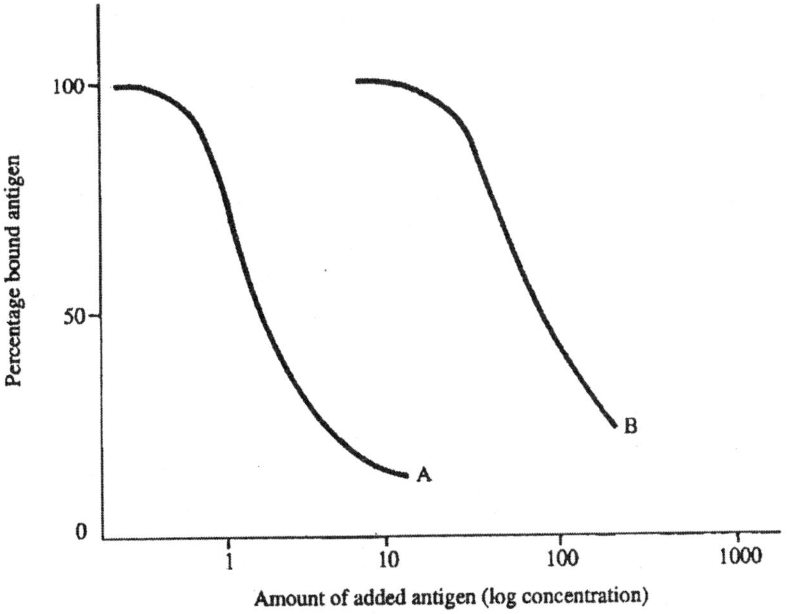

Fig. 10.10. Cross-reactivity. Compound B displaces 50% of the label at 100-fold greater concentration than compound A and is said to have a 1% cross-reactivity.

The Label

Immunoassays require a pure sample of either antibody or antigen for labelling with an appropriate molecule. Such a molecule should retain a high signal efficiency (*i.e.* be readily detected at low concentration) while its incorporation into the antigen or antibody should have no effect on their subsequent immunoreactivity.

Radioisotopes

The most commonly used isotopes are carbon-14, tritium (hydrogen-3) and iodine-125. Carbon-14 and tritium are both beta-emitting isotopes while iodine-125 emits both beta particles and gamma radiation. Carbon-14 and tritium are examples of internal labels since the radioactive atom replaces an existing atom within the antigen, while iodine-125 is described as an external label because it is usually necessary to attach the iodine covalently to the antigen.

There are advantages and disadvantages to each of these labels. The beta-emitting carbon-14 and tritium have the advantage that the labelled form of the molecules is identical to the unlabelled antigen, but suffer from the disadvantage that the efficiency of the measurement of the beta emission, which uses the technique of scintillation counting, is less than that of the gamma emissions associated with iodine-125. Equally, iodine-125 suffers from the problem that the covalent attachment of the isotope to the antigen often means that there is a significant structural difference between the unlabelled and the labelled antigen. The gain in signal measurement over the potential loss of immunoreactivity, however, is sufficient to make iodine-125 an isotope of choice for immunoassay systems. Iodine may be readily substituted onto the aromatic side-chain of the amino acid tyrosine by mild oxidation using a variety of agents; chloramine T, the enzyme lactoperoxidase *(EC* 1.11.1.7) and the sparingly soluble agent 'iodogen' (1,3,4,6-tetrachloro-3,6-diphenyl-glycouril) have all been used successfully to yield a stable and efficient label.

In applications where there is no suitable tyrosine residue available, a carrier molecule containing both a phenol or imidazole group suitable for iodination and an amine group suitable for coupling to a carboxylate group on the antigen may be used. After iodination the label is usually purified to remove damaged antigen and unreacted iodine and this may be conveniently accomplished using either gel permeation chromatography or HPLC.

Enzymes

Enzyme labels are usually associated with solid-phase antibodies in the technique known as enzyme-linked immunosorbent assay (ELISA). There are several variants of this technique employing both competitive and non-competitive systems. However it is best used in combination with two monoclonal antibodies in the 'two-site' format in which an excess of antibody is bound to a solid phase such as a test-tube or microtitre plate; the test antigen is then added and is largely sequestered by the antibody. After washing to remove the remaining biological material a second, enzyme-labelled, mon-oclonal antibody is added which recognizes a different epitope on the antigen (the second site), thus sandwiching the antigen between the enzyme-labelled and the solid-phase antibodies. The excess enzyme-antibody is washed away before addition of a suitable substrate and development of a colour which is measured either kinetically or as an end-point reaction. The enzymes commonly used as labels include alkaline phosphatase (EC 3.1.3.1), horseradish peroxidase (EC 1.11.1.7) and beta-galactosidase (EC 3.2.1.23).

1. Attachment of antibody to solid phase

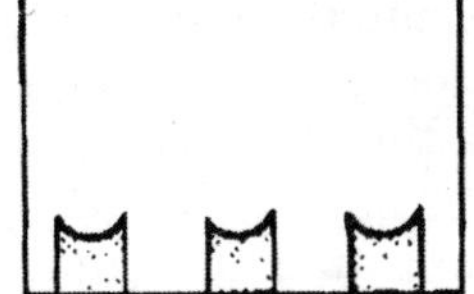

2. Wash

3. Incubate with sample containing antigen

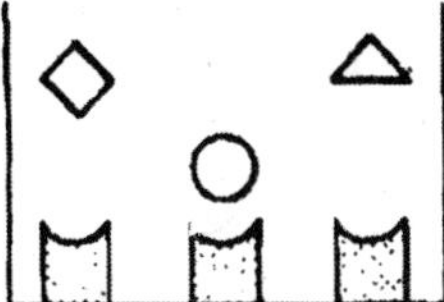

4. Wash

5. Incubate with antibody–enzyme conjugate

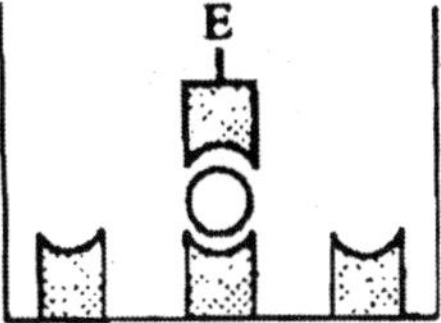

6. Wash

7. Incubate with enzyme substrate and measure product

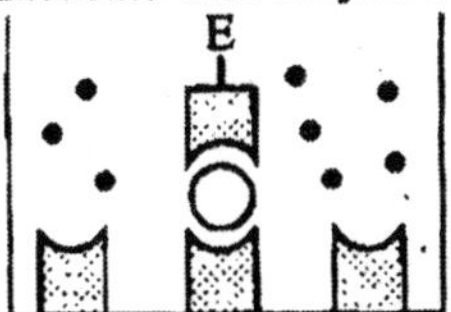

Fig. 10.11. The two-site assay employing two monoclonal antibodies directed against wo distinct epitopes.

The enzymes used should be capable of being covalently linked to the antigen or antibody without loss of either catalytic activity or immuno-reactivity. Glutaraldehyde may be used as a linking agent, while glycoprotein enzymes such as peroxidase may be linked via the carbohydrate group using periodic acid to form a reactive aldehyde group. The choice of enzyme is governed by the availability of substrate and the type of detector. Several substrates have been developed for use with horse-radish peroxidase: *o*-phenylene diamine (OPD), 2,2-azino-di(3-ethylbenzo-thiazoline-6-sulphonate) (ABTS) and 5, 5′-tetramethylbenzidine hydrochloride (TMB).

Enzymes may also be used to amplify the signal from the label, using cycling systems. Amplification of the signal from the label has the potential to increase the sensitivity of the immunoassay system.

Luminescent Labels

There are several different types of luminescence differing only in the source of energy used to excite the molecules to a higher energy state; radioluminescence occurs when energy is supplied from high energy particles; in chemiluminescence, energy is derived from a chemical reaction; in bioluminescence the excitation is performed by a biological molecule such as an enzyme; and in fluorescence or photoluminescence the excitation is derived from light energy. Each has been successfully used as a label system for immunoassay. In chemiluminescence immunoassay the antigen is tagged with a molecule such as luminol or an acridinium ester which emits light with a high quantum yield on oxidation.

Alternatively, the antigen may be labelled with a bioluminescent molecule such as luciferin, which emits light when oxidized by the enzyme luciferase. Successful immunoassays have been developed using photoluminescent labels such as fluorescein and rhodamine but there are significant drawbacks to the use of these compounds. A background signal can be generated by compounds present in the sample which themselves fluoresce when excited by light of the same wavelength as the fluorophore employed as the label.

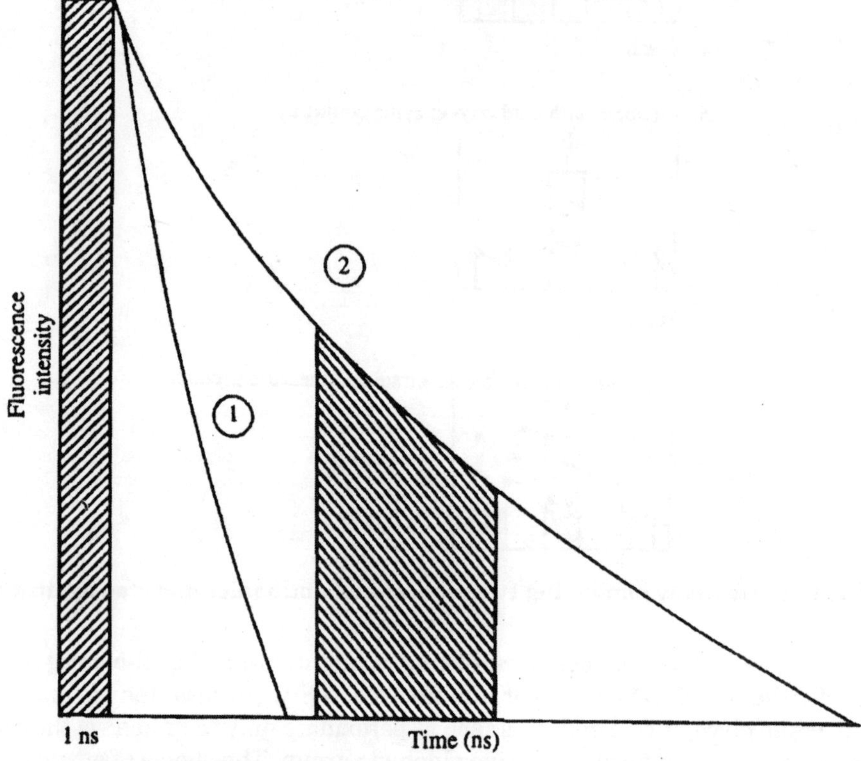

Fig. 10.12. Decay curve for europium chelates showing reading time after the decay of background fluorescence.

In addition, many biological molecules will act to decrease the emitted light by absorbing the energy, a process known as quenching. Biological molecules can also scatter the exciting light energy in such a way as to reduce the efficiency of excitation. Each of these effects has contributed to the limited usefulness of fluorescent labels in immunoassay: The problem of background fluorescence has been largely overcome with the introduction of time-resolved fluorimetry. This technique relies on the use of fluorophores with a long-lived fluorescence which can be measured after the background fluorescence has decayed.

The usual decay time for background fluorescence is about 10 ns while that of long-lived fluorophores such as chelates of the lanthanide metal europium is of the order of 103-106 ns. Instead of continuous excitation of the fluorophore, the exciting light is pulsed and readings made after the background emissions have decayed. Once a measurement has been taken, the fluorophore is pulsed again allowing the accumulation of signal from the label. Assays based on this principle have been shown to be both as precise and as sensitive as radioisotopic assays.

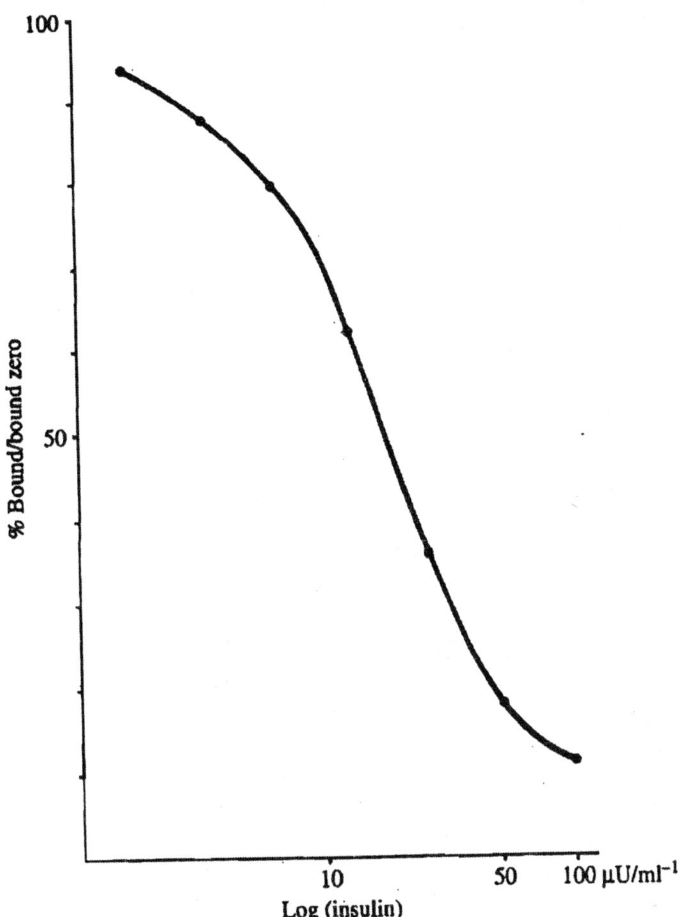

Fig. 10.13. Typical standard curve for competitive insulin assay.

Standard Materials

The quantitation of antigen will ultimately rely on comparison of test antigen responses with those of a series of standard solutions, and the type of standard solution will depend upon the nature of the assay. The matrix of the standard solution should resemble the sample matrix (blood, urine, saliva, etc.) as closely as possible and should of course be antigen free. This may be difficult to achieve in practice and only reputable suppliers should be used. Once the assay reagents have been optimized, a standard curve can be prepared.

It is customary to express the binding of the zero standard (B/B_0), although this is by no means the only way of expressing the final result. The responses of the unknown samples may now be noted on the curve and their concentrations conveniently read off.

Separation of Bound From Free Antigen

The whole quantitative basis of heterogeneous competitive immunoassay relies upon the physical separation of the bound and the free fractions in order that the relative proportions of each fraction may be assessed. Since the components are in solution, a variety of techniques have been developed in order to separate them. Some separation techniques rely on the physical removal of one of the fractions: charcoal will strongly adsorb the free fraction allowing its ready removal by centrifugation; the addition of dextran reduces the tendency of charcoal to 'strip' bound antigen from the complex; alternatively, the bound fraction may be precipitated by the addition of suitable concentrations of various protein precipitants such as alcohol, ammonium sulphate and polyethylene glycol (PEG). Amongst other techniques are those involving the immunoprecipitation of the bound fraction using a second antibody, which reacts with the proteins of the first antibody.

Table 10.4. Techniques used to Separate the Bound and Free Fractions in Immunoassay.

Principle	*Example*
Adsorption of free fraction	Dextran-coated charcoal
Precipitation of bound fraction	Ethanol, ammonium sulphate, PEG
Immunological	Use of second antibody directed against the primary antibody species
Solid phase	Using antibodies coupled to plastic; tubes, beads, micro-plates
	Using antibodies coupled to magnetic particles
Specific binding	Use of other specific binding properties; staphylococcal protein A, avidin-biotin

This second antibody may be produced by immunization: immunoglobulins from the species in which the primary antibodies were raised are injected into a different species. Such a separation procedure is often called a double antibody technique. The concentration of the second antibody is adjusted such that the soluble bound fraction is converted into an insoluble matrix which can be easily separated from the free antigen by centrifugation.

It is possible to accelerate the formation of the insoluble matrix by adding the second antibody in combination with a protein precipitant such as polyethylene glycol or ammonium sulphate (assisted second antibody technique). Many solid-phase systems have been developed in which either the primary or secondary antibody is immobilized onto surfaces such as plastics. Antibodies may be attached to plastic surfaces both by simple adsorption at high pH and by covalent linkage.

Various formats have been employed such as tubes, beads or microtitre plates. Once equilibrium has been reached the bound fraction is removed from the free by simple decantation of the tube contents, leaving the bound antigen attached to the solid phase. A novel use of this idea is the attachment of antibodies to magnetizable particles which, once the reaction is complete, allows the simple separation of bound from free fractions by placing the tubes above a strong magnetic field. When the particles have migrated to the bottom of the tube the free fraction may be decanted or aspirated away from the particles. Other substances that exhibit specific binding may be used to separate the free and the bound fractions: when attached to a solid phase the ability of staphylococcal protein A to bind to the FC fragment of certain isotypes of IgG can be utilized; the strong binding of the vitamin biotin to tetravalent avidin may also be employed.

Biotin may be readily incorporated into antibody molecules and these molecules may be subsequently captured by an avidin solid phase. Alternatively avidin may be used to provide a link between a biotinylated antibody and a biotinylated solid phase.

Homogeneous Assays

Homogeneous assays require no separation step in order to observe the ratio between the labelled and unlabelled antigen and the antibody. The most common of these systems is the

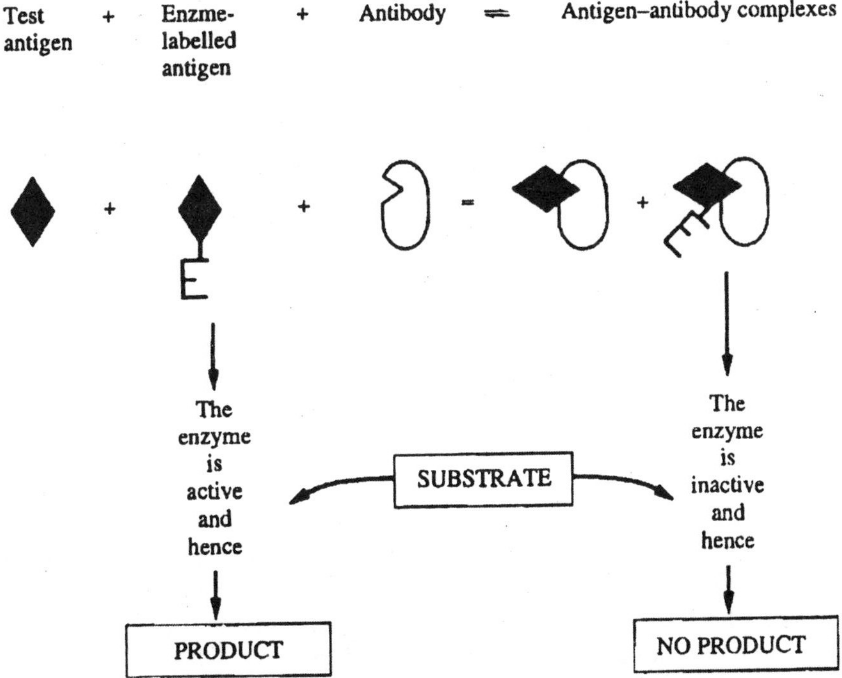

Fig. 10.14. Enzyme-multiplied immunoassay (EMIT). The three reactants, test (or standard) antigen, enzyme-labelled antigen and a limited amount of antibody are allowed to react and reach an equilibrium position. The unbound labelled antigen which remains is the only source of enzyme activity, the bound enzyme being inactivated. This free enzyme can be quantitated using a direct kinetic assay method and is proportional to the amount of unlabelled antigen originally present.

enzyme-multiplied immunoassay technique or EMIT, which is particularly suited to the measurement of small molecules (haptens) such as drugs. EMIT is a trade mark of the Syva Corporation of Palo Alto, California. Although it does not involve the separation of bound fraction from free it is nevertheless a competitive assay system.

The antigen is labelled with an enzyme in such a way that the enzyme retains its catalytic activity. When the antigen binds to the antibody the enzyme becomes inhibited, probably by an induced conformational change or by steric hindrance of the enzyme active site. This inhibition will be reduced by the presence of free antigen in the test sample competing for the antibody-binding sites. The more antigen that is present, the more activity is retained and the more coloured product can be formed.

The responses of the unknown are compared with the responses of standard doses of the antigen. The enzymes lysozyme (EC 3.2.1.17), malate dehydrogenase (EC 1.1.1.37) and glucose-6-phosphate dehydrogenase (EC 1.1.1.49) have all been used for this purpose. The dehydrogenase enzymes are particularly useful, owing to the additional ease of measuring their activity by monitoring the absorption of NADH at 340 nm. Other successful homogeneous assays have been developed such as the fluorescence polarization technique made available by Abbott Laboratories of North Chicago, USA. This competitive immunoassay technique relies upon the principle that, in solution, small molecules generally rotate more quickly than larger molecules.

The hapten label is fluorescent and excited by plane polarized light. If the hapten label is unbound then during the time between excitation and emission the molecule has rotated sufficiently to alter the plane of polarized light. Where the hapten label is bound, the larger complex does not rotate sufficiently during excitation and emission to alter the plane of polarized light and the emission may be measured. Thus the degree of binding of labelled hapten may be determined without separation of bound and free fractions. This technique is particularly well suited to the measurement of small molecules such as therapeutically administered drugs.

Membrane-based immunoassay devices

A particularly interesting advance in immunoassay technology has been the development of novel devices which contain all the necessary reagents, usually in a dry form. Many of these devices are designed for 'point-of-use' testing and are qualitative in nature, usually being assessed visually. Areas covered include pregnancy testing, drugs, microbiological antigens and environmental molecules such as pesticides and antibiotics. However a significant number are quantitative in design and involve dedicated instruments, some designed to be portable.

These immunoassay devices are directly comparable to the dry chemistry systems described earlier. The devices usually contain antibody immobilized on a membrane surface above a filter and absorbent pad. The sample is placed in the device and passes through the membrane into the absorbent pad. Where analyte is present in the sample it is sequestered by the antibody on the membrane and is then visualized by addition of a labelled second antibody. Labelled second antibody not bound to analyte also passes into the absorbent pad, sometimes with the aid of a wash solution.

The labelled molecule may be an enzyme, which would then require addition of a suitable substrate, or a label such as colloidal gold, which has the virtue of being visible without the aid of a second reagent. This type of approach is essentially non-competitive and usually requires the use of two monoclonal antibodies directed against two distinct epitopes on the analyte. Other devices have employed a two-stage competitive system in which analyte and labelled analyte compete for antibody in one part of the device. This is followed by transfer of the equilibrium mixture to a separate part of the device where membrane-immobilized antibody removes the unbound labelled material and allows the bound to go through the membrane into the absorbent pad.

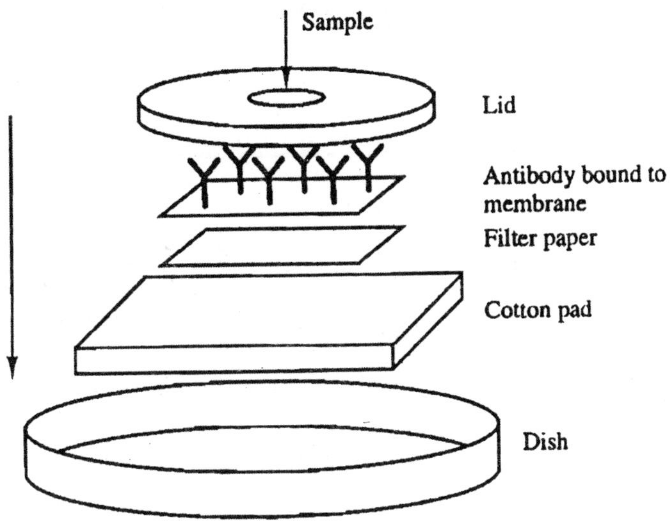

Fig. 10.15. Membrane-based device. Antibody is bound to the membrane and removes any test antigen as the sample is drawn through into the absorbent pad. A second labelled antibody is then applied and, if antigen is trapped in the membrane, this second antigen will also be held in the membrane and can be demonstrated by means of the label.

CHAPTER

11 Radioactivity

Atoms that have the same atomic number, and hence are the same element, but have different masses are known as isotopes. Radioisotopes, or more correctly, radionuclides spontaneously and continuously emit characteristic types of radiation. They are particularly useful in analytical biochemistry, the unique nature of the radiation providing the basis for many specific and sensitive laboratory methods.

Table 11.1. Physical Data of Some Frequently used Isotopes.

Element	*Symbol*	*Half-life*	*Beta emission*	*Gamma emission*
Calcium	^{45}Ca	165 d	+	–
Carbon	^{14}C	5760 a	+	–
Chlorine	^{36}Cl	3×10^5 a	+	–
Cobalt	^{60}Co	5.26 a	+	+
Hydrogen	^{3}H	12.2 a	+	–
Iodine	^{125}I	60 d	Electron capture	+
Iodine	^{131}I	8.04 d	+	+
Iron	^{59}Fe	45 d	+	+
Magnesium	^{28}MG	21.4 h	+	+
Nitrogen	^{13}N	600 s	Positron	+
Phosphorus	^{32}P	14.3 d	+	–
Potassium	^{40}K	10^9 a	Electron capture	+
Potassium	^{42}K	12.4 h	+	+
Sodium	^{22}Na	2.6 a	Positron	+
Sodium	^{24}Na	15 h	+	+
Sulphur	^{35}S	87.2 d	+	–

NATURE OF RADIOACTIVITY

The atomic nucleus is made up of *protons* and *neutrons*. The number of protons determines the atomic number and hence the identity of an element and is equal to the number of orbital electrons, a feature necessary to ensure the electrical neutrality of the atom. The atomic mass of the nucleus is made up by the additional neutrons that are present. Hence:

atomic number = the number of protons

atomic mass = the sum of the number of protons and neutrons

This information is normally shown as a *superscript* (*atomic mass*) and a *subscript* (*atomic number*) to the symbol for the element. Hence $^{14}_{6}C$ represents the isotope of carbon (atomic number 6) with the atomic mass of 14. In practice the subscript is often omitted because the atomic number is unique to the element that is represented by the appropriate letter (*e.g.*, C for carbon). For simplicity in the spoken form and often in the written form, isotopes are often referred to as carbon-14, phosphorus-32, etc. The stability of the atomic nucleus depends upon a critical balance between the repulsive and attractive forces involving the protons and neutrons. For the lighter elements, a neutron to proton ratio (N : P) of about 1 : 1 is required for the nucleus to be stable but with increasing atomic mass, the N : P ratio for a stable nucleus rises to a value of approximately 1.5 : 1. A nucle-us whose N : P ratio differs significantly from these values will undergo a nuclear reaction in order to restore the ratio and the element is said to be radioactive. There is, however, a maximum size above which any nucleus is unstable and most elements with atomic numbers greater than 82 are radioactive.

Types of Radioactivity

If a nucleus is too heavy and its atomic number exceeds 82, it may revert to a more stable arrangement by releasing both neutrons and protons. This is effected by the emission of an *alpha particle,* which contains two protons and two neutrons and is a helium nucleus, $^{4}_{2}He^{2+}$. Alpha particles are relatively large particles and are emitted with a limited number of energy levels. They carry a double positive charge and as a result attract electrons from the atoms of the material through which they pass, causing ionization effects. They have an extremely short range, even in air, and as a result present very little hazard as an external source of radiation but their effects within living cells or tissues can be serious. A nucleus that has an excess of neutrons will undergo neutron to proton transition, a process that may restore the N : P ratio but that requires the loss of an electron to convert the neutron to a positively charged proton, and as a result the atomic number increases by one.

The particle emitted is a high speed electron known as a negatron (β^-) and the atom is said to emit *beta radiation, e.g.*

$$^{14}_{6}C \rightarrow ^{14}_{7}N + \beta^-$$

Conversely, nuclei that contain an excess of protons undergo proton to neutron transition with the emission of a positively charged beta particle known as a positron (β^+) and with the reduction of the atomic number by one. A positron has only a very short existence, combining immediately with an electron of a nearby atom. The two particles disintegrate in the process with the emission of two gamma rays, *e.g.*

$$^{11}_{6}C \rightarrow ^{11}_{5}B + \beta^+$$

An alternative mechanism to positron emission, for the conversion of a proton to a neutron, involves a process known as *electron capture* (EC) in which the nucleus captures an orbital electron from an inner shell to restore the N : P ratio. Subsequently, an electron from another orbital falls into the vacancy left in the inner shell and the energy released in the process is emitted as an X-ray, the atomic number again being reduced by one, *e.g.*

$$^{125}_{53} \rightarrow I + e^- \rightarrow ^{125}_{52} \rightarrow Te + \text{radiation}$$

Beta emission from an atom shows a range of energy levels and although the maximum level is characteristic of the isotope, only a small proportion of the emissions may be of this energy. As a result, beta particles show a maximum range in a particular absorbing material with many emissions penetrating considerably less.

Penetration distances are about ten times greater than for alpha particles, values of 5-10 mm in aluminium being common. Because they are light, beta particles are easily deflected by other atoms and are less effective as ionizing agents than are alpha particles. Although a nucleus may have an N : P ratio in the stable range it is still possible for it to be in an unstable energetic state and to emit protons as electromagnetic radiation of extremely short wavelength known as *gamma rays.*

These have neither mass nor charge and as a result cannot cause direct ionization effects, but the energy associated with them is absorbed by an atom causing an electron to be ejected, which in turn produces secondary ionization effects. The atoms emitting gamma radiation suffer no change in either mass or atomic number although very few elements emit solely gamma radiation. Sodium-24 ($^{24}{}_{11}\rightarrow$Na) for example, emits beta radiation (negatron emission) and is converted to magnesium-24 ($^{24}{}_{11}\rightarrow$Mg). It also, however, emits gamma radiation in the process. In order to handle radioisotopes safely it is necessary, among other things, to define fairly carefully the penetrating power of the radiation emitted by any isotope.

Alpha particles, having only a relatively limited number of energy levels, are absorbed by contact with other atoms. The absorbing power of a material is referred to in terms of its equivalent thickness. The thickness required can be calculated by dividing the equivalent thickness by the density of the material. Beta radiation, because of its wide range of energies, shows an approximately linear relationship between the thickness of the material required to absorb radiation and the logarithm of the activity of the radiation.

The percentage of beta particles passing through a given thickness of absorbing material has a sigmoidal relationship to the logarithm of the maximum energy of emission. This information permits the selection of a particular thickness of absorbing material which will totally stop the beta radiation from one source and yet will allow the penetration of a relatively constant proportion of radiation from another source with a higher maximum emission energy. The absorption of gamma radiation is more difficult to assess and the most convenient means of expressing the absorbing power of a particular material is to quote the half-thickness.

Decay

The emission of radiation by an atom results in a change in the nature of the atom, which is said to have decayed. The rate at which a quantity of an isotope decays is proportional to the number of unstable atoms present and a graph of activity against time results in a typical exponential curve. For this reason the actual life span of a radioactive sample cannot be measured and a more meaningful expression of the rate of decay is provided by the half-life $t_{1/2}$. The half-life can be determined using the equation that describes the rate of radioactive decay:

$$\log_e \frac{N_t}{N_0} = -\lambda t$$

where N_t = the activity at time t,

N_0 = the activity at zero time,

λ = the radioactive decay constant,

t = time.

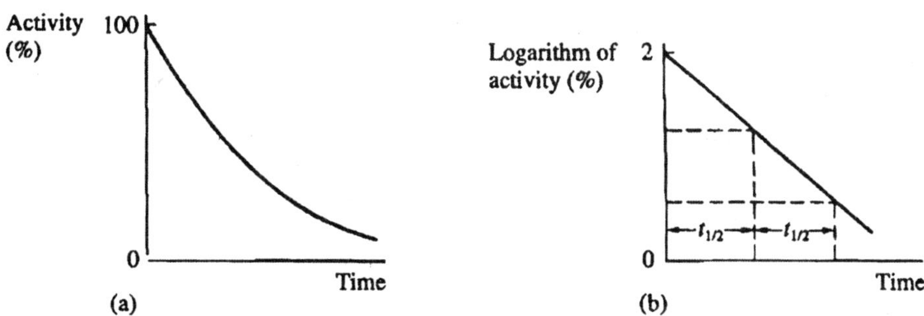

Fig. 11.1. Radioactive decay. The rate of radioactive decay is proportional to the number of unstable atoms present and although theoretically there should always be some activity left (*a*), in practice the activity does eventually fall to zero. A plot of the logarithm of the activity against time (*b*) results in a straight line from which the half-life can be determined.

In its linear form the equation is :

$$\log_e N_t = \log_e N_0 - \lambda t$$

and converting from natural logarithms it becomes :

$$\log_{10} N_t = \log_{10} N_0 - 0.4343\ \lambda t$$

This is the equation for a straight line with a slope of –0.4343 λ. If Nt is made equal to half N_0, then t will equal the half-life ($t_{1/2}$) and the equation becomes:

$$\log_{10} 1 = \log_{10} 2 - 0.4343\ \lambda_{1/2}$$

$$t_{1/2} = \frac{0.3010}{0.4343\lambda} = \frac{0.6931}{\lambda}$$

A plot of the logarithm of the activity of the sample ($\log_{10} N_t$) against time (t) is linear, and the half-life can be determined from the graph either by measuring the time interval between two readings of activity which vary by a factor of two or by using the value for the gradient of the line and substituting in the equation:

$$\text{Half-life} \quad (t_{1/2}) = \frac{0.3010}{\text{gradient}}$$

Units of Radioactivity

The curie unit (Ci) is based on the activity of 1 g of pure radium-226, which undergoes 3.7×10^{10} transformations per second. It is therefore defined as the quantity of a radioactive isotope which gives 3.7×10^{10} disintegrations per second. The SI unit of activity is the becquerel (Bq), which is equal to one nuclear transformation per second. Hence:

$$1\ \text{Ci} = 3.7 \times 10^{10}\ \text{Bq}$$

The specific activity of an isotope indicates the activity per unit mass or volume and is quoted as becquerels per gram (Bq g^{-1}) or millicuries per gram (mCi g^{-1}). A sample in which all the atoms of a particular element are radioactive is said to be carrier-free and is very difficult to achieve in practice.

Safety

Great care must always be exercised in handling radioisotopes. It is not only the powerful emitters that are dangerous but also weak emitters with long half-lives, *e.g.* tritium, carbon-14, which may be incorporated into the body and over a period of time can constitute a serious hazard.

There are many regulations regarding the handling and disposal of radioisotopes and these must be fully understood and observed when setting up a radioisotope facility. The hazards must be fully assessed and the laboratory must be equipped and approved for the intended applications. All manipulations must be assessed for specific hazards and standard operating procedures (SOPs) fully implemented.

There are, however, certain general principles of good practice which should constantly be kept in mind when working with radioisotopes.

1. All working surfaces and floors must be covered with an approved material and they must be regularly monitored for contamination.
2. Suitable protective clothing, including disposable gloves, must be worn.
3. Containers of a decontamination fluid must be readily accessible in which all contaminated glassware, etc. must be immersed after use.
4. Isotopes must be handled in an appropriate manner using forceps or handling devices if necessary. Adequate shielding must be used and film badges worn at all times.
5. Records of amounts and nature of all isotopes held in the laboratory must be kept, and used isotopes must be disposed of in the approved way.
6. After working, approved washing and decontamination procedures of both laboratory staff and equipment must be completed before the laboratory is vacated.

DETECTION AND MEASUREMENT OF RADIOACTIVITY

Radiation can be detected in several ways, all of which depend upon its direct or indirect ionization effects. The three methods most frequently used are the ionization of gases, the excitation of liquids or solids and the induction of chemical change. Radioactivity is measured as either the number of individual emissions in a unit time (differential measurements) or the total cumulative effect of all emissions in a given time (integral measurements). Generally, differential measurements are used for quantitative analytical work and integrating methods for dosimetry and autoradiography.

Regardless of the instrument or method of recording, any critical measurement of radioactivity demands that a background count (blank) must be done and the test value corrected for those emissions that are not due to the sample. Additionally, in order to get consistent results, the mean of replicate counts must be calculated and, where appropriate, statistical assessment must be undertaken because the fundamental basis of radioactive decay is a random process.

Geiger Counters

The ionizing effect of radiation on a gas may be detected if a potential difference is applied across two electrodes positioned in the gas. The electrons displaced as a result of ionization will be attracted to the anode and a current will then flow between the electrodes. This type of detector is based on the Geiger-Muller tube, which consists of a hollow metal cylinder (the cathode) within a glass envelope and a wire (the anode) running along the central axis of the cathode. The whole tube is filled with the counting gas, often a mixture of neon and argon, under relatively low pressure. Radiation entering the tube causes ionization of the detector

Fig. 11.2. Geiger–Muller tube. The tube is filled with an ionizable gas mixture, such as neon and argon, and a voltage applied across the electrodes. Ionization of the gas by incident radiation causes a current to flow between the electrodes.

gas and, as a result, a pulse of current flows between the electrodes, together with the emission of ultra-violet radiation. One problem with a Geiger-Muller tube is the fact that once it has fired, the ultraviolet radiation emitted causes further ionization of the gases (self-excitation).

It is necessary to quench this self-excitation and this is usually achieved by incorporating a halogen, *e.g.* bromine, into the gas mixture during manufacture. The bromine molecules absorb this extra energy and are split into their individual atoms. These atoms subsequently recombine when the counter is not being used and so regenerate the system. The end window of the tube must be thin enough to permit the weaker radiations to enter the tube (aluminium, 6-8 mg cm^{-2}; mica, 2 mg cm^{-2}) but even so alpha particles and very weak beta emissions are either completely or partially absorbed.

The emissions from the biologically important isotopes of tritium and carbon-14 fall into, this category and alternative detectors should be used for these isotopes. Gamma rays, being powerful, enter the tube very easily but cause little direct ionization. They do, however, produce emission of photoelectrons from the glass or metal walls of the tube and the internal electrodes. These photoelectrons then produce ionization of the gas with the resulting pulse of current. If the voltage applied to a Geiger-Muller tube is increased from zero, no response to radiation is detected despite the presence of a radioactive isotope until a minimum or starting voltage is reached. At this point a current is generated which increases rapidly to a plateau and finally at very high voltages (the breakdown potential) the tube goes into continuous discharge.

The counter should be operated at a voltage selected from the plateau region and will be severely damaged if voltages above the breakdown potential are applied for any significant period of time. A detector for liquid samples is a common variant of the Geiger-Muller tube and is designed either to be dipped into or to contain a liquid.

The main disadvantage of Geiger-Muller counters is the fact that after an ionization has occurred and the electrons have been discharged (a very rapid process) the resulting positive ions are discharged slowly and as a result prevent other electrons from reaching the anode. The period of time during which the detector is insensitive to further ionization is known as the dead-time and is often in the region of 500 μs. Correction factors are used to correct all readings for the counts lost during this time.

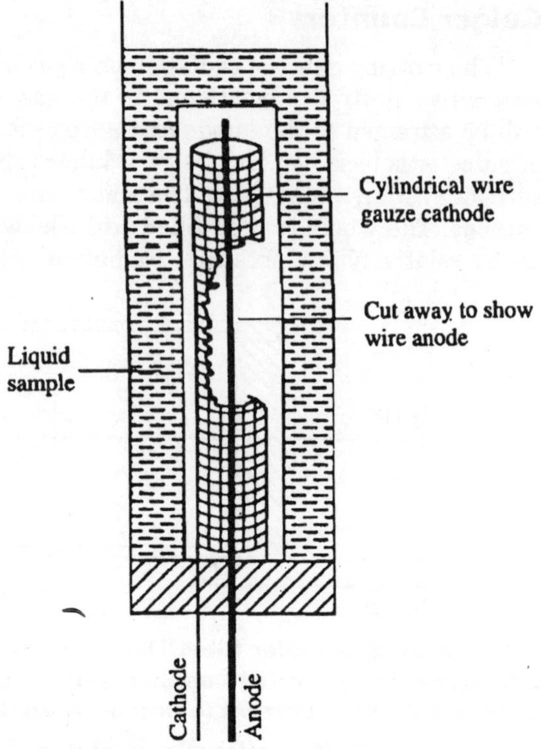

Fig. 11.3. A Geiger-Muller tube for measuring the activity of liquid samples.

Scintillation Counters

Some substances, known as fluors or scintillants, respond to the ionizing effects of alpha and beta particles by emitting flashes of light (or scintillations). While they do not respond directly to gamma rays, they do respond to the secondary ionization effects that gamma rays produce and, as a result, provide a valuable detection system for all emissions. A range of scintillants is available, many designed for maximum efficiency with specific isotopes.

Crystals of sodium iodide containing a small amount of thallous iodide are very efficient detectors of gamma radiation and instruments based on this design are often called gamma counters. The crystal is positioned on top of a photomultiplier which can detect the flashes of light and both are covered with a light-proof material that is thin enough to allow gamma radiation to pass into the crystal. (Fig. 11.4). When a sample is placed on the crystal, the pulses of current generated by the photomultiplier are counted electronically. Weak beta radiation and alpha particles often cannot penetrate the covering material but the use of a scintillant, which, together with the sample, will dissolve in a suitable solvent, enables a similar technique to be used. Liquid scintillation counters usually consist of two light-shielded photomultiplier tubes with the sample and a scintillant mixture contained in a glass vessel placed between them. Only those pulses that occur simultaneously from both photomultiplier tubes are due to scintillations and these are counted electronically, giving a detection system that can discriminate between genuine signals and random noise from the detectors. This technique is called coincidence counting.

A range of scintillants suitable for such work is available each showing particular light emission characteristics. They are dissolved in a suitable solvent, which should be miscible with the sample; toluene or xylene are frequently used for organic samples and dioxane for aqueous samples. The number of ionizations or scintillations detected by both types of detector is measured electronically and presented either as the total number of counts (a scaler) or as the number of pulses per minute (a count-rate meter).

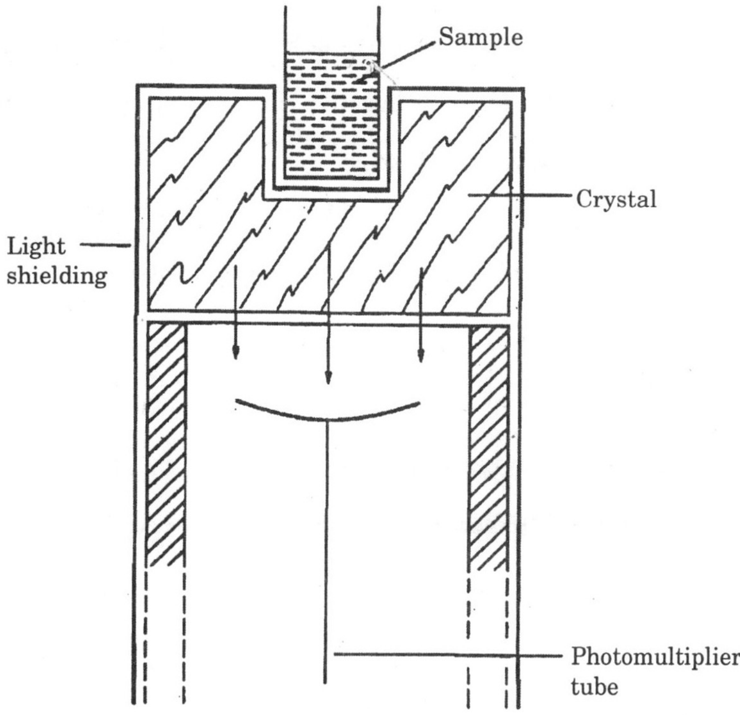

Fig. 11.4. A gamma counter. The scintillations from a crystal of fluor caused by gamma radiation are detected by a photomultiplier.

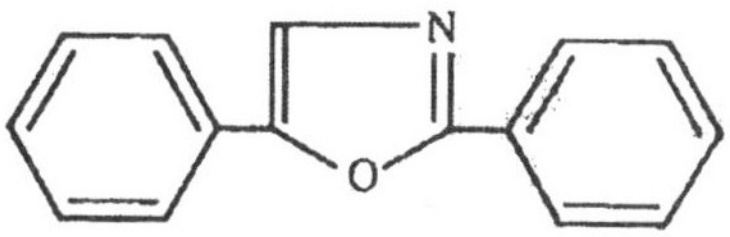

1-Phenyl-4-phenyloxazole
(PPO)

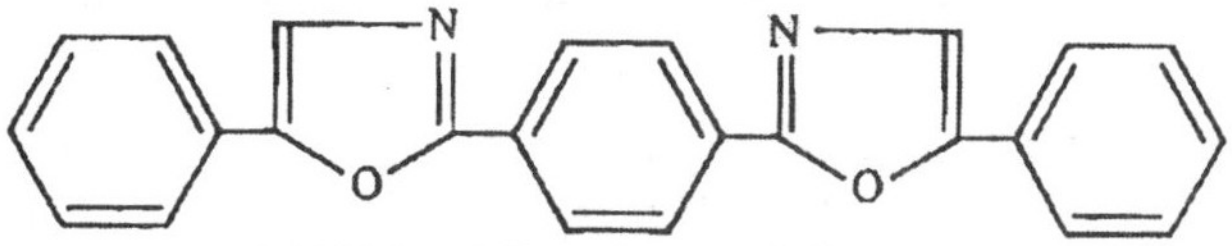

1,3-Di-2,5 (phenyloxazoyl) benzene
(POPOP)

Fig. 11.5. Scintillants. A range of organic scintillants is available with different solubility and emission characteristics. Scintillant 'cocktails' or mixtures contain a primary scintillant such as PPO and often contain a secondary scintillant which absorbs the radiation produced by the primary scintillant and re-emits it at a longer wavelength, e.g. POPOP.

Additionally, some counting devices (pulse height discriminators) are designed to record signals at specified voltages and can measure one isotope in the presence of another.

Autoradiography

Ionizing radiation has the same effect as light on photographic film and the extent of blackening of the film is related to the amount of radiation. Autoradiography is particularly useful for demonstrating the location of radioactive isotopes in tissues or chromatograms. The sample is placed on a photographic film which is protected from the light and allowed to remain in contact long enough for an adequate exposure.

The exposure time is dependent upon the intensity of the radiation and can usually only be determined by trial and error. It is possible, however, to predict an approximate exposure time from the fact that a total emission of 10^7 beta particles per square centimetre is often required.

Autoradiography is also a convenient way of monitoring the amount of radiation to which a worker has been exposed (dosimetry). If a small badge containing photographic film is worn continually and the photographic film developed after a set period of time, an estimate of the radiation received can be made from the degree of darkening of the film. No sophisticated instrumentation apart from photographic facilities are required for these types of application.

BIOCHEMICAL USES OF ISOTOPES

Tracers

Radioactive isotopes provide a very convenient way of monitoring the fate or metabolism of compounds that contain the isotopes. When used in this way, the isotope is described as a tracer and compounds into which the radioactive atom has been introduced are said to be labelled or tagged. The labelled molecules need only comprise a very small proportion of the total amount of the unlabelled radioactive substance because they act in the same way as the non-radioactive substance but can be detected very much more easily.

The varied applications of tracers in biochemistry range from studies of metabolism in whole animals or isolated organs to sensitive quantitative analytical techniques, such as radioimmunoassay. Phosphorus-32 is used in work with nucleic acids, particularly in DNA sequencing and hybridization techniques. In these instances the isotope is used as a means of visualizing DNA separations by autoradiographic techniques. While it is generally assumed that radioisotopes are metabolized in exactly the same way as the non-radioactive substances, occasionally some differences in reactivity are seen, due mainly to differences in the atomic weights of the two isotopes.

The effect is most obvious for elements with the lowest atomic number, where the difference in mass between isotopes is the greatest. For instance, the difference between hydrogen (mass 1) and tritium (mass 3) is more significant than the difference between iodine-127 and iodine-125. This isotopic effect is extremely slight in most cases and can be ignored unless critical studies of equilibrium constants and rates of reaction are involved. Ideally, molecules should be labelled by introducing a radioactive isotope in place of a normal atom, *e.g.* carbon-14 replacing a carbon-12 in a carbohydrate. This method of labelling involves the synthesis of the molecule either *in vivo* or *in vitro* and the use of enzymes often permits the isotope to be introduced in a particular position in the molecule.

The position of the labelled atom should be indicated wherever possible as for example in glucose-1-^{14}C. Alternatively, it may be necessary to label a molecule by introducing an additional atom and in this case it has to be assumed that its presence does not alter the reactivity or metabolism of the molecule. Proteins, for instance, are often labelled with an isotope of iodine. It is very important that any labels used are firmly attached to the molecule otherwise invalid results will be obtained.

The choice of an isotope for tracer studies requires an appreciation of not only the radiochemical properties of the element, but also the effects that they might have both biochemically and analytically. The isotope should have a half-life that is long enough for the analysis to be completed without any significant fall in its activity. Occasionally this might present a problem in that some elements only have radioactive isotopes with very short half-lives, *e.g.* fluorine-18 has a half-life of 111 min.

Conversely, isotopes with very long half-lives should not be used for *in vivo* studies because accumulation in the tissues of the recipient is unacceptable. The biochemically important elements of hydrogen, carbon and calcium are weak beta emitters and are difficult to detect using Geiger-Muller tubes. Scintillation counters are necessary for these weak emissions and are preferable for gamma emitters, while the Geiger-Muller detectors are suitable for the more powerful beta emitters. However, for autoradiography, alpha and weak beta emitters give the best resolution, while gamma emitters are usually unsuitable because they cause general fogging of the film.

Isotope Dilution Analysis

Isotope dilution analysis is the definitive method for the quantitative analysis of many compounds but is usually only carried out in specialist laboratories. If a known amount of an isotope with a known specific activity mixed with an unknown amount of the non-radioactive substance, the reduction in the specific activity can be used to determine the degree of dilution of the isotope and, hence, the amount of the non-radioactive isotope present. The isotope and the test substance must be thoroughly mixed before a representative sample of the mixture is purified and its specific activity determined.

Calculation

Isotope:	Amount	M
	Specific activity	S
Mixture:	Amount (unknown)	$M_x + M$
	Specific activity	S_X

The specific activity of the original radioisotope with a total activity *(A)* is:

$$S = \frac{A}{M}$$

The mixture with the non-radioactive isotope contains the same total activity but the amount of the substance is increased. Hence the specific activity is reduced to:

$$S_x = \frac{A}{M_x + M}$$

Substituting in this second equation an expression for the amount of activity *(A)* in the sample we have:

$$S_x = \frac{S \times M}{M_x + M}$$

Hence:

$$M_x = M\left(\frac{S}{S_x} - 1\right)$$

Radioactivation Analysis

Radioactivation analysis is used for the analysis of trace amounts of suitable elements. It is a technique that is not generally applicable to biological material although it has been used for the measurement of lead in hair and nails. It involves the simultaneous irradiation of the sample and a standard known mass of the same element to produce a radioactive isotope of the element. The activities of both the sample and the standard are then determined and, because their specific activities will be the same, it is possible to calculate the mass of the unknown sample. Activation analysis requires the use of a powerful source of neutrons as the activator and is suitable only for elements which form an isotope whose half-life is longer than the isotopes of other elements which may be produced.

Index

L

M

P

R